PHARMACOLOGY AND TOXICOLOGY

PHARMACOLOGY AND TOXICOLOGY

Editor

Volker Geiszler

scitus academics

Pharmacology and Toxicology

Edited by **Volker Geiszler**

This Book is Printed on Acid-Free Paper. ♾

ISBN: 9781681179292
Library of Congress Control Number: 2017948959

SCITUS Academics LLC,
616, Corporate Way, Suite 2, 4766,
Valley Cottage, NY 10989

www.scitusacademics.com

Printed in United States of America.

Preface

Veterinary pharmacology and toxicology are being increasingly recognized as important disciplines and were also rapidly changing and evolving in the mid-twentieth century. By the mid-1950s, pharmacology and toxicology were a highly active area of veterinary medicine; veterinary pharmacology and toxicology emerged as new disciplines, closely related, because they were involved in developing new compound, particularly new antimicrobials and antiparasitics to control infection diseases, and analgesic/non-steroidal anti-inflammatory drugs affecting different animal species. By the mid-1950s, veterinary toxicology was a highly active area of veterinary medicine. Toxicology was once the science of poisons and intoxication, but currently toxicology should be regarded as the "science of safety" because it is essential in ensuring safety of food, feed, drugs, pesticides, and industrial products used in daily life.

This volume 'Pharmacology and Toxicology' presents research studies in comparative pharmacology, clinical veterinary pharmacology and other aspects of pharmacology of interest to the veterinary profession. Biochemical and molecular interactions were discussed, and the tools of other disciplines were brought to focus upon the problems of domestic animal poisonings. It is primarily intended to students in pharmocology and veterinary nurses; it focuses on the practical application of knowledge and encourages a problem-solving approach by introducing a discussion of the potential uses of drugs and exploring a variety of therapies. It illustrates the more complex areas of pharmacology and aid comprehension of the general action mechanisms of the various drugs discussed. Complicated concepts are presented in a user-friendly way to maximize understanding. An understanding of mechanisms, potential benefits, and risks of antidotes is essential for clinicians who manage poisoned animals.

Table of Contents

CHAPTER 1

Perspectives in Veterinary Pharmacology and Toxicology

Arturo Anadón

Department of Toxicology and Pharmacology, Faculty of Veterinary Medicine, Universidad Complutense de Madrid, Madrid, Spain

Keywords: veterinary pharmacology and toxicology, antimicrobial resistance, contaminants and residues, environment, regulatory aspects, new challenges

Veterinary pharmacology and toxicology are being increasingly recognized as important disciplines and were also rapidly changing and evolving in the mid-twentieth century. By the mid-1950s, pharmacology and toxicology were a highly active area of veterinary medicine; veterinary pharmacology and toxicology emerged as new disciplines, closely related, because they were involved in developing new compound, particularly new antimicrobials and antiparasitics to control infection diseases, and analgesic/non-steroidal anti-inflammatory drugs affecting different animal species. The research and development cost of new antimicrobials that must fulfill the new regulatory requirements is considered high; in drug development projects, toxicology closely linked with pharmacology and pharmacokinetics is decisive in the success or failure of a new drug candidate. During the last years, the following new antimicrobial active substances have been authorized for use in veterinary medicine: difloxacin (belonging to the class fluoroquinolones, authorized in 1998 after marketing of enrofloxacin in 1987), valnemulin (a pleuromutilin, 1999), pirlimycin (a lincosamide, 2001), tulathromycin (a macrolide, 2003), tylvalosin (a macrolide, 2004), ceftiofur (a third-generation cephalosporin, 2005), cefovecin (a third-generation cephalosporin, 2006), gamithromycin (a macrolide, 2008), pradofloxacin (a fluoroquinolone, 2011), and tildipirosin (a macrolide, 2011). None of these antimicrobials belongs to a new class of antimicrobials. It is known that most veterinary drugs derive from either the human pharmaceutical industry (antibiotics, analgesic, antiobesity, anticancer, cardiovascular) or the plant protection products (insecticides, parasiticides). Like veterinary medicine, veterinary pharmacology and toxicology is not an easy matter because of multiplicity of animal species as well as breed, age, sex, and (patho)physiological status concerned that may respond differently to certain drugs, food, feed, pesticides, and industrial products as well as contaminants or toxins. The inter- and intraspecies differences as well the variability within and across animal populations are the essence of veterinary pharmacology and toxicology. It is know the species differences in the metabolism and excretory processes of different chemical compounds. In the 60s of the twentieth century, the extrapolation dose regimens from one species to another was a common practice; however, nowadays it is known that pharmacokinetics/toxicokinetics (PK/PD) differences remain for drugs, pesticides, contaminants, or toxins that is the cornerstone of veterinary pharmacology and toxicology. Moreover, the analytical techniques to measure plasma/tissue concentrations of xenobiotics are changed and the sensibility been improved. Analytical methods whether intended for use in pharmacokinetic (PK) and metabolism studies, residue depletion studies, or regulatory control programs for residues of veterinary drugs and pesticides, and contaminants, share a common subset of validation criteria. Performance characteristics to be determined for all method validation include specificity, accuracy, precision, limit of detection, limit of quantitation, susceptibility to interference, and information on method calibration. Practicability, applicability under normal laboratory conditions, and ruggedness are the additional criteria for the evaluation of regulatory methods. Residue programs consist of two principal activities: monitoring and surveillance. For monitoring drugs having maximum residue limits (MRLs), animal tissues, such as liver, kidney, muscle, milk, eggs, honey, or fat, are selected. Another group of samples/matrices frequently used are animal feed and drinking water as well as manure, urine, and hair. They are mostly used to monitor prohibited substances and can be taken prior to slaughtering. Analyzing hair has gained some popularity because it has been demonstrated that anabolic steroids can be detected in hair a long time after application of the drugs – that is, when residues cannot be detected any more in urine or manure. It has been suggested that the hair analysis is a strong means to control the illegal use of clenbuterol in animal production, and samples are available from living as well as slaughtered cattle. The thyroid gland is used to monitor the use of thyreostatics; the eye, vitreous humor, or retina (1, 2) have been suggested as tissues for residue screening, such as clenbuterol, ractopamine hydrochloride, and zilpaterol hydrochloride with β-adrenoceptor activity. Hair/feathers (both pigmented and non-pigmented) also show this accumulation effect; the concentration of clenbuterol in bovine brown or white hair was found to be about one-third that in black hair (3).

By the mid-1950s, veterinary toxicology was a highly active area of veterinary medicine. Toxicology was once the science of poisons and intoxication, but currently toxicology should be regarded as the "science of safety" because it is essential in ensuring safety of food, feed, drugs, pesticides, and industrial products used in daily life (4). Biochemical and molecular interactions were discussed, and the tools of other disciplines were brought to focus upon the problems of domestic animal poisonings. An understanding of mechanisms, potential benefits, and risks of antidotes is essential for clinicians who manage poisoned animals. Of the dozens of antidotes currently available, only a few are regularly used. These include activated charcoal, acetylcysteine, naloxone, sodium bicarbonate, atropine, flumazenil, therapeutic antibodies, and various vitamins. Even then, most are used in a minority of poisonings. There is little randomized trial evidence to support the use of most antidotes. Consequently, decisions about when to use them are often based on a mechanistic understanding of the poisoning and the expected influence of the antidote on the patient's clinical course. Importantly, most poisoned animal patients who reach veterinary clinic/hospital can recover with supportive care alone. In more serious poisonings, decisions regarding antidote use are generally guided by a risk/benefit assessment based on low quality evidence. With it, a new generation of veterinarian emerged. The developing veterinary toxicologist had to understand physiology, pathology, and chemistry, but with extensive knowledge of qualitative and quantitative analytical techniques. Judgment of clinical episodes and knowledge of metabolic and excretory processes were needed. Toxicology becomes intimately familiar with pharmacology and the molecular action of a wide variety of chemicals. Toxicology and pharmacology can be considered as the same discipline on two different ends of spectrum; pharmacology is the study of drugs used at doses to achieve therapeutic effects on an organism, while toxicology is the study of toxicants that produce harmful effect on an organism. Understanding treatments to be administered for specific intoxications was necessary. In conclusion, there are often confusing and confounding assortment of signs, lesions, and analytical results to reach rational interpretations and conclusions for the numerous problems being solved (5).

The veterinary medicine profession was initially focused on domestic animals, particularly those used for food consumption, and horse for transportation. With the growth of more specialized agriculture and production practices, the veterinary profession with its linkage to domestic livestock, stimulated growth of veterinary pharmacology and toxicology. Later on, the veterinary profession was concerned to a growing population of companion animals, including horses (used in equine activities relating to competitive sports and leisure activities), dogs and cats (as integral family members), and other animal species (i.e., will/exotic animals) (6). The antiparasitic, analgesics, antiobesity, cardiovascular, and anticancer drug market for dogs and cats has grown abundantly, which has been the result of the progress of the veterinary pharmacology and toxicology. In both fields of veterinary medicine (i.e., domestic and companion animals, and wild/exotic animals) was strengthened as observation-based medical practice was complemented and, ultimately, supplemented by science-based medicine. During this period, veterinary pharmacologists and toxicologists began to play an important role in veterinary medical diagnostic laboratories (6). With the strengthening of the science base of veterinary medicine, including the quality of the science in the veterinary medical curriculum, the emergence of the comparative medicine character of veterinary medicine became more apparent and was enhanced (7). Comparative Medicine is a distinct discipline of experimental medicine that uses animal models of human and animal disease in translational and biomedical research. Biomedical research included comparative pharmacology, toxicology, and pathology. These changes in the profession were accompanied by increased involvement of veterinarians in research on the species of traditional concern to the profession, domestic, and companion animals (8), and to participation in a broader range of biomedical research activities, involving use of the traditional laboratory animal species, driven largely by concern for human health (7). More and more veterinarians are now conducting pharmacological and toxicological research as their primary mission.

Pharmacology and toxicology are translational sciences. Pharmacology translates fundamental insights into drug action and fate into clinical therapy. Toxicology translates science, transferring knowledge from fundamental science into practical applications to safeguard animal health, human health, and the environment. Toxicology as translational science uses basic and clinical science to solve problems in toxicologically relevant fields. A high level of basic scientific research in toxicology is the prerequisite to assessing emerging risks in the process of developing drugs, chemicals, and materials (9). Nowadays, veterinary pharmacology and toxicology should be regarded as disciplines contributing to the paradigm "One World, One Health," reducing risk at the animal–human–ecosystems interface.

The development of veterinary pharmacology and toxicology is the same as that for humans, although between human and veterinary medicine, there are differences in size and economic factors (10). The evolution of veterinary pharmacology and toxicology occurred concurrently with evolution of its two roots; the profession of veterinary medicine and the science of pharmacology and toxicology. The veterinary pharmacologist and toxicologists recognize the necessity of considering the various animal species of concern as a main activity (6). Traditional indicators of target organ toxicity used in drug preclinical safety studies consist of a battery of clinical pathology parameters in blood and urine coupled with histopathologic examination of altered tissues, often measured in multidose toxicity studies. In recent years, the expansion of knowledge at the molecular and cellular level has provided the opportunity for considering the addition of a multitudinous of new endpoints to pharmacological and toxicological evaluations (11). Industry and regulatory agencies expect that new and better efficacy and toxicity biomarkers (the latter also known as safety biomarkers) will play an important role in improving key aspects of the drug development process. Successful translation of many biomarkers into clinical practice is now increasing. It is well known to the veterinary clinician, pharmacologist, and toxicologist and to physicians that biomarkers have been used in both human and veterinary medicine for centuries. In some cases, measurement of the biomarkers present in body fluids, urine, or exhaled breath serve as an indicator of exposure or, even, dose of a drug or toxicant. The application of genomics has already proved useful in fundamental research and for gaining a mechanistic understanding of drug activity (12).

New biomarkers of exposure will continue to be proposed. For each potential biomarker of exposure, it will be necessary to conduct experiments to validate the utility of the biomarker. A special challenge relates to recognizing the dynamics of the pharmacokinetics/toxicokinetics of various drugs and toxicants and establishment of quantitative relationships between exposure and dose at any particular time over the course of the intoxication.

The potential list of biomarkers for efficacy and toxic responses is seemingly endless. In all fields of medicine, from different kinds of cancer to various functional diseases of every organ system, new molecular markers are being identified on a regular basis. The challenge for pharmacologists and toxicologists is to consider, from among this array of opportunities, which biomarkers are sufficiently well validated with regard to their linkage to diseases and sufficiently reasonable in cost to warrant their use in exposure–response studies. This includes consideration of the new and highly sophisticated genomic tools. There is a special challenge in designing validation studies to make certain that the experimental design is directed toward identifying specific disease-related endpoints or toxicant-related effects rather than merely being another, albeit more sophisticated, marker of non-specific toxic effects. A serious issue in many previous validation studies has been the use of a single high exposure level and a few short-term observation times. Such studies are unable to evaluate exposure-related changes in biomarkers and may not be able to identify toxicant-specific changes. Most safety biomarkers have been used in safety assessment for many years without being formally qualified, and many of them show deficiencies with respect to sensitivity, specificity, and predictive value (11, 13). With that in mind, during the development of biomarkers, the following steps should be followed: (1) identification, (2) preclinical qualification, and (3) clinical qualification and diagnostic use (14).

Veterinary pharmacology and toxicology is in some countries a unique discipline, but in others is completely split as veterinary pharmacology and veterinary toxicology because of their development and complexity; the veterinary toxicology studies are not only focused in drugs but also addressed other subjects such as food and feed safety, plant protection products, and industrial chemicals among others.

VETERINARY PHARMACOLOGY

Veterinary pharmacology is defined as the study of the properties of drugs and all aspects of their interaction with living organisms. Drugs include any chemical agent (other than food) used in the treatment, cure, prevention or diagnosis of disease, or the control of physiological processes (10). The science of pharmacology draws on the knowledge and methods of many allied clinical and non-clinical disciplines, including chemistry, biochemistry, biology, physiology, pathology, toxicology, and medicine.

Veterinary pharmacology is an experimental science dealing with the properties of drugs and their effects on living systems. It has included study of sources of drugs (pharmacognosy), magnitude, and time course of the observed pharmacological effect on the body pharmacodynamics (PD), relationship between administered doses, the observed biological fluid/tissue concentrations of the drug, and time in the body pharmacokinetics (PK), use in the treatment of diseases (therapeutics), and poisoning effects (toxicology).

The residue definition for veterinary drugs in edible commodities of animal origin is obtained from metabolism studies conducted in target species and livestock animals. The metabolites, degradation products, and other transformation products are typically identified and quantified with methods based on the use of substances labeled with radioactive isotopes. Metabolites obtained in these studies are qualitatively compared with metabolites identified in laboratory animals, usually rats, to ensure that substances occurring in significant amounts in edible commodities have been included in the toxicological testing or to determine whether additional testing of individual metabolites is necessary. Metabolism studies in laboratory animals also serve to identify mammalian metabolism.

Drug–drug and drug–feed additive interactions deserve especial attention: a particular reference should be made to side effects and residue formation, and the likelihood of these interactions has to become part of the evaluation procedure. However, the most effective approach to prevent the occurrence of residues is certainly the rationally based selective use of veterinary drugs for which well-educated veterinarians are needed.

The form and the distribution of the residues that result from each authorized mode of application in each species should be determined, and the depletion of the residues from edible tissues or animal-derived foods should be studied. Total residue and metabolism study provides information to establish the appropriate marker residue (is the parent drug or any of its metabolites or a combination of any of these with a known relationship to the concentration of the total residue in each of the various edible tissues at the expected withdrawal time) and to determine the target tissue. A "marker residue" should be identified, which is usually the form of the drug (parent compound or metabolite) that is found at the highest concentration for the longest period in the target food. The tissue in which the highest residues are found is usually designated as a "target tissue" (represents the edible carcass from which residue depletes most slowly and is the edible tissue selected to monitor for the marker residue in the target animal) (15).

The extent to which such differences can be extrapolated to humans should be evaluated; for example, many sex differences in metabolism observed in rats do not occur in humans. The enzymes involved in the metabolism of foreign compounds represent the most important source of interspecies differences and human variability in the biodisposition of the compound and, for many cases, in the generation of toxic effects.

A further important role of such *in vitro* experiments is to investigate similarities and differences between humans and test species in the metabolism and effects of xenobiotics that may provide information critical to the extrapolations normally used in risk assessment.

Veterinary clinical pharmacology is a subset matter of the broad study of pharmacology and is devoted to the study of the clinical effects of drugs on animal patients with a goal of optimizing therapeutic dosage regimens. Knowledge of the PK and PD properties of drugs and their toxic effects is inherent in this veterinary discipline. Clinical pharmacology in the veterinary setting is the clinical discipline devoted to the optimal use of drugs in veterinary patients, maximizing their prophylactic or therapeutic benefits while ensuring that the adverse consequences of drug use are minimized.

Veterinary clinical pharmacology is a clinical science that integrates disease pathophysiology with fundamental concepts of pharmacology to provide a rational basis for drug therapy in animal patients (10). Brown (16) states that the goal of veterinary clinical pharmacology is to apply the principles of pharmacology to more successfully treat animal patients and to more rationally use medications in veterinary medical practice. A fundamental understanding of veterinary clinical pharmacology is essential for good clinicians. The demonstration of efficacy includes the following test phases: (1) description of the mode of action, (2) determination of dose and dosing interval(s), (3) dose confirmation trials, including persistent efficacy trials, and (4) where applicable, clinical field trials. Similarly, knowledge of the pharmacological action of drugs is meaningless unless one also has a basic understanding of the relevant physiology and pathophysiology of the system or tissue adversely affecting the health or welfare of the patient.

There are methods for conducting clinical efficacy studies for infectious diseases and non-infectious diseases. These approaches may include studies in a model of the disease, or alternate methods for evaluating the response to the therapeutic agent in order to get utility in the new animal drug development process (16). The choice of the clinical endpoint is critical and determines the study design. There are many compounds (e.g., antibacterials, anthelmintics), where efficacy endpoints are reasonably easy to identify (i.e., clinical and bacteriological response as determined by the use of appropriate clinical, postmortem, and bacteriological diagnostic methods, and the determination of mortality rate; decrease in worm count, decreased temperature). However, identification of efficacy endpoints for non-antibacterial or non-growth enhancement compounds is more difficult to determine. Veterinary pharmacology aids in the identification of appropriate endpoints, helping to make them more quantitative, reproducible, and representative of the clinical field situation. Whereas "therapeutics" is a term describing treatment of disease in general and includes use of drugs, surgery, radiation, behavioral modification, and other modalities. Demonstration of efficacy of veterinary drugs is usually dealing with general requirements for the assessment of efficacy of such products. Depending on the aim of the trial, it can be classed in one of the following three categories: (1) confirmatory, (2) exploratory, or (3) composite trial. Confirmatory trials can concern dose-determination trials, dose confirmation trials as well as controlled field trials.

For antimicrobials, the dose, the dosing interval, and the number of administrations of product should always be justified by considering the PK/PD relationship, if established, as well as the severity of the disease. In some cases, where the PK/PD relationship is well established using validated models, it may be possible to omit dose-determination studies and to evaluate in a clinical trial the efficacy of one or a very few regimens.

However, to be acceptable, the choice of the PK/PD parameter considered as best predictive of efficacy must be prospectively justified by independent data.

A prerequisite for clinical studies is to perform toxicological testing of new drugs with the objective to know potential undesirable effects of the drug candidates and drugs in the therapeutic range and above (17).

PHARMACOKINETICS/TOXICOKINETICS

Pharmacokinetics (PK) is the study of the characteristics of the time course and extent of drug exposure in individuals and populations and deals with the absorption, distribution, metabolism, and excretion (ADME) of drugs. PK is defined as the mathematical description of temporal changes in concentration of drugs within the body (18); such studies provide the experimental basis for drug dosage regimens in various animal species.

Bioequivalence is a clinical term referring to formulations of a drug with rates and extents of absorption that are sufficiently similar that there are not likely to be any clinically important differences with respect to either efficacy or safety. In order to demonstrate bioequivalence for systemically active drugs, a comparative PK study is generally undertaken and the similarity (defined by statistical and biological criteria) of Cmax (maximum concentration) and AUC (area under curve) of the formulations is assessed. For drugs not acting systemically, comparisons of clinical or other pharmacological endpoints may be necessary.

In the development of dosage forms, there are many factors that must be standardized in order to ensure that the biological availability of the active constituent is both predictable and reproducible. The following physicochemical and formulation-induced factors have been described as sources of potential bioequivalence between products.

Toxicokinetics (TK) relies on the same principles as PK, but there are differences in data availability and aims. The exposure environment is characterized by the concentration of the toxicant in the media, be it water, air, or feed, the quantities taken in, and the time course of the intake. Dose is the concentration, over time, of the toxicant in the various tissues of the subject (cow, human, rat). The characterization of the kinetics linking exposure with dose is referred to as TK (for a toxic agent) or PK (for a pharmaceutical). In actual practice, the term pharmacokinetics is frequently used when it would be more appropriate to use the term toxicokinetics. Toxicokinetics are used to describe the movement and disposition of the toxicant in the organism. This includes consideration of the route of entry: ingestion, inhalation, dermal, or purposeful administration by injection. A complete description of the toxicokinetics of a toxicant will take into account (1) the intensity and duration of the exposure, (2) the rate and amount of absorption of the toxicant from the site of entry, (3) the distribution of the toxicant within the body, (4) potential biotransformation to less, equal, or more toxic form, and (5) the rate of excretion by route (urine, feces, or exhalation). All of these aspects of toxicokinetics may be influenced by species differences in physiological and biochemical characteristics. Modern approaches to modeling toxicokinetics attempt to take account of both species differences and similarities in influencing the fate in the body of toxicants. It is also important to recognize that the exposure or dose level may influence the kinetics of a toxicant and its metabolite(s). This is an especially important consideration in extrapolating from laboratory studies that may be conducted at high doses to lower more environmentally relevant exposures/doses.

PHARMACOKINETIC/PHARMACODYNAMIC PREDICTED INDEXES

A PK/PD model is a mathematical description that provides clinically relevant information about the relationship between the pharmacokinetics and the pharmacological effect. There is a mounting body of evidence to support the use of PK/PD relationship for individual antibacterial agents in identifying those doses and dose intervals that are most likely to be efficacious and least likely to result in adverse effects and/or the selection of resistant organisms. This concept has implications for the safety and efficacy of antibacterial drugs (19). PK/PD modeling is a scientific tool to help developers select a rational dosage regimen for confirmatory clinical testing. A PK/PD model integrates the PK model (describing the relationship between dose, systemic drug concentrations, and time) with the PD model (describing the relationship between systemic drug concentration and the effect vs. time profile) and a statistical model (particularly, the intra- and interindividual variability of PK and/or PD origin). PK/PD concepts can be applied to individual dose optimization (20). Aspects of PK/PD relationship through concept, basic theory, and practical data reinforced on data analysis and modeling and simulation, and a chance to play an active role in analysis, interpretation, and discussion will be welcome. Use of the PK/PD relationship can be made to justify the dosages to be used in dose-determination studies. In some cases where the PK/PD relationship is well established using validated approaches, it may be possible to omit dose-determination studies and to confirm the efficacy of one or a very few dose regimens in clinical trials (dose confirmation and clinical field studies). To be acceptable, justification should be provided prospectively for the eligibility of such an approach (19). In conclusion, an effective antimicrobial should possess both PK and PD selectivity; it should distribute in the locus of the targeted pathogen, and it should have no PD impact on comensal microbiota of the gastrointestinal tract of the treated animal or on environmental ecosystems. Presently, we need to revise dosage regimens of the antimicrobial classes that are in the market and use preferentially narrow spectrum antimicrobials.

Veterinary drugs cover a broad range of chemical and class structures and usually undergo metabolism after any route of administration to an animal species. There are species differences in transformation and excretion on several drugs being the half-live values different. Modes of administration include injection, implantation, dermal application by spray or *pour-on*, and inclusion in feed or water, all of which may result in different rates of absorption, with possible differences in the tissue distribution and nature of the residues. Finally, as veterinary drugs are applied to a diverse group of target animals differing in species, breed, age, sex, and (patho)physiological status, the biological variance deserves to be given more attention not only in the developmental stage of new compounds but also at the stage of drug application by the practitioner (20).

VETERINARY TOXICOLOGY

Veterinary toxicology is a very complex subject as it deals with a wide variety of poisons. The field of toxicology is very broad, including the identification and characterization of a number of toxicants (pharmaceuticals, food and feed additives, natural toxins, consumer products, and specific chemicals), their physical and chemical properties, their fate in the body, and their biological effects. Currently, synthetic compounds constitute the largest number of chemicals that are frequently encountered in animal poisonings. Acute poisoning is a common presentation in emergency veterinary medicine. Moreover, veterinary toxicology is concerned with the treatment of disease conditions caused by poisons. Currently, there is a need to integrate veterinary medicine and specifically veterinary toxicology into environmental medicine.

The nature of the toxic responses depends not only on the toxicant but also on the route of exposure, the duration, and intensity of the exposure, and the characteristics of the exposed individual (i.e., species, gender, age, preexisting disease states, nutritional status, and prior exposure to the agent or related compounds) (6). The subject of veterinary toxicology is complicated greatly by the wide variations in responses of domestic, aquatic, wild, and zoo species to toxicants. Of course, there are many other factors that can be involved in the overall toxicity of a chemical.

Understanding the complete profile (especially mechanisms of toxicity) of each toxicant is the biggest challenge for today's veterinary toxicologists (21).

Veterinary toxicology is a multifaceted hybrid that draws on and contributes to the veterinary medical profession, the scientific field of toxicology and, broadly, to medical science. Some have characterized toxicology as a distinct scientific discipline. I view toxicology as an applied area of science addressing important societal issues by drawing on multiple scientific disciplines and professions (6). Veterinary toxicology had a much applied origin, involves the evaluation of toxicosis and deficiencies, identification and characterization of toxins and determination of their fate in the body, and the diagnosis and treatment of toxicoses in domestic animals and companion animals. However, the field has broadened to include concern for contaminants in human food products originating from animals and for contributing to the conduct and interpretation of safety/risk evaluations for pharmaceuticals, food additives, consumer products, and specific chemicals.

The recent worldwide melamine contamination in pet and swine feed indicates the relevancy of veterinary toxicology to current animal health and food safety, particularly on its toxicological aspects. Veterinary toxicology can be challenging because of the low frequency of cases observed in a practice setting. When a toxicosis occurs, it often involves a large number of animals and may also involve litigation (22).

Veterinary toxicologists who understand both normal and disease processes extending from the molecular level to the integrated mammalian organism and, indeed, populations, have an array of opportunities for making significant contributions to society (6).

In veterinary medicine, the discipline of clinical toxicology concerns the protection of companion, exotic/wildlife, and food-producing animals. Veterinary clinical toxicology links preventive and experimental toxicology with clinical veterinary medicine. Clinical animals are exposed to natural toxicants in their native environments, as well as to synthetic chemicals and drugs. The practicing veterinarian and the diagnostic or clinical veterinary toxicologist are often the first to respond to these problems, which may appear as a clinical syndrome of unknown cause. In veterinary medicine, understanding of sources of poisons, circumstances of exposure, diagnosis of the type of poisoning, treatment, and application of management or educational strategies to prevent poisoning is of much interest. The clinical diagnosis may be used initially to determine the organ systems affected, and the disturbances that must be controlled to save the animal's life and etiologic diagnosis is the most important diagnosis because it enables specific therapy and preventive measures to proceed. Clinical toxicology evaluation depends heavily on the determination of exposure and evidence for the contribution of interacting factors than can alter toxicity (23).

REGULATORY PHARMACOLOGY AND TOXICOLOGY

Regulatory pharmacology and toxicology is the study of efficacy and adverse effects of chemicals not just on humans but also on all living organisms, including natural toxins, animals, fungi, and insects. Referring to regulatory toxicology supports the development of standard protocols and new testing methods to continuously improve the scientific basis for decision-making processes. It is the duty of the industrial pharmacologist and toxicologist to aid in the application, interpretation, and circumstantial evaluation of laws, rules, and regulations. Extensive regulatory and monitoring systems have been established to ensure that chemical residues in food do not constitute an unacceptable health risk. Chemicals (i.e., veterinary drugs, medicated feed, or application of pesticides) are everywhere, in the food we eat and in the feed of animals. Regulatory authorities establish MRLs or tolerances and set withdrawal periods that ensure residues of the pharmacologically active substance will not exceed the MRL when the label instructions for the product are followed (15). Some chemicals are used to help people and animals, such as the chemicals in medicines. Surveillance programs by comparison, sample target-tissues from animals suspected of violative residues on the basis of clinical signs or herd history. Animals identified with violative residues of veterinary drugs or pesticides do not enter the food chain. Most people know that high doses of some chemicals used to help them in medicines can damage health. The same can occur in the animals (22).

Veterinary toxicologists try to find out (1) what the toxic effects of chemicals are; (2) where toxic effects occur; (3) how a chemical causes its toxic effects, and (4) what length of exposure causes a toxic effect. Veterinary toxicologists often find out about the effects of chemicals by collecting data from laboratory experiments on animals. They use their skills to predict what is likely to occur if humans come into contact with the same chemical. In laboratory studies, animals are often exposed to much higher levels of a chemical than would normally be experienced by humans from normal use of that chemical. Toxicologists have developed a number of ways of predicting the likely effects at lower exposure levels.

The previous strong emphasis on field observations was first complemented and then supplemented by experimentation. This, in turn, led to the current strong mechanistic orientation of toxicology. With this shift in toxicology came an increased awareness of the utility of a comparative medicine orientation in research directed primarily toward improving human health (7). With this comparative medicine orientation came increased opportunities for individuals educated in veterinary medicine, including veterinary pharmacology and toxicology, to contribute to general pharmacology and toxicology and biomedical science.

MEDICINAL PRODUCTS INTENDED FOR MINOR USE OR MINOR SPECIES

There is considerable concern in the animal health sector about the lack of authorized veterinary medicinal products for minor uses and for minor species (MUMS). A "minor use" within EU member states may be a major use in other parts of the world where climate, diseases and disease vectors, environment, and husbandry practices differ from those in the EU. All other animal species that are not considered major [i.e., *major food-producing species*: cattle (dairy and meat animals), sheep (meat animals), pigs, chickens (including laying hens), and salmon; *major companion animal species*: cats, dogs]. All other animal species, which are not considered major, are as a consequence, by default, classed as minor species. Minor use in a major species is generally considered as the use of veterinary medicinal products for the treatment of diseases that occur infrequently or occur in limited geographical areas and thus are indicated for a smaller market sector. The policy is intended to stimulate the development of new veterinary medicines for minor species and for diseases occurring infrequently or in limited geographical areas in major species that would otherwise not be developed in the current market conditions. For these reasons, the published guidelines are intended to reduce data requirements, where possible, for products classified as MUMS/limited market while still providing assurance of appropriate quality, safety, and efficacy and complying with the legislation in place and leading to an overall positive benefit–risk balance for the product (24). Studies on comparative metabolism may be of particular importance when assessing MRL recommendations for multiple species, particularly for the extension or extrapolation of MRLs to minor species. Where identical or very similar MRLs have been set for three major species from different animal classes (ruminants, monogastrics, and poultry), based on specific residue data, confirming a similar exposure situation of the consumer in relation to these species, it can be assumed that the exposure assessment and consequently the risk characterization on the basis of same/similar MRLs for further species beyond the animal classes concerned would be similar. The MRLs for chicken may be extended to turkeys based on similarity between the species (24).

TOXICITY TESTS

Toxicity test on laboratory animals are conducted to evaluate chemicals for their potential to cause cancer, birth defects, and other adverse health effects. Information from toxicity testing serves as the basis for regulatory decisions concerning toxic chemicals. Current test methods were developed incrementally over the past 50–60 years and are conducted using laboratory animals, particularly rats and mice. Test animals are often exposed to higher doses requiring assumptions about effects at lower doses or exposures. Often controversial uncertainty factors must be applied to account for differences between test animals and humans (25). The visional strategy for the replacement of animal testing by alternative approaches is based on the "3Rs principles" (Reduction, Replacement, and Refinement), which have been encouraged in particular in Europe. These principles have influenced the modern state and perspectives of the development of non-animal experimental methods. While full replacement has not been achieved yet for those toxicological endpoints, partial replacement strategies to reduce the number of animals used exist. The term "alternative of animal testing" is generally associated with the "3Rs principles," and was suggested due to animal welfare concerns (26).

In silico methods and *in vitro* systems are traditionally much used approaches to complement or partly replace animal tests in pharmacology and toxicology (26). The "integrated test strategies," which combine *in silico, in vitro,* and *in vivo* methods, allow the best predictions of toxicity.

Predominantly, predictive toxicological testing is derived from the findings to assess risks to animal, humans, and the environment. While traditional toxicology methods have relied heavily on animals, new high-throughput screening approaches to generate toxicological data are becoming increasingly available for the safety assessment of chemicals. Gene expression profiling lends itself to two highly topical areas of drug development: "predictive" toxicology and mechanism-based risk assessment. The last few years have seen a lot of progress being made in linking the profiles of gene expression induced by drugs with their toxicity (27).

Two recent reports prepared by Committees of the National Research Council provide insight into how some individuals envision the future of toxicity testing. The first NRC Report (28) addresses the application of toxicogenomic technologies in predictive toxicology and risk assessment. The second report provides a strategy and vision for toxicity testing in the twenty-first century (25). Both committees would have benefited from participation by veterinarians who understand the complexity of disease processes and toxicoses to compensate for the view of molecular and cell biologists who focus on the microlevel and systems biology and may overstate the role of mathematical models (6). The term systems biology (biological integration) is a realization that organisms do not consist of isolated subsets of genes, proteins, and metabolites. It uses an integrated approach to study and understand the function of biological systems, and how perturbations of such systems (29). The key elements in systems biology are bioinformatics and mathematics that enable data integration and interpretation. Genomics and systems biology had a impact on various aspects of pharmacology, toxicology, and therapeutics; so approaches based on genomics and systems biology are now being used in various stages of drug discovery and development (30).

The concept of systems biology has come to the forefront in recent years. The concept of a systems approach to normal biology and disease processes has been a fundamental concept undergirding both veterinary medicine and human biology.

PHARMACOGENETICS/TOXICOGENOMICS

The linkage between dose and adverse health outcome involves multiple mechanisms as various toxicants may potentially impact all the cells and organ systems of the body. Increasingly, scientists have attempted to model these relationships, which, in parallel to the nomenclature for the kinetic phase, are called toxicodynamics (TD) or PD models. It is obvious that multiple pathways may be involved in a toxicant producing disease and that knowledge of the individual steps will increase as knowledge of basic biological mechanisms increases. For example, the explosion of knowledge of basic biology at the level of the genome (genomics), proteins (proteomics), and metabolism (metabolomics) has provided a basis for exploring the mechanistic basis of toxicant-induced disease with a degree of refinement that could not even be envisioned even a short time ago. Genomics will also improve our understanding of comparative pharmacology and toxicology and thereby the rational use of animals and models (30).

The term genomics is used to encompass transcriptomics, proteomics, and metabolomics (or metabonomics) that describes the formation and fate of mRNA, proteins, and metabolites (30). The development of medium- to high-throughput methods in proteomics, genomics, and metabolomics, together with the revolution in bioinformatics, will lead to the accumulation of vast toxicological information, helping to elucidate bioactivity and mechanisms of toxicity in both *in vitro* and *in vivo* models (31). These approaches will also help in the development and establishment of new biomarkers of effect or exposure. To be more efficient, considerable improvement in the use of novel technical tools (omics-based technology) for effective safety assessment is needed. Omics analysis are now being accepted as essential methodological parts of systems or integrative biology, and it is possible to observe global changes in all RNAs (transcriptomics), proteins (proteomics), or metabolites (metabonomics, also called metabolomics, metabolic profiling, etc.) in cellular systems or tissues. This includes an array of new molecular biomarkers, which have received substantial attention. The omics will allow a comprehensive analysis of biological systems of many different animal species. Determination of the expression levels of multiple gene transcripts, proteins, and metabolites is now feasible and has been termed transcriptomics, proteomics, and metabolomics. Genomics can be used to identify new targets, since it is now possible to analyze changes in the transcription of >20,000 genes in various cell types and tissues in multiparameter experiments. There are two key approaches (1) the detection of cDNAs that correspond to the entire protein-encoding gene and (2) the use of RNAi to determine the relevance of mRNA levels in terms of protein expression and gene function. The transcriptomic data, and how they can be interpreted in the context of the pathology, and other biological data from a toxicology study are of much interest (30).

The toxicogenomics data have begun to be used more widely in the human health assessments of chemicals, and will likely be one of the crucial information sources for next-generation of risk assessment decisions. Given the general familiarity of the toxicologists with high-dimensional transcriptional profiling data and major traditional ways in which such data are analyzed, presented, and interpreted, this course is designed to demonstrate how the toxicogenomics data can become a key element in hazard identification, dose-response analysis, and selection of scientifically justifiable uncertainty factors. By superimposing the opportunities that are now afforded by both traditional and high-throughput genomics data onto the human health assessment paradigm, this course will be informative to the risk assessment practitioners and the toxicology research community and increases the scientific impact of the fundamental toxicology studies.

High-throughput screening (HTS) allows identifying desirable PK properties or compounds with the greatest *in vitro* potency and/or efficacy and/or least toxicity. Development of improved hESCH-based HTS assays for cardiotoxicity assessment; cardiac toxicity is a major concern in drug development.

Toxicogenomics is a mature field, which provides invaluable information on the molecular events preceding or accompanying toxicity; however, most traditional use of gene expression and other 'omic data in toxicology is largely the same as it was 10 years ago: the mode of action analysis, classification/prediction, and biomarker discovery. The toxicogenomics data have begun to be used more widely in the human health assessments of chemicals and will likely be one of the crucial information sources for next-generation of risk assessment decisions. Given the general familiarity of the toxicologists with high-dimensional transcriptional profiling data and major traditional ways in which such data are analyzed, presented, and interpreted, this course is designed to demonstrate how the toxicogenomics data can become a key element in hazard identification, dose-response analysis, and selection of scientifically justifiable uncertainty factors. By superimposing the opportunities that are now afforded by both traditional and high-throughput genomics data onto the human health assessment paradigm, this course will be informative to the risk assessment practitioners and the toxicology research community and increases the scientific impact of the fundamental toxicology studies. Transcriptional and posttranslational signals are known mechanisms, which promote efficient responses to DNA damage (30). Using HTS assays and small molecule analysis approaches, can be identified translational components of the DNA damage response, which operate through the modification of tRNA. Multiple research efforts are looking to use high- and medium-throughput *in vitro* screening data in identifying chemical hazards. For industrial and agricultural chemicals, research efforts in United States and Europe have characterized the *in vitro* biological activity of chemicals using multiple *in vitro* assays and technologies in order to predict *in vivo* toxicity and prioritize compounds for conventional toxicity testing. For pharmaceuticals, high-throughput *in vitro* screening assays have been integrated early into preclinical drug development to identify toxic compounds and guide medicinal chemistry efforts. In both cases, the application of *in vitro* toxicity screening for prioritization and hazard identification relies on its ability to accurately predict the results of *in vivo* laboratory animal studies and humans. More recently, preclinical testing has been tailored to the specific properties and mode of action of the new drug and may be adapted on the outcome of previous studies (32).

The current objectives of toxicoproteomics in diagnostic toxicology is to define molecular mechanisms of toxicity, screen for drug toxicities, and elucidate biomarkers or signature protein profiles in order to more accurately assess, predict, and diagnose toxicities.

The use of proteomics for diagnostic application and the application of microscopic analyses of feeds and animal ingesta for toxic components in the diagnostic process are presented.

VETERINARY MEDICINES AND ENVIRONMENTAL RISK

Advanced and developing societies are increasingly concerned with the impact of veterinary medicines on the environment (33). So, actually, an Environmental Risk Assessment (ERA) shall accompany an application for a marketing authorization for a medicinal product for veterinary use. The assessments are performed in two phases. In the first phase (Phase 1), the potential for environmental exposure is assessed according to the intended use of the veterinary medicine. The Phase 2 assessment is performed in two parts, Tier A and Tier B. Assessment will stop at Tier A if the product has been shown to present no risk. In Phase 1, the likelihood of exposure to the environment is assessed, and if the product does not meet certain criteria (e.g., if the soil PEC is greater then 100 μg/kg), then Phase 2 assessment is required. Phase 2 can be performed in two Tiers, in the first Tier, the likely impact of the substance is assessed using a range of standard tests, and the second Tier involves more detailed investigations into the compartment(s) of interest. A number of groups of veterinary medicines (i.e., fish farm medicines and anthelmintics) are known to be of environmental concern as well as the antimicrobials. After systemic administration, the substances can be metabolized and a mixture of the parent compound and metabolites may be excreted in the urine and the feces, developing in the case of antimicrobials, their selective pressure on microbiota harbored by waste, manure/slurry, and beyond in water and soil. Some 70% of the antimicrobials administered to food-producing animals are excreted as active substances into the environment; furthermore, some antimicrobials are stable in the environment for weeks or months (34). Residues from farms may contain antibiotics and antibiotic resistance genes that can contaminate natural environments. The clearest consequence of antibiotic release in natural environments is the selection of resistant bacteria. The same resistance genes found at clinical settings are currently disseminated among pristine ecosystems without any record of antibiotic contamination. For animals on pasture, the excreta will be released directly to soil, whereas for intensively reared animals, the main route of entry will be through slurry and manure spreading. Although there has long been concern over the use of chemical additives in livestock feed (e.g., potential consequences of insect-free manure), the effects on non-target dung insects and dung dispersal primarily become a concern with some drugs (e.g., macrocyclic lactones). For example, the introduction of doramectin into environment will be intermittent through the field application of sheep manure containing excreted drug residues or by direct defecation by sheep of manure containing doramectin residues. In order to assess the risks posed to the environment by veterinary medicines used to treat livestock, a number of models based on assumptions derived from experience with pesticides, industrial chemicals, and guidelines have been developed for regulatory agencies predicting concentrations of veterinary medicines in soil, groundwater, and surface waters. The potential impact of a veterinary medicine on the environment will be determined by a number of factors, including amount used, usage pattern, metabolism, persistence in manure and slurry, sorption and persistence in the environment, and ecotoxicity. Veterinary medicines can persist in soils for days to years, and studies have demonstrated that half-lives are influenced by a range of factors, including temperature, pH, and the presence of manure.

ANTIMICROBIALS AND ANTHELMINTICS RESISTANCE

Antimicrobial resistance (AMR) is the resistance of a microorganism to an antimicrobial drug to which it was previously sensitive. Resistant organisms (they include bacteria, viruses, and some parasites) are able to withstand attack by antimicrobial medicines, such as antibiotics, antivirals, and antimalarials, so that standard treatments become ineffective, and infections persist and may spread to others. Antibiotics have revolutionized the treatment of infectious diseases, but there are public health concerns that efficacy of antimicrobials in man is compromised by their use in animals. Antimicrobial resistence is a consequence of the use, particularly the misuse, of antimicrobial medicines and develops when a microorganism mutates or acquires a resistance gene. This issue is now a significant health problem, each year in the European Union (EU) alone, over 25,000 people die from infections caused by antibiotic-resistant bacteria. Infections due to these selected multidrug-resistant bacteria in the EU result in extra health-care costs and productivity losses of at least €1.5 billion each year. If the current trend is not altered, 300 million people worldwide are expected to die prematurely because of drug resistance over the next 35 years (35).

Antimicrobial resistance spreads through global tourism, transfer of patients between health-care facilities within and from outside the EU, and through trade in food and animals. It is an important global economic and a societal challenge that cannot be tackled by countries or public administrations alone. Therefore, the problem needs a comprehensive "One Health" approach to it. That means that a holistic, multi-sectorial approach involving many different sectors (public health, food safety, biosafety, environment, research and innovation, international cooperation, animal health and welfare as well as non-therapeutic use of antimicrobial substances) is needed to take care of this complex problem.

Antibiotic resistance is also a food safety problem: antibiotic use in feed – for treatment, disease prevention, or growth promotion – allows resistant bacteria and resistance genes to spread from food animals to humans through the food chain. The horizontal transfer of resistant bacteria from living animal to humans has been documented for methicillin-resistant *Staphylococcus aureus* (MRSA) investigated in pigs, cattle, horses, and domestic pets. Although most *S. aureus* infections are hospital acquired, transmission of MRSA from pigs to pig farmers and their families has been documented (36), but these transmissions concerning the livestock MRSA (ST398) seem to be low in humans. Livestock-acquired MRSA in contrast to the hospital-acquired MRSA (HA-MRSA) strains has a very low pathogenicity, and it is hardly transmitted horizontally between humans. The same can be indicated for *Escherichia coli* isolates expressing extended spectrum β-lactamases (ESBLs). The issue of MRSA in food animals is an emerging concern.

Recently, it has been discovered that there is a new mechanism of resistance in bacteria to colistin (caused by the *mcr-1* gene). The gene can be transferred between different types of bacteria, potentially causing a rapid development of resistance. The gene was first identified in bacteria (*Enterobacteriaceae*) in South China (37), and since then has also been found in the EU and other regions. In addition, colistin should be reclassified and added to Category 2 of the AMEG (Antimicrobial Advice ad hoc Expert Group of European Medicines Agency) classification system, which includes medicines reserved for treating infections in animals for which no effective alternative treatments exist. This category includes certain classes of antimicrobials listed by the World Health Organization as critically important in human health, especially fluoroquinolones and third- and fourth-generation cephalosporins, only be used in food animals when their use is justified and under specific restrictions. Responsible (prudent) use of antibiotics is an essential element in the fight against antibiotic resistance.

It is probable that one of the most effective measures to limit expansion of AMR is reducing the need for antibiotics in animal husbandry by improving animal health through biosecurity measures, disease prevention (including the introduction of effective vaccines, modulation of the innate immunocompetence, bacteriophage therapy), and good hygienic and management practices; and eliminating economic incentives that facilitate the inappropriate prescription of antibiotics. More targeted use of antimicrobials were still necessary to guard animal health (e.g., by the use of accurate diagnosis, evidence-based regional treatment guidelines, and correct dosing regimens). It is noticeable that it is necessary to provide improved summary product characteristics (SPC) guidance about the epidemiological circumstances under which this has shown to be effective and the extent of benefit demonstrated. Veterinary pharmacology contributes to make rational dosage schedule design and to minimize its emergence and spread.

There is now a need to develop veterinary medicines, which reduce the need for use of antimicrobials ("alternatives"), such as vaccines, and will facilitate the regulatory pathway for innovative products and novel therapies. Furthermore, there is now a need to develop new compounds having minimal impact on gut flora and on environmental bacterial ecosystems (34).

The development of nematode and trematode resistance to various groups of anthelmintics is a major problem. Compared with the development of AMR, resistance to anthelmintics in nematodes has been slow to develop under field conditions.

ANIMAL FEED CONTAMINATION

In the last years, the animal feed productions have had several contamination crises, such as transmissible bovine encephalitis caused by prions, unexpected sources of dioxin contamination of milk, meat, and eggs (38), and the presence of residues or unauthorized substances, such as melamine in milk, milk products, and animal feeds (39). Feed hygiene is another important subject to be highlighted because of contamination with *enterobacteriaceae* and anaerobes and the presence on environmental pollutants and contaminants, such as persistent organic pollutants (POPs), present in a broad variety of feed material and able to accumulate in animal tissues (i.e., dioxins and dioxin-like polychlorinated biphenyls and brominated flame retardants). Toxic heavy metals and other chemical elements and natural toxins, such as toxic plant metabolites and bacterial and fungal toxins, can be present in feed for animal consumption.

In recent years, increasing attention has been paid to the risk to consumers posed by chemical contaminants or residues in animal feed. This was caused by various cases of milk, eggs, or other animal products contaminated with environmental chemicals. For that, large number of emerging chemical contaminants in feed require the establishment of maximum limits (MLs).

The carry-over or cross-contamination from feed to food must be determined from animal experiments throughout the so-called transfer factors of chemicals from animal feed to animal products. The transfer factor is expressed as the concentration of the chemical in the animal food products (milligram/kilogram wet weight) divided by the concentration of the chemical in the animal feed (milligram/kilogram dry weight). Leeman et al. (40) have compiled data on transfer rates from a number of chemical contaminants in food products [i.e., pesticides, dioxins and furans, polychlorobiphenyls (PCBs), and polybrominated biphenyls (PBBs), heavy metals, mycotoxins, hormones, veterinary medicines, nitrosamines, and other compounds not belonging to one of the previous classes]. If a study on the carry-over is not available, a conservative worst-case estimate considers that all of the contaminant consumed is transferred to the animal product, e.g., the meat.

Carry-over or cross-contamination has been evaluated with some coccidiostats, particularly polyether ionophore. The health risk to non-target species resulting from the consumption of cross-contaminated feed with coccidiostats at levels of 2, 5, or 10% was evaluated recently by the European Food Safety Authority (EFSA) and a revision of the risk assessments is found in the paper of Dorne et al. (41).

To tackle these current and future challenges, we need to advance scientific knowledge through the application of novel approaches. Likewise, for an effective translation of scientific discoveries and applications in the field of veterinary pharmacology and toxicology, we need an outlet to enhance effective communication among scientists with different backgrounds and who share a common interest in veterinary pharmacology and toxicology. This was the drive for launching a veterinary pharmacology and toxicology specialty section in the journal *Frontiers in Veterinary Science*, an international, open-access, peer-reviewed, online journal. It gives me great pleasure to announce the inauguration of this new veterinary pharmacology and toxicology specialty section with the ethos of publishing the most impactful and interesting research discoveries in basic and fundamental areas of pharmacology and toxicology. The veterinary pharmacology and toxicology specialty section will publish original research articles and reviews of latest developments in specific subject areas and forward-looking perspectives.

As the founding Specialty Chief Editor for veterinary pharmacology and toxicology *Frontiers in Veterinary Science*, I have accepted the challenges and responsibilities of developing a premier professional journal in veterinary pharmacology and toxicology. The editorial team will strive to maintain excellence while improving the efficiency of the review process and taking advantage of the latest online technology. All editorial board members are leaders in their respective scientific areas and are willing to devote their time and expertise to build a journal of excellence to serve the veterinary pharmacology and toxicology research community, and life sciences in general, with the ultimate goal of improving animal and human lives. I do hope that the veterinary pharmacology and toxicology section will appeal to readers with veterinary, medical, and life science interests.

VETERINARY PHARMACOLOGY AND TOXICOLOGY IS A SPECIALTY SECTION OF FRONTIERS IN VETERINARY SCIENCE

Veterinary Pharmacology and Toxicology provides an international publication forum for scientists and clinicians investigating drugs and toxins in all animal species (livestock, poultry, game birds, avian, rabbits, companion animals, horses, wildlife, zoo animals, farmed fish and shellfish, bees, and laboratory animals). The section welcomes high quality original manuscripts in all topics of pharmacology and toxicology, including pharmacodynamics/toxicodynamics; pharmacogenetics/pharmacogenomics; toxicogenetics/toxicogenomics; pharmacokinetics/toxicokinetics; adverse effects of drugs, xenobiotics, plants and toxins; contaminants, pesticides, and drug residues in food animals and food products; clinical pharmacology and therapeutics; clinical toxicology; environmental toxicology; AMR; and regulatory pharmacology and toxicology. Veterinary pharmacology and toxicology welcomes submissions in traditional as well as new and emerging areas of research, encourages interdisciplinary studies, and seeks to advance global progress in these important disciplines.

I look forward to receiving your exciting research contributions in these areas. Your ideas and suggestions for making the journal highly impactful are welcomed and will be always valued.

AUTHOR CONTRIBUTIONS

The author confirms being the sole contributor of this work and approved it for publication.

CONFLICT OF INTEREST STATEMENT

The author declares that the research was conducted in the absence of any commercial or financial relationships that could be construed as a potential conflict of interest.

ACKNOWLEDGMENTS

The author thanks Mary Christopher for useful comments on the manuscript.

REFERENCES

1. Meyer HH, Rinke LM. The pharmacokinetics and residues of clenbuterol in veal calves. *J Anim Sci* (1991) 69:4538–44.
2. Brambilla GF, Agrimi U, Pierdominici E. Clenbuterol residues in vitreous humor and urine of calves. *Ital J Food Sci* (1991) 4:303–6.
3. Dursch I, Meyer HH, Karg H. Accumulation of the beta-agonist clenbuterol by pigmented tissues in rat eye and hair of veal calves. *J Anim Sci* (1995) 73:2050–53.
4. Casarett LJ. Toxicology. In: Casarett LJ, Doull J, editors. *The Basic Science of Poisons*. New York: Macmillan (1975). 3 p.
5. Oehme FW. Chapter 1. Veterinary toxicology: a historical perspective. 2nd ed. In: Gupta RC, editor. *Veterinary Toxicology*. San Diego, CA, USA: Elsevier/Academic Press (2012). p. 604–8.
6. McClellan RO. Chapter 2. Concepts in veterinary toxicology. 2nd ed. In: Gupta RC, editor. *Veterinary Toxicology*. San Diego, CA, USA: Elsevier/Academic Press (2012). p. 604–8.
7. Wilkinson L. *Animals and Disease: An Introduction to the History of Comparative Medicine*. New York: Cambridge University Press (2005).
8. Stahlheim OHV. *The Winning of Animal Health: 100 Years of Veterinary Medicine*. Ames, IA: Iowa State University Press (1994).
9. Gundert-Remy U, Barth H, Bürkle A, Degen GH, Landsiedel R. Toxicology: a discipline in need of academic anchoring – the point of view of the German Society of Toxicology. *Arch Toxicol* (2015) 89:1881–93.
10. Davis LE. Role of clinical pharmacology in veterinary medicine. In: Short CR, editor. *Proceding First Symposium of Veterinary Pharmacology and Therapeutics*. Baton Rougem, USA: Louisiana State Univiversity (1978). 147 p.
11. Anadón A, Martinez-Larrañaga MR, Castellano V. Chapter 34. Biomarkers of drug toxicity. In: Gupta RC, editor. *Biomarkers in Toxicology*. San Diego, CA, USA: Elsevier/Academic Press (2014). p. 593–607.
12. Anadón A, Castellano V, Martinez-Larrañaga MR. Chapter 55. Biomarkers in drug safety evaluation. In: Gupta RC, editor. *Biomarkers in Toxicology*. San Diego, CA, USA: Elsevier/Academic Press (2014). p. 923–45.
13. Boekelheide K, Schuppe-Koistinen I. SOT/EUROTOX debate: biomarkers from blood and urine will replace traditional histopathological evaluation to determine adverse responses. *Toxicol Sci* (2012) 129(2):249–55.
14. Marrer E, Dieterle F. Promises of biomarkers in drug development – A reality check. *Chem Biol Drug Des* (2007) 69(6):381–94.
15. Anadón A, Martinez-Larrañaga MR, Castellano V. Chapter 10. Regulatory aspects for the drugs and chemicals used in food producing animals. 2nd ed. In: Gupta RC, editor. *Veterinary Toxicology*. San Diego, CA, USA: Elsevier/Academic Press (2012). p. 135–57.
16. Brown SA. Perspectives in clinical veterinary pharmacology. *J Vet Pharmacol Ther* (1997) 20(Suppl 1):121–6.
17. Kinter LB, Valentin JP. Safety pharmacology and risk assessment. *Fundam Clin Pharmacol* (2002) 16(3):175–82.

18. Baggot DJ. *Principles of Drug Disposition in Domestic Animals the Basis of Veterinary Clinical Pharmacology*. Philadelphia, USA: Saunders (1977).
19. EMA. *Guideline for the Demonstration of Efficacy for Veterinary Medicinal Products Containing Antimicrobial Substances*. Committee for Medicinal Products for Veterinary Use (CVMP) (2016). EMA/CVMP/627/2001-Rev.1.
20. Toutain PL, Lees P. Integration and modelling of pharmacokinetic and pharmacodynamic data to optimise dosage regimens in veterinary medicine. *J Vet Pharmacol Ther* (2004) 27:467–77.
21. Gupta RC. Basic and clinical principles. 2nd ed. In: Gupta RC, editor. *Veterinary Toxicology*. San Diego, CA, USA: Elsevier/Academic Press (2012). p. 604–8.
22. Merck. *The Merck Veterinary Manual*. 10th ed. Whitehouse Station, NJ, USA: Merck & Co, Inc (2010).
23. Osweiler GD. *Toxicology. The National Veterinary Medical Series*. Media, PA, USA: Lippincott Williams & Wilkins (1996).
24. EMA. *Guideline on Quality Data Requirements for Veterinary Medicinal Products Intended for Minor Use or Minor Species (MUMS)/Limited Market*. (2016). EMA/CVMP/QWP/128710/2004-Rev.1.
25. NRC. *National Research Council of the National Academies. Toxicity Testing in the 21st Century: A Vision and a Strategy*. Washington, DC: National Academies Press (2007).
26. Anadón A, Martínez MA, Castellano V, Martínez-Larrañaga MR. The role of *in vitro* methods as alternatives to animals in toxicity testing. *Expert Opin Drug Metab Toxicol* (2014) 10(1):67–79.
27. Lord PG. Progress in applying genomics in drug development. *Toxicol Lett* (2004) 149:371–5.
28. National Research Council of the National Academies (NRC). *Committee on Applications of Toxicogenomic Technologies to Predictive Toxicology and Risk Assessment*. Washington, DC: National Academies Press (US) (2007).
29. Davidov E, Holland J, Marple E, Naylor S. Advancing drug discovery through systems biology. *Drug Discov Today* (2003) 8:175–83.
30. Witkamp RF. Genomics and systems biology – how relevant are the developments to veterinary pharmacology, toxicology and therapeutics? *J Vet Pharmacol Ther* (2005) 28:235–45.
31. Bhogal N, Grindon C, Combes R, Balls M. Toxicity testing: creating a revolution based on new technologies. *Trends Biotechnol* (2005) 23:299–307.
32. Pugsley MK, Toward R, Authire S, Gallacher DJ, Curtis MJ. Innovation in safety pharmacology testing. *J Pharmacol Toxicol Methods* (2011) 64(1):1–6.
33. Boxall AB, Fogg LA, Blackwell PA, Kay P, Pemberton EJ, Croxford A. Veterinary medicines in the environment. *Rev Environ Contam Toxicol* (2004) 180:1–91.
34. Lees P, Fink-Gremmels J, Toutain PL. Veterinary pharmacology: history, current status and future prospects. *J Vet Pharmacol Ther* (2013) 36:105–15.
35. ECDC, EMEA. *The Bacterial Challenge – Time to React a Call to Narrow the Gap between Multidrug-Resistant Bacteria in the EU and Development of New Antibacterial Agents*. Solna: ECDC & EMEA Joint Press Release (2009).
36. Voss A, Loeffen F, Bakker J, Klaassen C, Wulf M. Methicillin resistant *Staphylococcus aureus* in pig farming. *Emerg Infect Dis* (2005) 11:1965–6.
37. Liu YY, Wang Y, Walsh TR, Yi LX, Zhang R, Spencer J, et al. Emergence of plasmid-mediated colistin resistance mechanism MCR-1 in animals and human beings in China: microbiological and molecular biological study. *Lancet Infect Dis* (2015) 16:161–8.
38. Hoogenboom R, Zeilmaker M, Eijeren J, Kan K, Megelers M, Luykx D, et al. Kaolinic clay derived PCDD/Fs in the feed chain from a sorting process for potatoes. *Chemosphere* (2010) 78:99–105.
39. Qin Y, Lv X, Li J, Qi G, Diao Q, Liu G, et al. Assessment of melamine contamination in crop, soil and water in China and risks of melamine accumulation in animal tissues and products. *Environ Int* (2010) 36:446–52.

40. Leeman WR, Van der Berg KJ, Houben GF. Transfer of chemicals from feed to animal products. The use of transfer factors in risk assessment. *Food Addit Contam* (2007) 24:1–13.

41. Dorne JL, Fernandez-Cruz ML, Bertelsen U, Renshaw DW, Peltonen K, Anadón A, et al. Risk assessment of coccidostatics during feed cross-contamination: animal and human health aspects. *Toxicol Appl Pharmacol* (2013) 270(1):196–208.

CHAPTER 2

Potential Role of Pet Cats As a Sentinel Species for Human Exposure to Flame Retardants

Luis A. Henríquez-Hernández[1], Elena Carretón[2], María Camacho[1], José Alberto Montoya-Alonso[2], Luis D. Boada[1,3], Verónica Bernal Martín[1], Yaiza Falcón Cordón[2], Soraya Falcón Cordón[2], Manuel Zumbado[1] and Octavio P. Luzardo[1,3]*

[1]*Toxicology Unit, Research Institute of Biomedical and Health Sciences (IUIBS), Universidad de Las Palmas de Gran Canaria, Las Palmas, Spain*

[2]*Internal Medicine Service, Faculty of Veterinary Medicine, Research Institute of Biomedical and Health Sciences (IUIBS), Universidad de Las Palmas de Gran Canaria, Las Palmas, Spain*

[3]*Spanish Biomedical Research Centre in Physiopathology of Obesity and Nutrition (CIBERObn), Las Palmas, Spain*

ABSTRACT

Flame retardants are a wide group of chemicals used by the industry to avoid combustion of materials. These substances are commonly found in plastics, electronic equipment, fabrics, and in many other everyday articles. Subsequently, ubiquitous environmental contamination by these common chemical is frequently reported. In the present study, we have evaluated the level of exposure to polychlorinated biphenyls (PCBs), brominated diphenyl ethers (BDEs), and organophosphorous flame retardants (OPFRs) in pet cats through the analysis of their serum. We also analyzed the level exposure to such chemicals in a series of 20 cat owners, trying to disclose the role of pet cats as sentinel species of human exposure to FRs. Our results showed that PCBs, banned 40 years ago, showed the lowest levels of exposure, followed by BDEs—banned recently. Congeners PCB-138 and PCB-180 were detected in ≥50% of the series, while BDE-47 was detected in near 90% of the pet cats. On the other hand, the highest levels were that of OPFRs, whose pattern of detection was similar to that observed in humans, thus suggesting a potential role of cats as a sentinel species for human exposure to these currently used FRs. Six out of 11 OPFRs determined [2-ethylhexyldiphenyl phosphate, tributylphosphate, triisobutylphosphate, triphenylphosphate, tris (2-chloroethyl) phosphate, and tris (2-chloroisopropyl) phosphate] were detected in 100% of the samples. It will be interesting to perform future studied aimed to elucidating the potential toxicological effects of these highly detected chemicals both, in cats and humans.

Keywords: persistent organic pollutants, polychlorinated biphenyls, organophosphorus compounds, brominated flame retardants, pets, biomonitoring of pollutants

INTRODUCTION

Flame retardants (FRs) are chemical additives in materials to prevent combustion and to delay the spread of fire after ignition (1). There are two types of FRs, the additive FRs, which are mixed with the base material and are mainly of mineral origin, and reactive FRs, which are bonded to the base material, and are usually organohalogenated or organophosphorus compounds (2). Many halogenated chemicals, such as some brominated flame retardants (BFRs) and polychlorinated biphenyls (PCBs), are environmentally persistent, bioaccumulative, and/or toxic in the environment, as well as toxic to animals and humans. Phosphate flame retardants (PFRs) have been used commercially since the 1970s in a variety of products as possible substitutes for bromine-containing formulations (3), and currently these are the most used worldwide.

Polychlorinated biphenyls are polyhalogenated aromatic hydrocarbons that have been widely used in a variety of applications: closed systems such as electrical capacitors and transformers, paints, rubbers, glues, wire coatings, insulating tapes, thermal paper, fungicides and other pesticides, or fertilizers among many others (4). These compounds are volatile and evaporate easily from water surfaces and their movement through the atmosphere, resulting in widespread dispersal into the environment (5). Moreover, due to their chemical characteristics they accumulate (and biomagnify) in the food chain and the ingestion of PCBs with food in animals and human being has been reported to be relevant, leading to bioaccumulation throughout their lives (6). PCB production decreased and eventually ceased in the 1970s (7). In most Western countries (including Spain), PCBs were banned in the late 1980s (8, 9). However, thousands of tons are still releasing from obsolete electric and electronic equipment (10, 11), and detectable levels of such compounds are nowadays found in serum of humans (12) and other animals (13).

Brominated flame retardants (BFRs) are industrial chemicals produced to fire-protect construction materials, indoor decorations, furniture, textiles, electronics, and electrical appliances (14). One of the major classes of BFRs are the polybrominated diphenyl ethers (poly-BDEs) (15). Unlike PCBs, for which it has been established that dietary intake is the most important route of exposure, BFRs also have indoor dust as a significant source of exposure via inhalation and hand-to-mouth activities, of special concern for small children (16) and cats (17). Despite the fact that the Stockholm Convention on Persistent Organic Pollutants listed some BDEs (tetra-, penta-, hexa-, and hepta-BDEs) in Annex A—elimination—they are often detected in the environment (13). As PCBs, BDEs are considered endocrine disrupters (18) and have been associated with the development of diseases in humans and also in other species such as pet cats (19).

Phosphate flame retardants can be divided in three main groups: inorganic PFRs, organic PFRs (including organophosphate esters, phosphonates, and phosphinates), and halogenated PFRs (20). Organophosphorous flame retardants (OPFRs) are currently considered as more environmentally friendly and safer than the brominated-based flame retardants. As a consequence, PFRs are being increasingly employed in consumer products. However, chemicals such as tris(2-chloroethyl)phosphate (TCEP) or tris(1,3-dichloro-2-propyl)phosphate (TDCPP) are considered dangerous for the environment (21–23). Although PFRs have been present in industrial formulations since 40 years ago, information about their environmental fate or their effects in humans and biota are still scarce.

The aim of the present study was to assess the serum levels of PCBs, BDEs, and OPFRs in pet cats from Gran Canaria (Canary Islands, Spain). Since cat pets have been closely associated with the indoor environment, and besides due to their grooming habits, cats have been classically used to evaluate the level of human exposure to BDEs through the indoor dust, serum levels of such compounds were also determined in a series of humans that share environment with them, trying to evaluate the correlation among species, and to disclose whether pet cats may be sentinels of human contamination by these pollutants.

MATERIALS AND METHODS

Sampling

Cat blood samples were prospectively collected between October and November 2016. The blood was extracted through cephalic vein puncture. Twenty-two pet cats visiting the Veterinary Hospital of the Faculty of Veterinary of the University of Las Palmas de Gran Canaria (Canary Islands, Spain) for routine care were included in this study. We employed heparinized tubes for the blood extraction, and serum was immediately obtained by centrifugation. Serum samples were kept frozen at −20°C until sample preparation for chemical analysis.

Human blood samples (n = 20) were collected from volunteers, which fulfilled the condition of being cat owners (although not the owners of these cats), and being residents in Gran Canaria. Blood collection of these volunteers took place between October and December 2016 in the Insular Hospital (Gran Canaria, Canary Islands, Spain). All human volunteers provided their written informed consent to participate in this study. The cat owners also signed an informed consent of the participation of their pet cats in this study. We obtained the approval from the Research and Ethics Committee of our Institution (Universidad de Las Palmas de Gran Canaria, Canary Islands, Spain).

Reagents and Analytes of Interest

A total of 37 chemical pollutants were determined all cat samples. The 11 OPFRs were also determined in human serum samples. We have considered also sums of groups of contaminants as follows: Sum of marker PCBs, sum of those congeners considered as markers of environmental contamination for PCBs (M-PCBs, IUPAC congeners #28, 52, 101, 138, 153, and 180); sum of dioxin-like PCBs (DL-PCBs), sum of the 12 dioxin like PCBs measured (IUPAC congeners #77, 81, 105, 114, 118, 123, 126, 156, 157, 167, 169, and 189); sum of BDEs, sum of congeners #28, 47, 85, 99, 100, 153, 154, and 183; sum of OPFRs, sum of 2-ethylhexyldiphenyl phosphate, tri(2-ethylhexyl) phosphate, tributylphosphate (TBP), triethylphosphate, tiisobutylphosphate, triphenylphosphate (TPhP), tris((2-chloro-1-chloromethyl)ethyl)phosphate, tris (2-butoxyethyl)phosphate (TBOEP), TCEP, tris (2-chloroisopropyl)phosphate, and tricresyl phosphate.

We prepared stock solutions of each compound in cyclohexane (1 mg/mL), which were stored at −20°C. Diluted solutions from 0.03 to 20 ng/mL were used for calibration curves (10 points) in cyclohexane containing 1% olive oil as analyte protectant (especially for organophosphorous compounds), as described previously (23, 24). PCB 202 and diazinon-d10 were employed as internal standards. All the standards were neat compounds and were acquired from Dr. Ehrenstorfer Reference Materials (Augsburg, Germany). All the organic solvents were of mass spectrometry grade (Merck, Darmstadt, Germany), and the ultrapure water was produced in the laboratory using a Milli-Q Gradient A10 apparatus (Millipore, Molshein, France). The rest of the reagents employed in this work were acquired from Sigma-Aldrich (St. Louis, USA).

Extraction and Clean-up Procedure

One-milliliter aliquots of plasma samples were subjected to solid-phase extraction as previously reported for animal serum samples (25, 26). Briefly, samples were mixed with 1 mL of water/n-propanol (prepared in a proportion 85/15, v/v) and applied to 200 mg (3 mL) Chromabond® C18ec columns (Macherey-Nagel, Germany). These solid-phase extraction columns were mounted in a manifold (Waters Corporation, USA), which was operated under vacuum. Fist, the columns were cleaned and conditioned using 2 mL methanol and 2 mL of water/n-propanol at a constant flow rate of 1.5 mL/min. After the application of serum samples, the columns were washed two times with water/n-propanol (500 μL), and the columns were allow to dry leaving them during 30 min under vacuum. The elution of the compounds of interest was performed with two volumes of cyclohexane (250 μL each). Finally, the solvent of the extract were evaporated to dryness under a nitrogen stream. The dry residues were then re-solubilized in 200 μL of cyclohexane (27) and employed in the subsequent chromatographic analyses. Analyte recoveries were good, as all were in the range 89–107%.

Procedure of Chemical Analysis

We employed a GC System 7890B equipped with a 7693 Autosampler (Agilent Technologies, Palo Alto, CA, USA) for gas chromatographic separations. Two fused silica ultra-inert capillary columns Agilent J&WHP-5MS (Crosslinked 5% phenylmethylpolysiloxane, Agilent Technologies) each with a length of 15 m, 0.25 mm i.d., and a film thickness of 0.25 μm were connected in series and used as the stationary phase. Both columns were connected by a Purged Ultimate Union (PUU; Agilent Technologies). Helium (99.999%) at a constant flow rate of 1.0 mL/min for column 1 was used as the carrier gas. We employed the back-flushing technique during the chromatographic separations. In this technique the low volatility analytes and matrix components that are retained at the head of the column because of their low mobility under the chromatographic conditions used are eliminated through a purge valve. This is done by reversing the helium flow in the first chromatographic column once the last analyte of interest has been eluted from it. The oven temperature program was programmed as follows: (a) 60°C held for 1 min; (b) increase to 170°C at a rate of 40°C/min; (c) increase to 310°C at a rate of 10°C/min to 310°C; (d) 3 min hold time; and (e) cool down to 60°C. Each chromatographic analysis lasted 20.75 min. Injector and transfer line were set at 280°C. Standards and samples were injected (1 µl) in the splitless mode using a 4-mm ultrainert liner with glass wool (Agilent Technologies).

The detection of the analytes was performed using a Triple Quad 7010 mass spectrometer (Agilent Technologies, Palo Alto, CA, USA). We employed chlorpyrifos-methyl (Rt = 9.143 min) as the time reference to adjust the retention times of the analytes in this method. For this we employed the Retention Time Locking feature of the Agilent equipment used. All the analytical details of this method have been previously published (23). Nitrogen (99.99%) was used as the collision gas. Collision gas flow was set at 1.5 mL/min. The QqQ mass spectrometer was operated under the previously described conditions (28). The quantification was done using ten-point calibration curves. These calibration curves were constructed using a least-squares linear regression from the injection of standard solutions (1% olive oil), ranging from 0.03 to 20 ng/mL. The limits of quantification varied among compounds and ranged from 0.03 to 0.15 ng/mL (23, 28). The results of this study have been expressed in nanogram contaminant per gram of lipid weight (ng/g lw) for the normalization of the results.

Quality Assurance and Quality Control (QA/QC)

Each sample was analyzed in duplicate and the means were used for the calculations. Each 16 vials (8 samples) we included 3 controls: (a) a reagent blank consisting of cyclohexane with 1% olive oil; (b) a vial consisting of cyclohexane containing 1% olive oil, but also containing a concentration of 2 ng/mL of each of the analytes of interest; and (c) a procedural blank consisting of a sample of lyophilized human serum (Medidrug Basis Line, Medichem, Germany), which was reconstituted and fortified at a concentration of 10 ng/mL of each analyte under study and extracted using same procedure described above for the samples. The results of the analyses were considered acceptable since the concentration of the analytes of the procedural blank were in all cases within 15% of the deviation from the theoretical value.

Statistical Analysis

For the management of the database and statistical analyses, we employed PASW Statistics v 19.0 (SPSS Inc., Chicago, IL, USA). We checked the normality of the data the Kolmogorov–Smirnov test. As the distributions of the pollutants lacked normality and homoscedasticity, we used non-parametric tests (the Mann–Whitney and Kruskal–Wallis tests) in the subsequent analyses. The relationships between the categorical variables were examined by means of the chi-square test. The results were reported as mean ± SD and median. Probability levels of less than 0.05 (two-tailed) were considered statistically significant.

RESULTS

Serum Levels of PCBs in Pet Cats and Humans

A total of 18 PCBs—6 M-PCBs and 12 DL-PCBs—were measured in the serum of 22 pet cats and 22 humans included in the study. A total of six compounds were detected in the series of cats (33.3%), although four of them were detected in ≤10% of pet cats (Table 1). In the case of humans, only five congeners were detected, and three were detected in more than 70% of samples. In the case of cats, only PCB-138 (detected in 50% of the series) and PCB-180 (detected in 18.2% of the series)—both considered as M-PCBs—were measured in a significant number of subjects. In fact, the highest concentrations were observed for these two congeners (138 and 153), which accounted for almost half of the measured total PCBs concentration, followed by PCBs 101 and 180, and all of them are not dioxin-like PCBs (M-PCBs). Median of sum M-PCBs was 0.35 ng/g fat, similar to the median of sum Total-PCBs. Thus, the contribution of DL-PCBs to the body burden was scarce in cats (Table 1). The concentrations detected in human serum were significantly higher for PCB-138, PCB-153, PCB-180, and for the sum of M-PCBs and total PCBs. The median value of the total sum of PCBs was around 50 times higher in human serum than in cat serum.

Table 1. Concentrations of polychlorinated biphenyls (ng/g fat) in the whole series of cats (n = 22) and humans (n = 20).

Congener	Pet cats					Humans					P
	n	(%)	Mean	SD	Median	n	(%)	Mean	SD	Median	
PCB-28	0	0	–	–	–	0	0	–	–	–	–
PCB-52	0	0	–	–	–	0	0	–	–	–	–
PCB-77	0	0	–	–	–	0	0	–	–	–	–
PCB-81	0	0	–	–	–	0	0	–	–	–	–
PCB-101	2	9.1	0.96	3.52	0.00	0	0	–	–	–	–
PCB-105	0	0	–	–	–	0	0	–	–	–	–
PCB-114	0	0	–	–	–	0	0	–	–	–	–
PCB-118	1	4.5	0.16	0.75	0.00	1	5	0.04	0.12	0.00	n.s.
PCB-123	0	0	–	–	–	0	0	–	–	–	–
PCB-126	1	4.5	0.44	2.07	0.00	1	5	0.03	0.09	0.00	n.s.
PCB-138	11	50.0	8.66	33.27	0.35	18	90	4.82	4.56	3.45	0.0224*
PCB-153	2	9.1	8.95	38.27	0.00	14	70	7.66	7.58	6.05	0.0013**
PCB-156	0	0	–	–	–	0	0	–	–	–	–
PCB-157	0	0	–	–	–	0	0	–	–	–	–
PCB-167	1	4.5	0.24	1.11	0.00	0	0	–	–	–	–
PCB-169	0	0	–	–	–	0	0	–	–	–	–
PCB-180	4	18.2	0.67	2.19	0.00	16	80	9.44	9.21	6.45	<0.0001****
PCB-189	0	0	–	–	–	0	0	–	–	–	–
∑M-PCBs	11	50.0	20.08	78.31	0.35	18	90	21.92	21.14	15.95	<0.0001****
∑DL-PCBs	2	9.1	0.84	3.23	0.00	2	10	0.23	0.71	0.00	n.s.
∑Tot-PCBs	11	50.0	20.08	78.31	0.35	18	90	22.15	21.62	15.95	<0.0001****

*$P < 0.05$; **$P < 0.01$; ****$P < 0.001$.

PCBs, polychlorinated byphenils.

∑M-PCBs: sum of marker PCBs (IUPAC congeners #28, 52, 101, 138, 153, and 180).

∑DL-PCBs: sum of dioxin-like PCBs (IUPAC congeners #77, 81, 105, 114, 118, 123, 126, 156, 157, 167, 169, and 189).

∑Tot-PCBs: sum of all 18 PCBs cited above.

Serum Levels of BDEs in Pet Cats and Humans

A total of 8 BDEs—tri-BDEs (congener #28), tetra-BDEs (congener #47), penta-BDEs (congeners #85, 99, and 100), hexa-BDEs (congeners #153 and 154), and hepta-BDEs (congener #183)—were measured in the serum of 22 pet cats and 20 humans included in the study. In cats, a total of five compounds were detected (62.5%), although four of them were detected in ≤10% of pet cats (Table 2). Only two of the compounds detected in cat serum were also detected in human serum samples. BDE-153 was detected in 10% of human samples and was not detected in cats. Only BDE-47 was measured in a significant number of subjects (n = 19, 86.4% in cats; n = 10, 50% in humans). Median values of BDE-47 were 3.75 ng/g fat in cat serum, and 0.90 ng/g fat in human serum, and this congener was the main contributor to the total body burden of BDEs in both species (4.45 and 2.2 ng/g fat in cats and humans, respectively, Table 2). The median value of the total sum of BDEs was around half in human serum than in cat serum.

Table 2. Concentrations of brominated diphenyl ethers (ng/g fat) in the whole series of cats (*n* = 22) and humans (*n* = 20).

Congener	Pet cats					Humans					*P*
	n	(%)	Mean	SD	Median	*n*	(%)	Mean	SD	Median	
BDE-28	0	0	–	–	–	0	0	–	–	–	–
BDE-47	19	86.4	4.86	4.36	3.75	10	50	1.34	1.42	0.9	0.0002***
BDE-85	2	9.1	0.08	0.25	0.00	0	0	–	–	–	–
BDE-99	1	4.5	0.09	0.41	0.00	0	0	–	–	–	–
BDE-100	0	0	–	–	–	0	0	–	–	–	–
BDE-153	0	0	–	–	–	2	10	0.28	0.86	0.00	–
BDE-154	1	4.5	0.18	0.83	0.00	0	0	–	–	–	–
BDE-183	2	9.1	0.26	1.00	0.00	0	0	–	–	–	–
ΣBDEs	23	95.5	5.48	4.35	4.45	13	65	1.62	1.41	2.2	<0.0001****

****P* < 0.005; *****P* < 0.001.
BDEs, brominated diphenyl ethers.

Serum Levels of OPFRs in Pet Cats and Humans

To our knowledge, this is the first time describing serum levels of OPFRs in pet cats. A total of 11 compounds were measured in the serum of 22 pet cats and 20 humans included in the study. A total of six compounds were detected in 100% of the samples, and other one was detected in 82% of cats and 90% of humans (TBOEP). Only one compound, tris ((2-chloro-1-chloromethyl)ethyl) phosphate, was not detected in any cat (Table 3). However, it is striking the case of tricresyl phosphate, which was detected in 77.3% of cats and in none of the human samples. 2-Ethylhexyldiphenyl phosphate and tris (2-chloroisopropyl) phosphate were the pollutants detected at the highest concentrations in both species. On the other hand, triethylphosphate, detected in 50% of pet cats, was the pollutant with lowest concentration (median concentrations of 0.60 ng/g fat in cats and 0.00 ng/g fat in humans). Sum OPFR in cats was 909.25 ng/g fat and 877.72 ng/g fat in humans. These results of cats and humans were not significantly different. The median values of the sum OPFRs reported here is at least 200 times higher than serum levels of BDEs and almost 2,500 times higher than median values of PCBs.

When we compared the concentrations of OPFRs found in the serum of cats with those found in human serum we found a similar profile of distribution (Figure 1) between both species. The levels in humans were slightly lower in human serum than in cat serum, and for two of these compounds the differences were statistically significant (Table 3). However, as shown in Figure 2, the grouped bar graph of the averaged concentrations found in both species show that the proportions were similar for the majority of these compounds, except for tricresyl phosphate, which was found only in cat samples.

Table 3. Concentrations of organophosphorous flame retardants (ng/g fat) in the whole series of cats (n = 22) and humans (n = 20).

Congener	Pet cats					Humans					P
	n	(%)	Mean	SD	Median	n	(%)	Mean	SD	Median	
2-Ethylhexyldiphenyl phosphate	22	100	617.64	273.30	535.20	20	100	177.6	567.3	425.8	0.0008***
Tri(2-ethylhexyl) phosphate	6	27.3	0.53	1.34	0.00	10	50	1.15	1.49	0.4	n.s.
Tributylphosphate	22	100	93.13	62.46	72.80	20	100	58.26	19.06	64.80	n.s.
Triethylphosphate	11	50.0	2.96	6.93	0.60	6	30	1.54	3.22	0.00	n.s.
Triisobutylphosphate	22	100	77.05	72.92	48.50	20	100	48.66	21.32	47.70	n.s.
Triphenylphosphate	22	100	29.58	15.39	24.65	20	100	18.52	6.83	22.67	n.s.
Tris ((2-chloro-1-chloromethyl)ethyl)phosphate	0	0	–	–	–	0	0	–	–	–	–
Tris (2-butoxyethyl)phosphate	18	81.8	53.08	42.00	44.90	18	90	106.3	149.8	56.41	n.s.
Tris (2-chloroethyl)phosphate	22	100	12.10	17.43	7.15	20	100	3.12	1.27	3.69	0.0003***
Tris (2-chloroisopropyl)phosphate	22	100	150.40	111.97	114.45	20	100	83.65	28.94	93.97	n.s.
Tricresyl phosphate	17	77.3	13.41	12.91	10.40	0	0	–	–	–	–
∑OPFRs	22	100	1049.88	558.93	909.25	22	100	712.12	304.8	877.72	n.s.

***$P < 0.005$.
OPFRs, organophosphorous flame retardants.

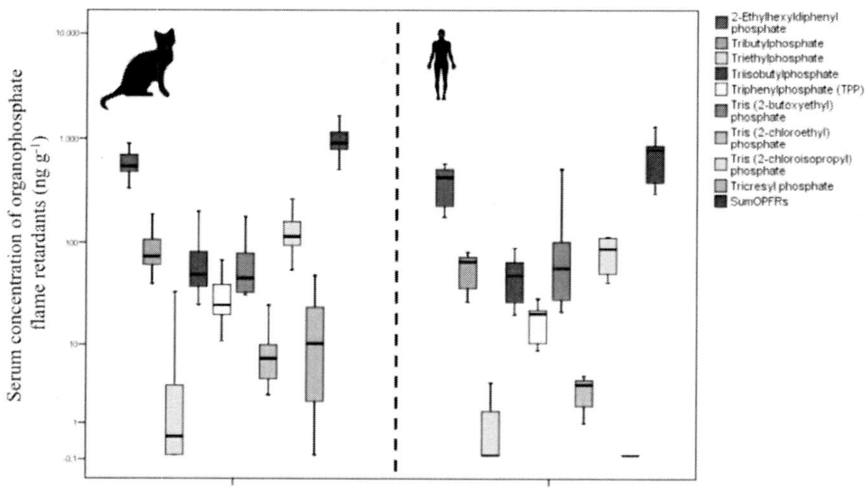

FIGURE 1. BOX PLOT SHOWING THE SERUM LEVELS OF THE OPFRS DETECTED IN ≥50% OF THE SAMPLES, AMONG CATS (N = 22) AND HUMANS (N = 20). The lines connect the medians, the boxes cover the 25th to 75th percentiles, and the minimal and maximal values are shown by the ends of the bars.

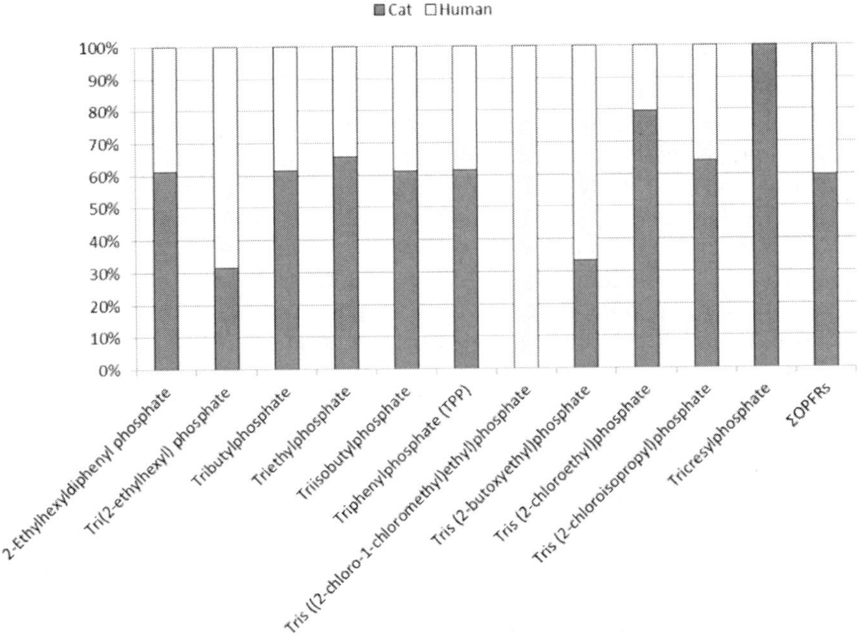

FIGURE 2. STACKED BAR GRAPH SHOWING THE PROPORTIONALITY OF THE AMOUNT OF EACH OPFR BETWEEN CAT AND HUMAN SERUM (EXPRESSED AS PERCENTAGE).

DISCUSSION

Serum levels of different types of flame retardants—including PCBs, BDEs, and OPFRs—were determined in a cohort of pet cats aimed to describe the exposure of such animals to these compounds, especially regarding to OPFRs, which have been less studied in this species. The same compounds were determined in the serum of 20 pet cat owners (although they were not the owners of the cats included in this study, due to logistical limitations), with the aim of comparing the concentrations found, and to disclosing whether pet cats may serve as sentinels of human exposure to these pollutants.

Regarding the group of legacy flame retardants—PCBs—only PCB 138 and PCB 180 were detected in a significant number of subjects and at very low concentrations. The profile of contamination by these legacy flame retardants is very similar between cats and humans, but the concentrations found in the latter are several times higher. Besides, PCB congeners were also more frequently detected in human beings than in pet cats.

The profile of PCB contamination we found in this study is very similar to those reported in other previous studies in cats. However, the median values reported here were much lower than those in previous studies. Thus, in a cohort of 20 cats from Pakistan, PCB congeners (101, 118, 138, and 153) were detected in ≥70% of the series (29); and, while in that study a median value for ∑PCBs of 36 ng/g lipid weight was reported, in this study we have found a median value which is 100 times lower (0.35 ng/g lipid weight). It is probable that in Pakistan a more recent and continued exposure to obsolete electric and electronic equipment (EEE) and the waste they generate (WEEE) have existed. Until mid 1980s, EEE used to contain high amounts of PCBs, as they were employed as flame retardants in those devices. It has been reported that in certain areas (especially urban areas) of developing countries, exposure to PCBs is high due to the exposure to WEEE (11). Also in previous studies in cats from France (13) or USA (19), the cats also exhibited higher serum levels of PCBs, and also a higher variety of congeners detected. However, it has been reported that a geographical gradient in the level of PCB contamination exists, and the pattern varies greatly from region to region (30). Thus, the apparent discrepancy in the level of PCB contamination in cats from the Canary Islands with those from other parts of the world has a parallelism with the values observed in humans from this archipelago, in which also lower values of PCBs than those in other populations have been reported (12). It should be noted that the economy of the Canary Islands is based fundamentally on a few economic sectors: tourism and, to a much lesser extent, farming and fishing, and that other economic sectors, such as traditional polluting industries, have a limited presence in the Islands (31), contributing to explain the low levels of PCBs detected in human and pet cats. In any case, serum levels of ∑PCBs were the lowest among all the FRs included in this study, which is logical because, although these are very persistent chemical, they were banned for 40 years. This result is consistent with those published by other authors worldwide (12, 13, 29). Although the comparison between cats and humans indicate that the profile of contamination in very similar between both species, pet cats exhibited much lower concentrations of those. Thus, pet cats do not seem to be good sentinels of the human exposure to this group of pollutants, as previously reported also for pet dogs (28, 32, 33).

With respect to the second most abundant group of flame retardants in cat serum—the BDEs—it is important to note that, similarly to what occur with PCBs, most of them have been banned, although more recently than PCBs. Due to this ban, different authors have reported a decreasing trend in the environmental and serum levels along the last decade. Currently, the only compound of the BDE family that is still approved for use is BDE-209 (www.pops.int), and it has to be taken into account that unfortunately this chemical was not included in this study, because it can not be determined in our laboratory due to technical limitations. In any case, and even without having included the BDE 209, our results are very similar to those reported in cats from Pakistan (median ∑BDEs = 6.1 ng/g fat) (29), USA—Georgia, Massachusetts, and North Carolina—(median ∑BDEs = 5.9 ng/g fat) (17), but lower than those reported in Sweden (median ∑BDEs = 24.1 ng/g fat) (34), and than in those cats from California, in which the authors reported extraordinarily high levels of BDEs (median ∑BDEs = 2904 ng/g fat), mainly due to the high levels of BDE-99 detected in those animals (19). While diet, mainly through fish intake, is considered the main route of exposure for PCBs (23, 35–37), the ingestion of dust from the environment is considered the most important route of exposure for BDEs in cats (38). This may explain why, although the profile of contamination between humans and cats are very similar, the concentrations of congeners are slightly (and significantly) higher in cats than in humans. Thus, due to their grooming habits, cats have been classically used to evaluate the level of human exposure to BDEs through the indoor dust (39). In this sense, a positive correlation between serum cat levels and dust has been consistently reported in the literature (34), although the serum levels tend to be several times higher in cats than in humans. Compared to humans, ∑BDE levels in cats are often 20- to 100-fold greater than median levels in US adults (17, 38), but the correlation of levels and the pattern of contamination (congener distribution and proportions) between both species has been reported to be very good. In our study this pattern is maintained, and the concentrations found are much more similar between both species than those reported in other studies (median value of sum BDEs is only double in cats than in humans of this study). Hence, pet cats have been proposed as sentinels to better assess human exposure and adverse health outcomes related to low level but chronic BDE exposure (17). Our study is consistent with previous ones and seems to reinforce the hypothesis that pet cats are good sentinels for human exposure to this group of flame retardants.

Finally, with regard to the group of flame retardants most abundant in cat and human serum, the OPFRs, it should be noted that to date there are very few studies that have determined them in blood or serum of higher vertebrates, and as far as we know this is the first report in cats and, as well as one of the few available in humans. Despite this limitation, our results agree with the scarce data available in humans, which have reported a high frequency of detection (70–100%) of TBOEP, TPhP, TCEP, and TBP (40). High amounts of OPFRs have been detected in dust from houses (41), schools (42), and other everyday places (43). According to our results, when comparing cats to humans, the profile of exposure was comparable among both species (Figures 1 and 2). Although some compounds, similarly to what occurs with BDEs, were found at higher concentrations in serum of pet cats than in human serum, in general terms both patterns of exposure were virtually overlapped (Figure 1), suggesting that cats may play a potential role of cats as sentinels of the human exposure to OPFRs. Although the concentrations of some of these compounds were significantly different between cats and humans, the differences in concentration were relatively small, and when we summed the concentrations of all of the OPFRs the significance disappeared. However, it is worth to mention the case of TCP, because this compound was detected at a very high concentration in cats, but it was completely absent in humans (Figure 2). It is not easy to find an explanation to this discrepancy, because of the lack of supporting bibliography. It is worth to note, however, that the reports on environmental levels of OPFRs are extremely variable, not only regarding the concentrations, but also in the compounds mixture. In the case of TCP, it has been reported that this compound is frequently associated with urban areas, where a high disposal of WEEE exists (44). However, due to the wide range of current applications of OFPRs and the wide variety of mixtures employed, reports for TCP and other chemicals may vary from undetected to account for more than 50% of the ∑OPFR (45, 46). This could explain the disparities observed among studies (especially regarding to amount of chemicals detected, but also the composition of the mixtures). Taking this into account, a plausible hypothesis to explain the discrepancy between cats and cat owners of this study is the fact that, due to logistical reasons, we could not recruit serum from the cats and their owners. In this case, the humans recruited for this study were cat owners but not the owners of these cats, and therefore the samples of both species may not be directly compared. However, despite this limitation, we think that the findings are of interest due to the global contamination of OPFRs and the daily exposure to these compounds. Besides, our results allow to suggest that cats are good sentinels, not only of human exposure to BDEs, but also to OPFRs. Moreover, pet cat and human exposures to OPFRs are of concern since they may pose a risk for health [i.e., TCEP is classified as carcinogen (category 3) (47)].

CONCLUSION

Serum levels of different flame retardants were measured in serum of pet cats and cat owners. Our results showed that PCBs, banned 40 years ago, showed the lowest levels of exposure in both species, although the levels in humans were almost 50 times higher. The frequencies of detection and levels were followed by BDEs, which were banned much more recently. In the case of the compounds within this group the profile of contamination between humans and cats were very similar, and also the concentrations were closer, although higher in cats. Finally, OPFRs were frequently detected, and at the highest concentrations in our series of cats and humans and both, the pattern of detection and concentrations were similar between both species. Our study indicates that, unlike what occurs for PCBS, pet cats are good sentinel of human exposure to BDEs and OPFRs. It seems that further research is needed to elucidating the biotransformation routes of cats and humans, as well as studies aimed to disclose, which is toxicological relevance of such a high exposure to potentially toxic chemicals.

ETHICS STATEMENT

The study was approved by the ethics committee of the University of Las Palmas de Gran Canaria. Informed consent was obtained from human participants. Informed consent was obtained from animal owners.

AUTHOR CONTRIBUTIONS

LH-H: laboratory analysis of PCBs, first draft preparation, and manuscript approval. EC: sample obtention, clinical examination of cats, and manuscript revision and approval. MC: laboratory analysis of BDEs, first draft preparation, and manuscript revision and approval. JM-A: sample obtention, clinical examination of cats, and manuscript revision and approval. LB: collaboration in manuscript preparation, design of figures, and manuscript revision and approval. VBM: sample preparation and manuscript revision and approval. YFC: sample preparation and manuscript revision and approval. SFC: sample preparation and manuscript revision and approval. MZ: laboratory analysis of OPFRs and manuscript revision and approval. OL: study design and supervision and manuscript preparation and approval.

CONFLICT OF INTEREST STATEMENT

The authors declare that the research was conducted in the absence of any commercial or financial relationships that could be construed as a potential conflict of interest.

ACKNOWLEDGMENTS

The authors thank Mrs. María de los Reyes Suárez Hanna for her technical assistance.

REFERENCES

1. Kemmlein S, Hahn O, Jann O. Emissions of organophosphate and brominated flame retardants from selected consumer products and building materials. *Atmos Environ* (2003) 37(39–40):5485–93.
2. EFRA. *Frequently Asked Questions*. Brussels: EFRA - The European Flame Retardants Association (2007).
3. Horrocks AR, Davies PJ, Kandola BK, Alderson A. The potential for volatile phosphorus-containing flame retardants in textile back-coatings. *J Fire Sci* (2007) 25:523–40.
4. Safe S. Polychlorinated biphenyls (PCBs), dibenzo-p-dioxins (PCDDs), dibenzofurans (PCDFs), and related compounds: environmental and mechanistic considerations which support the development of toxic equivalency factors (TEFs). *Crit Rev Toxicol* (1990) 21(1):51–88.
5. Headrick ML, Hollinger K, Lovell RA, Matheson JC. PBBs, PCBs, and dioxins in food animals, their public health implications. *Vet Clin North Am Food Anim Pract* (1999) 15(1):109–31, ix–x.
6. Safe SH. Polychlorinated biphenyls (PCBs): environmental impact, biochemical and toxic responses, and implications for risk assessment. *Crit Rev Toxicol* (1994) 24(2):87–149.
7. Tanabe S. PCB problems in the future: foresight from current knowledge. *Environ Pollut* (1988) 50(1–2):5–28.
8. Kimbrough RD. Human health effects of polychlorinated biphenyls (PCBs) and polybrominated biphenyls (PBBs). *Annu Rev Pharmacol Toxicol* (1987) 27:87–111.
9. Kimbrough RD. Polychlorinated biphenyls (PCBs) and human health: an update. *Crit Rev Toxicol* (1995) 25(2):133–63.
10. García-Álvarez N, Martín V, Fernández A, Almunia J, Xuriach A, Arbelo M, et al. Levels and profiles of POPs (organochlorine pesticides, PCBs, and PAHs) in free-ranging common bottlenose dolphins of the Canary Islands, Spain. *Sci Total Environ* (2014) 493:22–31.
11. Luzardo OP, Boada LD, Carranza C, Ruiz-Suarez N, Henriquez-Hernandez LA, Valeron PF, et al. Socioeconomic development as a determinant of the levels of organochlorine pesticides and PCBs in the inhabitants of Western and Central African countries. *Sci Total Environ* (2014) 49(7–498):97–105.
12. Henriquez-Hernandez LA, Luzardo OP, Almeida-Gonzalez M, Alvarez-Leon EE, Serra-Majem L, Zumbado M, et al. Background levels of polychlorinated biphenyls in the population of the Canary Islands (Spain). *Environ Res* (2011) 111(1):10–6.
13. Braouezec C, Enriquez B, Blanchard M, Chevreuil M, Teil MJ. Cat serum contamination by phthalates, PCBs, and PBDEs versus food and indoor air. *Environ Sci Pollut Res Int* (2016) 23(10):9574–84.
14. Alaee M, Arias P, Sjodin A, Bergman A. An overview of commercially used brominated flame retardants, their applications, their use patterns in different countries/regions and possible modes of release. *Environ Int* (2003) 29(6):683–9.
15. EFSA. Scientific opinion on polybrominated diphenyl ethers (PBDEs) in Food. *EFSA J* (2011) 9(5):2156.

16. Lunder S, Hovander L, Athanassiadis I, Bergman A. Significantly higher polybrominated diphenyl ether levels in young U.S. children than in their mothers. *Environ Sci Technol* (2010) 44(13):5256–62.
17. Dye JA, Venier M, Zhu L, Ward CR, Hites RA, Birnbaum LS. Elevated PBDE levels in pet cats: sentinels for humans? *Environ Sci Technol* (2007) 41(18):6350–6.
18. UNEP/WHO. *The State-of-the-Science of Endocrine Disrupting Chemicals – 2012*. Geneva: WHO (2013).
19. Guo W, Gardner S, Yen S, Petreas M, Park JS. Temporal changes of PBDE levels in California house cats and a link to cat hyperthyroidism. *Environ Sci Technol* (2016) 50(3):1510–8.
20. Andresen JA, Grundmann A, Bester K. Organophosphorus flame retardants and plasticisers in surface waters. *Sci Total Environ* (2004) 332(1–3):155–66.
21. Cequier E, Sakhi AK, Marce RM, Becher G, Thomsen C. Human exposure pathways to organophosphate triesters – a biomonitoring study of mother-child pairs. *Environ Int* (2015) 75:159–65. doi:10.1016/j.envint.2014.11.009
22. Kosarac I, Kubwabo C, Foster WG. Quantitative determination of nine urinary metabolites of organophosphate flame retardants using solid phase extraction and ultra performance liquid chromatography coupled to tandem mass spectrometry (UPLC-MS/MS). *J Chromatogr B Analyt Technol Biomed Life Sci* (2016) 1014:24–30.
23. Henriquez-Hernandez LA, Montero D, Camacho M, Gines R, Boada LD, Ramirez Bordon B, et al. Comparative analysis of selected semi-persistent and emerging pollutants in wild-caught fish and aquaculture associated fish using Bogue (*Boops boops*) as sentinel species. *Sci Total Environ* (2017) 581–582:199–208.
24. Lynam K, Smith D. *Ultra Inert (UI) Wool Liner Performance Using an Agilent J&W DB-35ms UI Column with and without an Analyte Protectant for Organophosphorus (OP) Pesticides*. (2011). Available from: https://www.agilent.com/cs/library/applications/5990-8235EN.pdf
25. Kakuschke A, Valentine-Thon E, Griesel S, Gandrass J, Perez Luzardo O, Dominguez Boada L, et al. First health and pollution study on harbor seals (*Phoca vitulina*) living in the German Elbe estuary. *Mar Pollut Bull* (2010) 60(11):2079–86.
26. Camacho M, Orós J, Henríquez-Hernández LA, Valerón PF, Boada LD, Zaccaroni A, et al. Influence of the rehabilitation of injured loggerhead turtles (*Caretta caretta*) on their blood levels of environmental organic pollutants and elements. *Sci Total Environ* (2014) 487:436–42. doi:10.1016/j.scitotenv.2014.04.062
27. Kouzayha A, Al Iskandarani M, Mokh S, Rabaa AR, Budzinski H, Jaber F. Optimization of a solid-phase extraction method using centrifugation for the determination of 16 polycyclic aromatic hydrocarbons in water. *J Agric Food Chem* (2011) 59(14):7592–600.
28. Henriquez-Hernandez LA, Carreton E, Camacho M, Montoya-Alonso JA, Boada LD, Valeron PF, et al. Influence of parasitism in dogs on their serum levels of persistent organochlorine compounds and polycyclic aromatic hydrocarbons. *Sci Total Environ* (2016) 562:128–35.
29. Ali N, Malik RN, Mehdi T, Eqani SA, Javeed A, Neels H, et al. Organohalogenated contaminants (OHCs) in the serum and hair of pet cats and dogs: biosentinels of indoor pollution. *Sci Total Environ* (2013) 449:29–36.
30. Henriquez-Hernandez LA, Luzardo OP, Pérez-Arellano JL, Carranza C, Sanchez NJ, Almeida-Gonzalez M, et al. Different pattern of contamination by legacy POPs in two populations from the same geographical area but with completely different lifestyles: Canary Islands (Spain) vs. Morocco. *Sci Total Environ* (2016) 541:51–7.
31. Zumbado M, Goethals M, Alvarez-Leon EE, Luzardo OP, Cabrera F, Serra-Majem L, et al. Inadvertent exposure to organochlorine pesticides DDT and derivatives in people from the Canary Islands (Spain). *Sci Total Environ* (2005) 339(1–3):49–62.
32. Ruiz-Suárez N, Rial C, Boada LD, Henríquez-Hernández LA, Valeron PF, Camacho M, et al. Are pet dogs good sentinels of human exposure to environmental polycyclic aromatic hydrocarbons, organochlorine pesticides and polychlorinated biphenyls? *J Appl Anim Res* (2016) 44:143–55.

33. Henriquez-Hernandez LA, Carreton E, Camacho M, Montoya-Alonso JA, Boada LD, Valeron PF, et al. The heartworm (*Dirofilaria immitis*) seems to be able to metabolize organochlorine pesticides and polychlorinated biphenyls: a case-control study in dogs. *Sci Total Environ* (2017) 575:1445–52.
34. Norrgran Engdahl J, Bignert A, Jones B, Athanassiadis I, Bergman A, Weiss JM. Cats' internal exposure to selected brominated flame retardants and organochlorines correlated to house dust and cat food. *Environ Sci Technol* (2017) 51(5):3012–20.
35. Bocio A, Domingo JL, Falcó G, Llobet JM. Concentrations of PCDD/PCDFs and PCBs in fish and seafood from the Catalan (Spain) market: estimated human intake. *Environ Int* (2007) 33(2):170–5.
36. Rodriguez-Hernandez A, Camacho M, Henriquez-Hernandez LA, Boada LD, Ruiz-Suarez N, Valeron PF, et al. Assessment of human health hazards associated with the dietary exposure to organic and inorganic contaminants through the consumption of fishery products in Spain. *Sci Total Environ* (2016) 557–558:808–18.
37. Rodriguez-Hernandez A, Camacho M, Henriquez-Hernandez LA, Boada LD, Valeron PF, Zaccaroni A, et al. Comparative study of the intake of toxic persistent and semi persistent pollutants through the consumption of fish and seafood from two modes of production (wild-caught and farmed). *Sci Total Environ* (2017) 575:919–31.
38. Guo W, Park JS, Wang Y, Gardner S, Baek C, Petreas M, et al. High polybrominated diphenyl ether levels in California house cats: house dust a primary source? *Environ Toxicol Chem* (2012) 31(2):301–6.
39. Chow K, Hearn LK, Zuber M, Beatty JA, Mueller JF, Barrs VR. Evaluation of polybrominated diphenyl ethers (PBDEs) in matched cat sera and house dust samples: investigation of a potential link between PBDEs and spontaneous feline hyperthyroidism. *Environ Res* (2015) 136:173–9.
40. Li P, Jin J, Wang Y, Hu J, Xu M, Sun Y, et al. Concentrations of organophosphorus, polybromobenzene, and polybrominated diphenyl ether flame retardants in human serum, and relationships between concentrations and donor ages. *Chemosphere* (2017) 171:654–60.
41. Mizouchi S, Ichiba M, Takigami H, Kajiwara N, Takamuku T, Miyajima T, et al. Exposure assessment of organophosphorus and organobromine flame retardants via indoor dust from elementary schools and domestic houses. *Chemosphere* (2015) 123:17–25.
42. Cristale J, Hurtado A, Gomez-Canela C, Lacorte S. Occurrence and sources of brominated and organophosphorus flame retardants in dust from different indoor environments in Barcelona, Spain. *Environ Res* (2016) 149:66–76.
43. Tokumura M, Hatayama R, Tatsu K, Naito T, Takeda T, Raknuzzaman M, et al. Organophosphate flame retardants in the indoor air and dust in cars in Japan. *Environ Monit Assess* (2017) 189(2):48.
44. He CT, Zheng J, Qiao L, Chen SJ, Yang JZ, Yuan JG, et al. Occurrence of organophosphorus flame retardants in indoor dust in multiple microenvironments of southern China and implications for human exposure. *Chemosphere* (2015) 133:47–52.
45. Stapleton HM, Klosterhaus S, Eagle S, Fuh J, Meeker JD, Blum A, et al. Detection of organophosphate flame retardants in furniture foam and U.S. house dust. *Environ Sci Technol* (2009) 43(19):7490–5.
46. Ali N, Dirtu AC, Van den Eede N, Goosey E, Harrad S, Neels H, et al. Occurrence of alternative flame retardants in indoor dust from New Zealand: indoor sources and human exposure assessment. *Chemosphere* (2012) 88(11):1276–82.
47. ECHA. *TRIS (2-Chloroethyl) Phosphate, TCEP. Summary Risk Assessment Report.* Dortmund: European Chemicals Agency (2008).

CHAPTER 3

The Effect of Supplementing Air-Dried *Moringa stenopetala* Leaf to Natural Grass Hay on Feed Intake and Growth Performances of Arsi-Bale Goats

Aberra Melesse [1,*], Degnet H/Meskel [2], Sandip Banerjee [1], Aster Abebe [1] and Amsalu Sisay [1]

[1]*School of Animal and Range Sciences, Hawassa University, P.O. Box 05, Hawassa, Ethiopia*

[2]*Wachamo University, Hossana, Ethiopia*

ABSTRACT

The most constraining factor in goat production in the tropics is underfeeding mainly attributed to limitations of feed both in quantity and quality. This study was conducted to assess the effect of supplementing different levels of air-dried *Moringa stenopetala* leaf (MSL) as a protein source on nutrient intake and growth performances of Arsi-Bale male goats. A total of 24 yearling goats with average initial body weight of 13.6 ± 0.25 kg were used in the study. The goats were blocked by live weight into four groups (*n* = 6 per group) and the groups were then randomly allocated into four supplemented treatments. All goats received a basal diet of natural grass hay *ad libitum* and 340 g/head/day concentrate. The treatments were the control diet with no supplementation (Treatment 1, T1) and diets supplemented with MSL at a rate of 120 g/head/day (Treatment 2, T2), 170 g/head/day (Treatment 3, T3) and 220 g/head/day (Treatment 4, T4). The duration of the experiment was 75 days. The results indicated that the average daily feed intake was ($p < 0.001$) higher in goats supplemented with T3 and T4 diets. The total dry matter, organic matter, and crude protein intakes of goats fed with T3 and T4 supplementations were ($p < 0.001$) also higher than those reared in T1 and T2 diets. Goats reared in T3 and T4 diets had lower ($p < 0.05$) feed conversion ratio than those fed with T1 and T2 diets. The final body weight in goats reared in T3 and T4 diets was 18.2 kg and 18.5 kg, respectively, being ($p < 0.05$) higher than those of T1 (15.8 kg) and T2 (16.3 kg). The average daily weight gain in goats fed with T3 and T4 diets was 111 and 114 g/goat/day, respectively, which was ($p < 0.05$) higher than those reared in the control (T1) (54.0 g/goat/day) and T2 (58.1 g/goat/day) diets. It can thus be concluded that goats reared at high level of MSL supplementation (T3 and T4) had better nutrient intake, feed conversion efficiency and growth performances, suggesting its potential as a good protein supplement to natural grass hay at the farmer's management level.

Keywords: Arsi-Bale goats; natural grass hay; *Moringa stenopetala* leaf; nutrient intake; supplementation; weight gain

1. INTRODUCTION

Ethiopia has the largest livestock population among African countries and has different agro-ecological zones ranging from semi-arid to arid environments which are suitable for goat production. There are approximately 570 breeds and types of goats in the world of which 89 are found in Africa [1]. The present estimated population of goats in Ethiopia is 29.3 million [2]. Goats play important roles in communities by improving the livelihood of resource-challenged farmers by creating alternative employment opportunities, enhancing family income by sale of live animals, skin, manure *etc.* Apart from contributing to human nutrition in form of meat and milk, they also act as an income buffer to the risks associated with erratic climatic changes [3]. In addition, demand for goat's meat is on the rise throughout the world, particularly in Ethiopia, due to increased human population coupled with income growth particularly in the cities. Regardless of their good attributes, production of goats remains low due to diseases, poor genotype, management, and nutrition; nevertheless, there is ample scope for improving the productivity of goats [4].

One of the most constraining factors in goat production in the tropics is underfeeding, which is mainly attributed to limitations of feed in both quantity and quality. The situation is exacerbated during the dry season when natural pastures usually dry out and are overgrazed, resulting in low contents of protein and energy in the fodder. As a result, large flocks of productive livestock cannot be maintained on such low quality feeds to meet the basic maintenance requirements of the animals. It is therefore important to supplement the meager available fodder with some amount of concentrates in order to improve intake and digestibility of such poor quality feed resources. However, the use of such supplements is usually limited under smallholder livestock production systems due to their inaccessibility and high cost.

As a result, in order to mitigate the problems associated with the lack of dietary protein, there is a need to look for some alternative but cheap sources of protein. Previous works have reported that leaves of multipurpose trees growing in the backyards of the farmers can be used as a cheap source protein for supplementing livestock [5]. In this regard, leaves from the Moringa tree may serve as a multipurpose protein source for livestock. The uniqueness of this tree is that it does not shed its leaves during the dry season when other sources of fodder become scarce [6]. It is a multipurpose tree cultivated both for human food and animal feed and is found in abundance in certain parts of Southern Ethiopia. Recent studies conducted by [7,8] and [9] have indicated that the leaves of *Moringa stenopetala* are rich in protein (28.2%–36.1%) and contain substantial amounts of essential amino acids as well as macro and trace minerals [10].

There are several studies reporting on the use of the fresh foliage of *Moringa* spp. as a source of livestock feed. According to the report of [11], supplementation of fresh *Moringa oleifera* leaves has resulted in the improvement of milk yield in crossbred cattle. Promising results have been obtained from inclusion of air-dried *M. stenopetala* leaf supplementation to the basal diet of sheep [12] and growing chicken [8,13]. However, information on the feeding value of *M. stenopetala* leaf on the performance of goats in the tropical countries, particularly in Ethiopian, is lacking. Thus, this research was conducted to assess the effect of supplementing air-dried *M. stenopetla* leaf meal to natural grass hay on feed intake and body weight performances of Arsi-Bale goats.

2. MATERIALS AND METHODS

2.1. Experimental Site

The experiment was carried out at the sheep and goat farm of School of Animal and Range Sciences, Hawassa University (Ethiopia), which lies geographically between 7°5′ N latitude and 38°29′ E longitude at an altitude of 1700 m above sea level. The average rainfall ranges from 800 to 1100 mm. The mean minimum and maximum temperatures in the study area are 13.5 °C and 27.6 °C, respectively [14].

2.2. Preparation of Experimental Feeds

Fresh *Moringa stenopetala* leaves were collected from the trees owned by the farmers in Mirab Abaya (district town) which is situated near Arbaminch city. The district is situated between 6°4′ N latitude and 37°34′ E longitude with the mean altitude of 1220 m above the sea level. The fresh leaves were harvested from the available trees regardless of the age of the tree. The leaves were trimmed from its twigs and then shade dried on a plastic sheet. During the drying process, regular turning of leaves was done to ensure uniform drying. The air-dried Moringa leaves were finally transported to the experimental site and ground into coarse powder which hereafter is referred to as air-dried *Moringa stenopetala* leaf (MSL). The powder was packed in bags of 100 kg and stored until used. Natural grass hay was bought from a nearby private farm and hand chopped into the size of 3 to 5 cm for ease of feeding. Samples of MSL and various ingredients were subjected to chemical analysis before being used in the formulation of experimental diets.

2.3. Experimental Animals and Their Management

In this experiment, 24, about one-year-old (age determined by dentition) Arsi-Bale intact male goats were used with an average initial body weight of 13.6 ± 0.25 kg. Goats were purchased from local market and quarantined for a fortnight, during which they were treated with 250 mg of albendazole (Chengdu Qiankun Veterinary Pharmaceuticals Co. Ltd., Chengdu, China), administered through drenching gun for de-worming the animals. Moreover, 1.5 mL Oxytetracycline (Chengdu Qiankun Veterinary Pharmaceuticals Co. Ltd., Chengdu, China) was provided intravenously for three days to treat the animals from coughing caused due to transportation from the market to the experimental station. The barns were also sprayed with Diazinone (Chengdu Qiankun Veterinary pharmaceuticals col. Ltd., Chengdu, China) to take care of any ecto-parasite infestation.

2.4. Experimental Design and Treatment Diets

The feeding trial was a completely randomized design (CRD) consisting of one control and three supplemental treatment diets with six goats within each treatment. The trial was conducted for 75 days exclusive of 15 days of adaptation period. All the experimental goats had *ad libitum* access to natural grass hay and water. They were also provided with a concentrate at a rate of 2.5% of their body weight; this was done according to the recommendation of [15]. Consequently, as the average initial body weight of goat was 13.6 kg, the calculated concentrate supplied per goat per day was 340 g. The diets therefore contained 340 g concentrate only (treatment, T1), 340 g concentrate plus 35% MSL supplementation (120 g) (treatment 2, T2), 340 g concentrate plus 50% MSL supplementation (170 g) (treatment 3, T3), and 340 g concentrate plus 65% MSL supplementation (220 g) (treatment 4, T4). Accordingly, the total concentrate/MSL mixture offered to T1, T2, T3, and T4 treatment groups were 340, 460, 510, and 560 g/head day, respectively. The concentrate was a mixture of 50% wheat bran, 35% maize, 14% Noug (*Guizotia abyssinica*) seed cake, and 1% salt. Natural grass hay and the concentrate/MSL mixture were offered separately.

2.5. Data Collection

At the end of the quarantine period, the goats were weighed early in the morning (prior to being offered any feed) for two consecutive days and the body weight was averaged. Then the goats were blocked into control and supplemented groups (n = 6 per group) using initial body weight and the groups were then randomly allocated into four supplemented treatments. Animals were ear tagged and housed in individual pens and each animal was provided with feeder and watering trough.

Animals were offered the concentrate/MSL mixture twice a day at 8:00 a.m. and 1:00 p.m. in equal portions throughout the experimental period. Samples of daily feeds offered and refused were collected, measured and pooled over experimental period for each feed and animal and stored in plastic bags. Representative samples of the offered and refused feeds for the different treatments were then taken for chemical analysis. The daily average feed intake of hay and concentrate/ MSL mixture was determined separately by the difference between the amounts of feed offered and refused and then averaged. Body weight of each goat was recorded every fortnight after overnight fasting to determine body weight change. Average individual daily weight gain was calculated as the difference between final body weight and initial body weight of the goat divided by the experimental days.

2.6. Chemical Analysis of the Feed

The analysis of ash and moisture contents was conducted using the method of [16]. The dry matter content of the feed was determined by drying the samples at 105 °C overnight in a force air oven. Ash was determined by combusting the samples at 550 °C for 5 h in a muffle furnace. Nitrogen (N) was determined using the Keldjhal method and then the crude protein (CP) was calculated as N × 6.25 [16]. The acid detergent fiber (ADF) and neutral detergent fiber (NDF) contents were analyzed using the method of [17] in an ANKOM® 200 Fiber Analyzer (ANKOM Technology Corp., Fairport, NY, USA). All samples were analyzed in duplicates at Animal Nutrition Laboratory of Hawassa University.

2.7. Statistical Analysis

The collected data were subjected to the General Linear Models of SAS (SAS, 2010, version 9.3, SAS Institute Inc., Cary, NC, USA). When significant differences were observed among treatment means, they were separated by Duncan Multiple Range Test. Comparisons with $p < 0.05$ were considered significant and all statements of statistical differences were based on this level unless noted otherwise. The following linear model summarizes the statistics employed to analyze the data:

$$Yij = \mu + Ai + Aj + eij$$

where *Yij* is the observed *j* variables in the *i*th supplement, μ is the overall mean of the observed variable, *Ai* is the effect due to *i*th supplement levels (*i* = 0, 120, 170, 220 g/head/day), and *eij* is the random residual error

3. RESULTS

3.1. Chemical Composition of Experimental Feed

Chemical composition of the experimental feeds is presented in Table 1. The crude protein (CP) content of the *M. stenopetala* leaf was the highest followed by that of the Noug seed cake; however the lowest value was observed for the hay offered. The dry matter (DM) content of the feed was more or less similar across the feed ingredients. The organic matter (OM) content was lowest for the *M. stenopetala* leaf while it was highest for maize. The results also indicated that the ether extract (EE) values were highest for Noug seed cake and lowest for hay. The neutral detergent fiber (NDF) values were highest for hay and the lowest for *M. stenopetala* leaf.

Table 1. Proximate composition of ingredients of the concentrate, natural grass hay, and air-dried *Moringa stenopetala* leaf.

Nutrients (%, DM Basis)	Ingredients of the Concentrate Mix			Natural Grass Hay	MSL
	Maize	Wheat Bran	Noug Seed Cake		
Dry matter	93.3	93.6	93.9	95.6	94.4
Organic matter	98.1	88.5	90.1	94.0	88.0
Crude protein	6.50	16.3	25.0	5.10	29.5
Ether extract	9.60	5.42	11.5	2.86	5.90
NDF	10.0	36.5	32.5	61.2	17.8
ADF	4.90	17.5	21.5	25.7	16.5

MSL: air-dried *Moringa stenopetala* leaf meal; NDF: neutral detergent fiber; ADF: acid detergent fiber; DM: Dry matter.

3.2. Nutrient Contents of the Experimental Treatments

The results from the Table 2 indicate that the CP content was highest in T4 diet supplemented with 220 g/head/day MSL and lowest in non-supplemented control diet (T1). The DM content of the feed was similar across the treatments while the organic matter (OM) and ether extract (EE) content was lowest in the control (T1) and the highest values were observed in the goats supplemented with 220 g/head/day MSL diet (T4) while the reverse was true for ash, acid detergent fiber (ADF), and NDF contents.

3.3. Nutrient Intakes

As presented in Table 3, goats reared in T3 and T4 diets had ($p < 0.01$) higher intakes of CP, DM, OM, NDF, and ADF for natural grass hay than those fed with T1 and T2. For concentrate/MSL mixtures, the highest and lowest DM, OM, and CP intakes were observed in goats reared in T4 and T1, respectively. Conversely, the NDF and ADF intakes for concentrate/MSL was highest in goats reared in the control diet (T1), but lowest in those fed with the T3 diet. The total intakes of DM, OM, and CP were ($p < 0.001$) higher in goats fed with T4 than those reared in other treatment diets. These parameters were also ($p < 0.001$) higher in goats reared in T3 than those in T2 and T1 diets and goats fed with T2 had significantly higher DM, OM, and CP intakes than those of T1. On the other hand, goats reared in the control diet (T1) had ($p < 0.001$) higher ADF intake than those supplemented with different levels of MSL. The goats fed with T2 diet had ($p < 0.001$) higher NDF intake than other treatments.

Table 2. The average proximate compositions (%) of treatment diets composed of natural grass hay, concentrate supplemented with different levels of air-dried *Moringa stenopetala* leaf.

Treatments	DM	Ash	OM	CP	EE	NDF	ADF
T1	92.3	6.20	88.0	16.0	5.7	69.3	56.1
T2	92.7	7.10	92.0	16.7	6.10	45.5	38.3
T3	93.6	8.01	92.9	17.2	8.70	36.5	23.8
T4	94.8	8.03	93.8	18.7	9.50	33.6	22.5

DM, dray matter; OM, organic matter; CP, crude protein; EE, ether extract; NDF, neutral detergent fiber; ADF, acid detergent fiber; T1, no supplementation; T2, 120 g/head/day supplementation with air-dried *Moringa stenopetala* leaf; T3, 170 g/head/day supplementation with air-dried *Moringa stenopetala* leaf; T4, 220 g/head/day supplementation with air-dried *Moringa stenopetala* leaf.

3.4. Body Weight Change and Feed Conversion Efficiency

While there were no significant differences between the supplemented groups at the beginning of the experiment, significant differences in body weight were recorded across the different treatment diets at the later stages of the experiment (Table 4). Accordingly, the non-supplemented goats (T1) had the lowest final body weight while those reared on T4 supplementation had the highest. Goats supplemented with T3 and T4 diets had significantly higher values for the final body weight, total and daily weight gain than those reared in T1 and T2 diets. The values also indicated that there was no significant difference between those receiving T3 and T4 supplemented diets. Moreover, growth parameters of goats reared in T1 did not differ from those of T2, indicating two distinctly truncated categories.

As shown in Table 4, the feed intake values were ($p < 0.001$) higher in goats supplemented with increasing levels of MSL (T3 and T4) than those of the non-supplemented (T1) and less supplemented (T2) groups. No significant differences were noted between goats supplemented with T3 and T4 levels and those fed with the T1 and T2 diets. The FCR value was lower in goats reared in T3 and T4, ($p < 0.05$) than those fed with T1 and T2 diets. However, FCR was similar ($p > 0.05$) between goats reared in T1 and T2 diets; and between those in T3 and T4 diets.

Table 3. The effects of hay supplemented with different levels of air-dried *Moringa stenopetala* leaf on nutrient intakes of Arsi-Bale goats.

Parameters (g/head/day)	T1	T2	T3	T4	SE	P
Hay intake						
Dry matter	291 b	272 b	372 a	404 a	22.3	**
Organic matter	287 b	268 b	366 a	397 a	21.9	**
Crude protein	15.5 b	14.7 b	19.8 a	21.3 a	1.21	**
NDF	187 b	174 b	239 a	259 a	14.3	**
ADF	78.3 b	73.3 b	100 a	108 a	6.00	**
Supplemental intake						
Dry matter	281 d	321 c	370 b	420 a	3.73	***
Organic matter	267 d	319 c	367 b	417 a	3.64	***
Crude protein	48.7 d	57.8 c	68.0 b	83.2 a	0.689	***
NDF	211 a	158 b	144 d	150 c	1.604	***
ADF	170 a	133 b	94.5 d	100 c	1.	***
Total intake						
Dry matter	572 c	594 c	742 b	825 a	23.7	***
Organic matter	544 c	587 c	733 b	814 a	23.4	***
Crude protein	64.3 d	72.4 c	87.8 b	105 a	1.55	***
NDF	397 a	332 b	383 a	408 a	14.9	**
ADF	249 a	206 b	194 b	208 b	6.355	***

[a,b,c,d] Means within the same row with different superscript letters are significantly different ($p < 0.05$); ** $p < 0.01$; *** $p < 0.001$; SE: standard error of the mean; NDF: neural detergent fiber; ADF: acid detergent fiber; T1: no supplementation; T2: 120 g/head/day supplementation with air-dried *Moringa stenopetala* leaf; T3: 170 g/head/day supplementation with air-dried *Moringa stenopetala* leaf; T4: 220 g/head/day supplementation with air-dried *Moringa stenopetala* leaf.

Table 4. The effects of supplementation of different levels of air-dried *Moringa stenopetala* leaf on feed intake and growth performances of Arsi-Bale goats.

Growth Performance	T1	T2	T3	T4	SE	P
Initial body weight (kg/head)	13.5	13.8	13.6	13.7	0.481	NS
Final body weight (kg/head)	15.8 b	16.3 b	18.2 a	18.5 a	0.544	**
Total weight gain (kg/head)	2.27 b	2.44 b	4.65 a	4.80 a	0.439	***
Average daily gain (g/head)	54.0 b	58.1 b	111 a	114 a	10.4	***
Average daily feed intake (g/head)	609 b	632 b	785 a	841 a	24.8	***
FCR (g feed/g gain)	12.6 a	11.5 a	7.69 b	7.41 b	1.49	*

[a,b] Means within the same row with different superscript letters are significantly different ($p < 0.05$); * $p < 0.05$; ** $p < 0.01$; *** $p < 0.001$; SE = standard error of the mean; FCR: feed conversion ratio; NS: Not significantly different ($p > 0.05$); T1: no supplementation; T2: 120 g/head/day supplementation with air-dried *Moringa stenopetala* leaf; T3: 170 g/head/day supplementation with air-dried *Moringa stenopetala* leaf; T4: 220 g/head/day supplementation with air-dried *Moringa stenopetala* leaf.

4. DISCUSSION

4.1. Nutrient Composition of M. stenopetala and Experimental Feed

The CP values of MSL as assessed in the current study are similar to those reported by [7,8,10,13]. The CP values are also similar to the findings of [18] for *M. oleifera* leaves and [10] for *M. stenopetala* leaf meal. However, the CP content in *M. stenopetala* and *M. oleifera* leaves as reported by [19,20] are lower than those obtained from the present study while high values were reported by [9], which may be attributed to the maturity stage of leaves as the young leaves usually have higher protein values than the older ones. The CP content in the Moringa leaves in the present study indicated that they can be considered good protein supplements to low quality forages in the smallholder livestock production system. The EE values in the present study are similar to those reported by [21,22] for the leaves of *M. stenopetala* and *M. oleifera*, respectively. However, lower EE values were reported by [8] for *M. stenopetala* leaf. This could be due to differences in Moringa species, maturity stage of leaves, and type of soils in which the plans were grown.

The contents of NDF and ADF in MSL in the present study are comparable to the observations of [8] for the same Moringa species; but were higher than those of *M. oleifera* leaf meal reported by [22,23] The ash content in MSL as observed in this study is similar to the observations of [8,19] for *M. stenopetala* leaves. The ash content was also similar to those reported by [10,24] for *M. oleifera* leaves. However, the ash content of MSL in the present study was higher than those reported by [7,21,25]; and lower than those of [22] for *M. oleifera* leaves. Similarly, [18,26] have reported high contents of ash (14.1% and 14.8%, respectively) in the leaves of *M. oleifera.*

The comparatively high NDF and ADF contents of the natural grass hay in the present study indicate that it is of a poor nutritional quality and is not adequate to support maintenance requirements of goats as the CP content is well below the minimum level (7%) required for the microbial function [27]. The CP, DM, and OM contents of the hay used in the current study were consistent with the reports of [28]. The CP, NDF, and ADF values for wheat bran and Noug seed cake were found to be similar to those reported by [29]. The average contents of DM and ADF in Noug seed cake were comparable with the reports of [30]. However, the values for CP and ash in Noug seed cake reported by [30] were lower than observed in the present findings which might be due to the maturity stage of the Noug seed and type of soil in which the plants were grown.

The CP, EE, and NDF values of wheat bran as assessed in the current study were comparable with the observations of [31].

4.2. Nutrient Intake

Feed intake is an important parameter in determining the nutritive value of animal's feed as it account for nutrient supply. In the present study, there is significant difference in intakes of DM, OM and CP and these parameters increased with increasing supplementations level of MSL which is consistent with the findings of [11]. Sánchez [11] observed increased intake in cows supplemented with *M. oleifera* leaves to a basal diet of grass hay. Similar results have also been reported by [12] and [32] who found that supplementation of sheep with *M. stenopetala* fed with a basal diet of grass hay improved intake of DM, OM, and CP. The increased DM, OM, and CP intake with increasing levels of *M. stenopetala* supplementation may possibly be due to the lower fiber and higher CP contents of the leaves thereby enhancing the palatability of the feed. This could lead to an increase in microbial population and efficiency, thereby facilitating the rate of breakdown of the digesta which eventually leads to increment in feed intake. The low total nutrient intakes observed in non-supplemented goats might be due to the low level of CP in the hay.

4.3. Body Weight and Gain

There was improvement in final body weight, total weight gain, average daily weight gain, and feed conversion efficiency in the treatment groups with high levels of MSL supplementations (T3 and T4). The findings are consistent to those of [5] and [12] who reported increased body weight values in sheep receiving supplement containing *M. stenopetala* leaf meals. Studies by [33] also indicated linear increase in body weight of sheep supplemented with *Gliricdia sepium* leaf meal when fed a basal diet of elephant grass and Rhodes grass.

The average daily weight gain as were observed among the goats reared on T3 and T4 diets were in the range of values reported by [34] in sheep fed on a basal diet of Rhodes grass hay and those of [35] where the *M. oleifera* leaf was used as a supplementary feed for goats. The average daily weight gains are also in accordance with the reports of [36], who supplemented M. oleifera leaf meal to goats. Increment in body weight was also reported in goats fed with cassava leaf meal as compared to those reared on grass hay alone [37].

The difference in feed conversion ratios among the treatment groups are comparable with the observations of [38], who reported significantly higher values of efficiency in rabbits fed with *M. oleifera* leaf meal than those of the control diet. In this study, goats supplemented with high levels of MSL (170 and 220 g/day and goat) consumed less feed but gain more weight as compared to those reared in the control and less supplemented (120 g/head/day) groups. This implies that due to the year round availability and easy access of *M. stenopetala* tree by smallholder farmers, the leaf can be used as a very useful protein supplement to small ruminants, particularly during the dry season, where most of local feed resources become scarce.

5. CONCLUSIONS

The supplementation of *Moringa stenopetala leaf* to natural grass hay has considerably improved the nutrient intake and body weight gain of Aris-Bale goats as compared to those of the non-supplemented goats. Therefore, supplementation of low quality feed resources such as natural grass hay with Moringa leaves is a viable option for improving the productivity of local goats under smallholder farmer conditions where conventional protein supplements are beyond the reach of local farmers.

ACKNOWLEDGMENTS

This research project was fully supported by the research fund granted by the Hawssa University, Research and Development Directorate Office (now Vice President for Research and Technology Transfer) for which the authors are highly grateful. The support received from Tadesse Bekore in analyzing the feed nutrient composition is highly acknowledged.

AUTHOR CONTRIBUTIONS

The first author prepared the research proposal and solicited fund. He together with the third author closely monitored and supervised the second author who has been mainly involved in data collection and organization as a part of his Master thesis. Moreover, the first author analyzed the data using appropriate statistical packages and prepared the manuscript. The forth and third authors assisted the first author during the preparation of the research proposal.

CONFLICTS OF INTEREST

The authors declare no conflict of interest.

REFERENCES

1. Gala, S. Biodiversity in goats. Small Rumin. Res. 2005, 60, 75–81.
2. Central Statistical Agency; Federal Democratic Republic of Ethiopia. Agricultural Sample Survey, Report on Livestock and Livestock Characteristics; CSA: Addis Ababa, Ethiopia, 2014; volume 2, pp. 12–25.
3. Rumosa-Gwaze, F.G.; Chimonyo, M.; Dzama, K. Communal goat production in Southern Africa: A review. Trop. Anim. Health Prod. 2009, 41, 1157–1168.
4. Simela, L.; Merkel, R. The contribution of chevon from Africa to global meat production. Meat Sci. 2008, 80, 101–109.
5. Manaye, T.; Tolera, A.; Zewdu, T. Feed intake, digestibility and body weight gain of sheep fed Napier grass mixed with different levels of Sesbania sesban. Livest. Sci. 2009, 122, 24–29.
6. Thurber, M.D.; Fahey, J.W. Adoption of Moringa oleifera to combat under-nutrition viewed through the lens of the "Diffusion of Innovations" theory. Ecol. Food Nutr. 2009, 48, 212–225.
7. Melesse, A.; Bulang, M.; Kluth, H. Evaluating the nutritive values and in vitro degradability characteristics of leaves, seeds and seedpods from Moringa stenopetala. J. Sci. Food Agric. 2009, 89, 281–287.
8. Melesse, A.; Tiruneh, W.; Negesse, T. Effects of feeding Moringa stenopetala leaf meal on nutrient intake and growth performance of Rhode Island Red chicks under tropical climate. Trop. Subtrop. Agroecosyst. 2011, 14, 485–492.
9. Negesse, T.; Makkar, H.P.S.; Becker, K. Nutritive value of some non-conventional feed resources of Ethiopia determined by chemical analyses and an in vitro gas method. Anim. Feed Sci. Technol. 2009, 154, 204–217.
10. Melesse, A.; Steingass, H.; Boguhn, J.; Schollenberger, M.; Rodehutscord, M. Effects of elevation and season on nutrient composition of leaves and green pods of Moringa stenopetala and Moringa oleifera. Agroforest. Syst. 2012, 86, 505–518.
11. Sánchez, R.; Spröndly, E.; Ledin, I. Effect of feeding different level of foliage of Moringa oliefera to creole dairy cows on intake, digestibility, milk production and composition. Livest. Sci. 2006, 101, 24–31.
12. Gebregiorgis, F.; Negesse, T.; Nurfeta, A. Feed intake and utilization in sheep fed graded levels of dried moringa (Moringa stenopetala) leaf as a supplement to Rhodes grass hay. Trop. Anim. Health Prod. 2011, 44, 511–517.
13. Melesse, A.; Getye, Y.; Berihun, K. Effect of feeding different levels of Moringa stenopetala leaf meal on growth performance, carcass traits and some serum biochemical parameters of Koekoek chickens. Livest. Sci. 2013, 157, 498–505.
14. Banerjee, S.; Melesse, A.; Dotamo, E.; Berihun, K.; Beyan, M. Effect of feeding different dietary protein levels with Iso-Caloric ration on nutrients intake and growth performances of dual-purpose koekoeck chicken breeds. Int. J. Appl Poult. Res. 2013, 2, 27–32. Available online: http://ijscience.com/Journal_of_Applied_

Poultry_Research/pdf-files/Issue-2-2013/27-32.pdf (accessed on 27 November 2015).
15. McDonald, P.; Edwards, R.A.; Greenhalgh, J.F.D.; Morgan, C.A.; Sinclair, L.A.; Wilkinson, R.G. Animal Nutrition, 7th ed.; Pearson Education Limited: Edinburgh, UK, 2010; p. 692.
16. AOAC (Association Official Analytical Chemists). Official Methods of Analysis, 16th ed.; AOAC Inc.: Arlington, VA, USA, 1995; p. 1298.
17. Van Soest, P.J.; Robertson, J.B.; Lewis, B.A. Methods of dietary fiber, neutral detergent fiber and non-starch polysaccharides in relation to animal nutrition. J. Dairy Sci. 1991, 74, 3583–3597.
18. Kakengi, A.M.V.; Kaijage, J.T.; Sarwatt, S.V.; Mutayoba, S.K.; Shem, M.N.; Fujihara, T. Effect of Moringa oleifera leaf meal as a substitute for sunflower seed meal on performance of laying hens in Tanzania. Livest. Res. Rural Dev. 2007, 19, 120. Available online: http://www.lrrd.org/lrrd19/8/kake19120.htm (accessed on 27 November 2015).
19. Dechasa, J.; Sonder, K.; Alemayehu, L.; Mekonen, Y.; Anjulo, A. Leaf yield and nutritive value of Moringa stenopetala and Moringa oleifera accessions: Its potential role in food security in constrained dry farming agro forestry system. In Proceedings of the Moringa and Other Highly Nutritious Plant Resources: Strategies, Standards and Market for Better Impact on Nutrition in Africa, Accra, Ghana, 16–18 November 2006; pp. 1–14.
20. Yang, R.Y.; Tsou, S.C.S.; Lee, T.C.; Chang, L.C.; Kuo, G.; Lai, P.Y. Moringa a novel plant rich in antioxidants bio-available iron and nutrients. Am. Chem. Soc. Symp. Ser. 2006, 925, 224–239.
21. Odeyinka, S.M.; Oyedele, O.J.; Adeleke, T.O.; Odedire, J.A. Reproductive performance of rabbits fed Moringa oleifera as a replacement for Centrosema Pubescens. In Proceedings of the 9th World Rabbit Congress, Verona, Italy, 10–13 June 2008; pp. 411–416.
22. Asaolu, V.O.; Jarju, A.K.; Joof, A.; Odeyinka, S.M.; Manne, J.; Darboe, D.; Jallow, M. Moringa Oleifera Horticulture-Livestock Integration Approach to Improving Incomes and Livelihoods of Women Farmers: The Gambian Experience. Available online: http://hdl.handle.net/10625/44856 (accessed on 15 July 2015).
23. Olugbemi, T.S.; Mutayoba, S.K.; Lekule, F.P. Moringa oleifera leaf meal as a hypocholesterolemic agent in laying hen diets. Livest. Res. Rural Dev. 2010, 22, 84. Available online: http://www.lrrd.org/lrrd22/4/olug22084.htm (accessed on 27 November 2015).
24. Olugbemi, T.S.; Mutayoba, S.K.; Lekule, F.P. Effect of Moringa (Moringa oleifera) inclusion in Cassava based diets fed to broiler chickens. Int. J. Poult. Sci. 2010, 9, 363–367.
25. Nouala, F.S.; Akinbamijo, O.O.; Adewumi, A.; Hoffmann, E.; Muetzel, S.; Becker, K. The influence of Moringa oleifera leaves as substitute to conventional concentrate on the in vitro gas production and digestibility of groundnut hay. Livest. Res. Rural Dev. 2006, 18, 121. Available online: http://www.lrrd.org/lrrd18/9/noua18121.htm (accessed on 27 November 2015).
26. Babiker, M.S. Chemical composition of some non-conventional and local feed resources for poultry in Sudan. Int. J. Poult. Sci. 2012, 11, 283–287.
27. Van Soest, P.J. Nutritional Ecology of the Ruminant, 2nd ed.; Cornell University Press: Ithaca, Greece, 1994.
28. Banerjee, S.; Demo, K.; Abebe, A. Some serum biochemical and carcass traits of Arsi bale rams reared on graded levels of Millettia ferruginea leaf meal. World Appl. Sci. J. 2013, 28, 532–539.
29. Seyoum, B.; Zinash, S. The Composition of Ethiopian Feedstuffs; IAR: Addis Ababa, Ethiopian, 1989.
30. Gizachew, L. Crude Protein and Mineral Status of Forages Grown on Pelvic Vertisol of Ginchi, Central Highlands of Ethiopia. Ph.D. Thesis, University of the Free State, Bloemfontein, South Africa, November 2002.

31. Eshetie, T. Compatiblity of Quality Protein Maize and Sesame Seed Meal as a Substitute for Synthetic Amino Acids in Broiler Rations. Master's Thesis, Alemaya University of Agriculture, East Harerge, Ethiopia, March 2005.
32. Foster, J.L.; Adesogan, A.T.; Carter, J.N.; Blount, A.R.; Myer, R.O.; Phata, K. Intake, digestibility and N retention by sheep supplemented with warm-season legume hays or soybean meal. J. Anim. Sci. 2009, 87, 2891–2898.
33. Mpairwe, D.R.; Sabiiti, E.N.; Mugerwa, J.S. Effect of dried Gliricidia sepium leaf supplement on feed intake, digestibility and nitrogen retention in sheep fed dried KW4 elephant grass (Pennisetum purpureum) ad libitum. Agroforest. Syst. 1998, 41, 139–150.
34. Nurfeta, A. Feed intake, digestibility, nitrogen utilization and body weight change of sheep consuming wheat straw supplemented with local agricultural and agro-industrial by-products. Trop. Anim. Health Prod. 2010, 42, 815–824.
35. Aregheore, E.M. Intake and digestibility of Moringa oleifera-batiki grass mixtures by growing goats. Small Rumin. Res.2002, 46, 23–28.
36. Moyo, B.; Masika, P.J.; Muchenje, V. Effect of supplementing crossbred Xhosa lop-eared goat castrates with Moringa oleifera leaves on growth performance, carcass and non-carcass characteristics. Trop. Anim. Health Prod. 2012, 44, 801–809.
37. Tilahun, S.; Animut, G.; Urge, M. Effects of supplementing cassava leaf meal, brewers' dried grain and their mixture on body weight change and carcass traits of local goats fed urea treated tef straw. J. Livest. Sci. 2013, 4, 31–43.
38. Nuhu, F. Effect of Moringa Leaf Meal (MOLM) on Nutrient Digestibility, Growth, Carcass and Blood Indices of Weaner Rabbits. Ph.D. Thesis, Kwame Nkrumah University of Science and Technology, Kumasi, Ghana, February 2010.

CHAPTER 4

Phosphorus Bioavailability: A Key Aspect for Conserving this Critical Animal Feed Resource with Reference to Broiler Nutrition

Xiuhua Li *, Dagong Zhang, Tsung Y. Yang † and Wayne L. Bryden

Poultry Science Unit, School of Agriculture and Food Sciences, The University of Queensland, Gatton Qld 4343, Australia

ABSTRACT

Phosphorus (P) is an essential element, and the majority of animal feed phosphate is derived from phosphate rock that is a non-renewable resource. Current global P reserves may be depleted in 50–100 years. This poses the challenge of securing future P supply for the global animal feed industries. Currently, nutritionists formulate diets with substantial safety margins to guarantee that animals do not become P deficient. Excessive dietary P concentrations increase, not only the cost of diets, but also P excretion and pollution of the environment. We contend that understanding P bioavailability is central to the sustainable use of this mineral in animal agriculture. Poultry accounts for approximately 50% of animal feed phosphate consumption worldwide and for this reason we use the meat chicken or broiler as a case study to explore the nuances of P bioavailability. We conclude that, to tackle the challenge of dietary P bioavailability, cooperative research on a global scale is needed to standardise measurement procedures in order to produce a robust and reliable database which can be used by nutritionists to formulate diets to meet the bird's P requirements precisely. Achievement of this goal will assist endeavours to sustain the global supply of phosphorus.

Keywords: phosphorus; requirements; available P; bioavailability; poultry; animal nutrition; rock phosphate; global feed supply

1. INTRODUCTION

Phosphorus (P) is an essential element for all forms of life, from single to multicellular organisms. It is required for normal muscle growth and egg formation, is an important component of nucleic acids, the genetic code, and phospholipids, and is also a co-factor of many enzyme systems. Phosphorus plays a vital role in maintaining osmotic and acid-base balance, energy metabolism, amino acid metabolism and protein synthesis. Phosphorus was named "life's bottleneck" by chemist and science writer Isaac Asimov [1], who claims: "Life can multiply until all the phosphorus has gone, and then there is an inexorable halt which nothing can prevent," also writing, "We may be able to substitute nuclear power for coal, and plastic for wood, and yeast for meat, and friendliness for isolation—but for P there is neither substitute nor replacement."

The main ingredients of non-ruminant diets are cereal grains and their by-products. These diets usually require supplementation with inorganic P, because of the low concentrations, and low availability of total P in cereal feed ingredients. Meat and bone meal has relatively high bioavailable P content but is banned from use in animal feed in European countries. This policy increases the demand for inorganic feed phosphates [2]. The majority of feed phosphate is derived from phosphate rock that is a non-renewable resource and becoming increasingly scarce and expensive [3]. Current global P reserves may be depleted in 50–100 years [4,5]. This poses the challenge of securing future P supply for the international and national animal feed industries. Poultry account for approximately 50% of animal feed phosphate consumption worldwide [6]. Feed phosphates make a significant contribution to the P content of any broiler diet; generally providing as much as 60% of the bird's P requirements [7]. There is limited information on biologically determined P values in feed ingredients for poultry [8,9,10]. Therefore, nutritionists have to formulate diets with substantial safety margins to guarantee that birds do not become P deficient. Greater dietary P concentrations increase the cost of poultry production, and P excretion into manure. Excessive amounts of P in manure contribute to the pollution of lakes, streams and wetlands leading to surface water eutrophication. Eutrophication of fresh water globally is accelerated by P accumulation [11,12].

All these concerns have attracted much global attention and stimulated the re-examination of the use of inorganic P in animal diets. This issue is pertinent to all non-ruminant animals that require dietary P supplementation, especially poultry (broilers, layers and breeders) and pigs. In this review, the focus is on meat chickens or broilers as much of the P used in animal nutrition is in poultry diets. In an effort to optimise P use by poultry, the P requirements of the modern broiler have been re-examined [13,14,15,16,17,18] and defining P bioavailability in feedstuffs has commenced [2,19]. Mention of bioavailability, which encompasses digestion, absorption and utilisation of P, evokes much debate and is the thrust of this review.

In many instances, the terms "availability" and "bioavailability" are interchangeable when referring to nutrients. However, to avoid confusion in this review, we use the term bioavailability to encompass the digestion, absorption and utilisation of P, as the term "available P" has a specific connotation in P nutrition as described below. We contend that bioavailability is central to the sustainable use of P in animal agriculture. A thorough understanding of P bioavailability will permit the development of strategies to enhance efficiency of P utilisation, reduce wastage of P resources and reduce environmental pollution by P.

2. PHOSPHORUS METABOLISM

It is important when discussing phosphorus requirements and bioavailability to appreciate the metabolism of phosphorus (Figure 1) and to remember the close association that phosphorus homeostasis has with the metabolism of two other nutrients; calcium (Ca) and vitamin D. The major features of these interactions are briefly reviewed here and the interested reader is referred to detailed reviews on the topic [20,21]. The major store of Ca and P is the skeleton where the cationic and anionic forms, respectively, of these minerals connect to form hydroxyapatite, which confers rigidity on the bone matrix [2]. Bone is continually turned over and to ensure that the biological demands for both Ca and P are met, metabolism of these essential minerals is tightly coordinated.

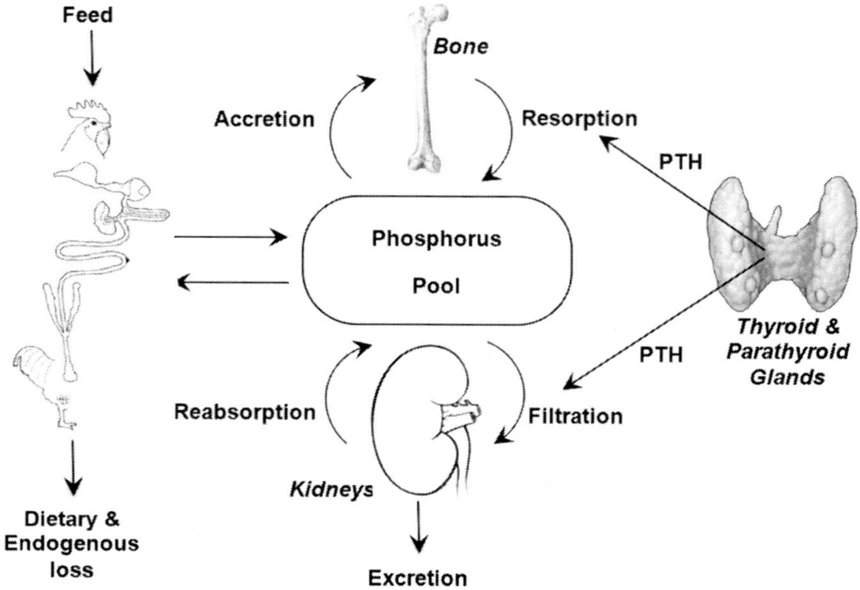

Figure 1. Phosphorus metabolism. Coordination of three organ systems, intestine, kidneys and bone, is central to Ca and P homeostasis. Reduced circulating concentrations of Ca causes increased secretion of parathyroid hormone (PTH) by the parathyroid glands. PTH increases bone resorption and renal Ca reabsorption but increased renal excretion of P. In the kidneys, PTH also stimulates formation of the hormonal form of vitamin D_3 which increases intestinal absorption of Ca and, to a lesser extent, the absorption of P. Elevated blood Ca concentrations as a result of these processes creates a negative feedback loop to ensure homeostasis. If hypercalcaemia occurs, calcitonin is secreted by the thyroid gland to inhibit bone resorption.

Utilisation of Ca and P is modulated by the relative amounts of each in the diet [2]. However, the quantities of each available for their respective metabolic functions is determined by efficiencies of intestinal absorption, glomerular filtration, renal tubular reabsorption, rates of transfer from blood to bone, and intestinal endogenous losses [20]. The efficiencies of these processes in the different tissues are modulated by several hormones, chiefly parathyroid hormone (PTH) and the hormonal form of vitamin D_3 (1,25 dihydroxycholecalciferol; calcitriol) [20,21]. The control of P metabolism differs from that of Ca, with less control in the gut and greater control in the kidneys. Moreover, there is increasing evidence that dietary protein and carbohydrate affect intestinal metabolism of Ca and P [21]. Together, the factors that determine intestinal absorption and kidney excretion along with endocrine regulation of absorption and reabsorption in the kidneys facilitate P homeostasis.

3. PHOSPHORUS REQUIREMENTS

Research on the P requirement of broilers has been the subject of numerous investigations for many decades; however, the minimum requirement for this nutrient has still not been established with any certainty. The published results are variable as different breeds of birds, age of birds, feed ingredients (with or without phytase supplementation), Ca and P sources were used in different studies. In some instances, the differences observed may reflect true differences in requirements, not merely variation resulting from experimental conditions. To further complicate interpretation of results, different criteria, including body weight gain, feed conversion ratio (FCR), tibia breaking strength, tibia ash, or toe ash contents have been used to assess responses to P (Table 1).

Table 1. Summary of available P requirements (g/kg diet) of broilers.

	AP (g/k) (Bird Age, Day)		
Criteria of Assessment	1–21	21–42	Reference
Carcass P analysis	3.7 (day 0–10), 3.0 (day 10–30)	2.4 (day 30–40)	[2]
Bone breaking strength	3.9	3.0	[13]
FCR		1.63	[22]
FCR	3.3–3.9		[23]
Growth and tibia ash	3.2–2.8 (day 18–32)	2.4–1.9 (day 32–42), 1.1 (day 43–49)	[24,25]
Growth, bone ash and processing losses	4.5	4.0	[26]
Weight gain	3.5 (day 0–14)	3.0 (day 15–49)	[15,18]
Weight gain	3.2–3.4		[27]
Weight gain		1.86	[22]
Weight gain	3.2–3.5		[23]
Weight gain	2.8–2.9		[28]
Tibia or toe ash	4.5	3.5	[29]
Tibia ash		3.3	[22]
Tibia ash	3.5–3.9		[23]
Tibia ash	3.7–3.9	3.1 (day 49)	[30]

Evidence in the literature clearly demonstrates that the P requirement of broilers is much lower than NRC [29] recommendations and the values currently used by the industry [2,7,13,15,17,18,22,23,24,25,27,28,31]. The greatest challenge in applying a lower dietary P strategy in practice is to decide what dietary concentrations of P to use in a formulation. This is due to a lack of information on biologically determined P values in feed ingredients for poultry and an agreed P evaluation system [8,9,10].

4. EVALUATION OF PROCEDURES THAT DETERMINE BIOAVAILABLE PHOSPHORUS

There are many approaches for determining P bioavailability of feed and feed ingredients. Some methods are biological and rely on *in vivo* measures including growth, bone strength, bone ash content and blood parameters. Chemical procedures have also been developed to estimate P bioavailability. Most published values using both biological and chemical procedures are qualitative rather than quantitative. Recently, there has been an international move to establish ileal digestible P as the method of choice for determining P bioavailability. An overview of the different approaches for the determination P bioavailability is given below.

4.1. Biologically Determined Values

Initial attempts to quantitate the availability of P for chicks from various phosphate sources used tibia bone ash as the response factor. The method was based on relative percentage of tibia bone ash of chicks fed a test supplement to that of chicks fed a reference standard, thus obtaining relative bioavailability P values [32]. A positive linear relationship was found between tibia bone ash percentage and dietary P intake. A similar relationship was found with toe ash percentages [33,34,35]. Hoshii and Yoshida [36] demonstrated that P bioavailability determined with a slope ratio assay using toe ash was closely associated with the values determined by carcass P retention, suggesting that changes in toe ash concentration reflect changes in net P utilisation.

In the 1980s and '90s, the bioavailability of P was routinely estimated by non-linear bioassays (asymptotic or sigmoidal curves) of body weight and tibia bone or toe ash, using a graded range of P supplements [37,38,39,40,41]. Others [42,43] determined P bioavailability using responses in bone ash content, bone breaking strength, body weight gain or blood inorganic P concentration of birds fed a test P source compared with the response of birds to a standard reference P source. Bioassays based on growth, blood parameters, tibia bone or toe ash percentage and tibia bone breaking strength can only provide a relative biological reference value or ranking [42,43] and do not measure true or apparent

P bioavailability *per se*. The values do not define the quantity of bioavailable P and therefore have limited value when formulating diets [44].

Early attempts to determine true or apparent P retention by collecting excreta in balance studies with chicks were unsuccessful [45,46]. Later, Simons and Versteegh [47] were successful using a balance assay based on excreta collection, when birds were fed a basal diet, nearly free of P (5 g/kg Ca), to which test phosphates were added at low concentrations of approximately 2 g/kg AP. At this level, there is negligible urinary P excretion [47].

4.2. Chemically Determined Values

The most common P evaluation system (NRC, [29,48]) is based on the relationship of P with phytate; non-phytate P (NPP) and phytate P (PP). Phytate is the bound form in which P is found in plants; it is discussed in detail elsewhere (see [49,50,51,52,53]). The NPP system has the advantage of being based on the chemical analysis of feed ingredients. Non-phytate P is the P in the feed or ingredient that is not bound as part of phytic acid. It can be determined by subtracting analysed PP from analysed total P [14]. With this approach, it is assumed that NPP, together with P from non-plant sources are completely available to poultry and non-ruminant animals; assuming that PP is not available to non-ruminant animals. If PP in plant feedstuffs is not analysed, it is assumed that NPP is 30% of total plant P and the remaining 70% is PP. Non-phytate P has been referred to as "AP" or used interchangeably with "AP." These assumptions based on simple differentiation from chemical analysis have been adopted in diet formulation worldwide, since publication by the NRC in 1977.

The NPP system has limitations. Chemical analysis determines how much P is present, but does not indicate to what degree, if any, P is utilised by animals and birds. In other words, chemical analysis does not provide information on P bioavailability of the ingredient or diet and thus any changes in PP and P bioavailability of an ingredient will not be reflected in this system. Numerous studies have found that NPP is not 100% bioavailable and up to 80% of PP can be bioavailable to birds [8,54,55]. In fact, no elements are completely absorbed and utilised by poultry [56].

Phytate has been found in the excreta of the chicks fed a phytate-free (casein-gelatin) diet, whereas none was found in the excreta of germ-free chicks fed the same diet. It appears that phytate may be synthesised by the microflora in the avian digestive tract [57]. Therefore, it is possible that the net microbial hydrolysis of phytate occurring during transit of digesta through gastrointestinal tract may be partly offset by microbial synthesis [58]. This makes determination of dietary NPP requirement by total collection or indigestible marker and analysing NPP in excreta unreliable since some NPP may have been incorporated into phytate. Other chemical or *in vitro* approaches have attempted to relate the solubility of P supplements to P bioavailability determined *in vivo* and have been unsuccessful [59].

4.3. Ileal Digestibility as an Estimate of Phosphorus Bioavailability

In an attempt to overcome the difficulties associated with current approaches to P bioavailability determinations, the Nutrition Working Group of the European Federation of Branches of World's Poultry Science Association (hereafter Rodehutscord and WPSA, [60]) has developed a protocol to determine ileal P digestibility by regression analysis. Two to three diets are formulated for each test P source. A low P basal diet is used, and a minimum of two levels of the P source under test are added to the test diet. At the highest level of inclusion, P supply must not exceed the requirement. With such conditions, the AP of the P source under test can be determined from the slopes of linear regression equations (see [60] for details). The protocol is currently being evaluated in an international collaborative ring test. Collinearity appears to exist between two variables used in the regression analysis and may lead to unreliable and unstable estimates of regression coefficients [61].

Finally, none of the methods outlined above measure P bioavailability *per se* but rather provide an estimate. Accurate measurements of bioavailability would require very expensive and laborious experiments, including the use of radioactive isotopes. What is required is a robust procedure that will provide reliable and repeatable estimates of P bioavailability. It would appear that precaecal P digestibility is the method of choice. A similar situation exists with amino acid availability where precaecal or ileal digestibility is used to estimate availability. Detailed discussions that describe amino acid digestibility assays are available [62].

5. TERMS USED TO DESCRIBE BIOAVAILABLE PHOSPHORUS

Over the years, a number of terms have been suggested to define certain aspects of P utilisation as described by Rodehutscord and WPSA [60] and outlined below. These terms have largely arisen to reflect differing approaches to the determination of bioavailable P.

Digestible P is the portion of dietary total P that is not recovered in faeces. The determination of digestible P requires *in vivo* studies with quantitative determination of dietary P intake and P excretion in faeces. Alternatively, digestible P can be determined by using an indigestible marker. However, as bird excreta consists of faeces and urine, determination of digestible P requires colostomised birds to exclude urinary P content.

Retainable P (RP) indicates the proportion of dietary total P that is retained in the body and can be determined quantitatively following measurement of P intake and P excretion in faeces [60]. It is also called total tract P digestibility and is easy to measure. However, as urine is included in the determination of digestible P, this term is not an appropriate description of the measurement in poultry. Importantly, if a P intake is lower than the requirement, the P content of the urine can be ignored. In these circumstances, the determination of RP should be similar to the determination of digestible P.

Available P (AP) is the quantity of P that is absorbed from the diet by the animal. This definition of AP is different from previous interpretations [29,60] where it has been equated to NPP and RP. In the currently acceptable definition of AP [60], it is determined by including an indigestible marker in the feed and is the proportion of total dietary P that is not recovered in the terminal ileum. It is also called precaecal digestible P or ileal P digestibility [60]. The major advantage of this method is that if P intake exceeds requirements, urinary excretion of P does not confound the results.

There is now general agreement that ileal digestible P is the preferred approach for determining AP as a quantitative estimate of P bioavailability. Rodehutscord and WPSA [60] recommend that retainable P values not be used in the future to avoid confusion. Retainable P reflects both P digestibility and post-absorptive utilisation since bird excreta includes faeces and urine. Dietary supply of P relative to requirement and body status affects total tract P utilisation [63]. In several studies of individual feed ingredients and mixed diets, the RP in total tract is substantially different from AP in the ileum (Table 2). Based on these and other publications that demonstrate a similar trend, both utilisation of P or Ca in feeds and requirement in birds should be expressed on an ileal digestible basis [63]. Furthermore, AP can be used to measure the marginal level of P supplied in animal diets to meet the P requirement of the animal [60].

Table 2. Comparisons of retainable P (RP) with available P (AP) in the ileum of poultry [63].

Ingredient	RP (%)	AP (%)	Reference
Low-phosphorus mixed diet	43	29	[64] [1]
Low-phytate soybean meal	77	94	[65] [2]
Conventional soybean meal	60	94	[65] [2]
Low-phosphorus mixed diet	59	52	[66] [1]
Low-phosphorus mixed diet	45	30	[67] [1]
Canola meal	39	66	[63] [2]

[1] Expressed as apparent digestibility; [2] Expressed as true digestibility derived from regression analysis.

6. PHOSPHORUS SOURCES AND AVAILABILITY

Sources of P in poultry diets include feed ingredients of both plant and animal origin and also inorganic forms of the mineral. These sources differ in P bioavailability.

6.1. Plant Feedstuffs

Plant feedstuffs include cereal grains, oil seeds and the respective by-products. Phytate is the principal storage form of P in many plant tissues. As commercial broiler diets are primarily comprised of feedstuffs of plant origin, most of the P present in these ingredients is in the form of phytate P (PP) [68]. The concentration of PP within the same ingredient can vary considerably. Barrier-Guillot *et al.* [69] found that the concentration of PP ranged from 0.092% to 0.268% dry matter (DM) in wheat and depended on soil fertility, plant variety, and stage of plant maturity at harvest.

The bioavailability of PP for broilers ranges from 0% to 80% (Table 3) [2] depending on diets, bird age and metabolic adaptation. The degree to which PP is utilised by birds will depend, to a large extent, upon the hydrolysis of PP in the gastrointestinal tract. Moreover, the bioavailability of PP in different feed ingredients ranges from 0% to 60% [8,47,67], while the bioavailability of total P within ingredients varies considerably (15% to 75%; Li *et al.*, unpublished data [70]). There are many factors affecting PP utilisation: phytase activity of dietary ingredients, dietary concentrations of fibre, Ca, P and vitamin D_3, age of bird and genotype [58,71].

Table 3. Total P (TP) (g/kg), retainable P (RP) (% of TP), non-phytate P (NPP) (g/kg), phytate P (PP) (g/kg), PP (% of TP) and calculated PP degradability (%) of some of the feedstuffs in three-week-old broilers [2].

Feedstuffs	TP (g/kg)	RP (% TP) *	NPP (g/kg)	PP (g/kg)	PP (% TP)	PP Degradability (%) **
Beans	4.9	52	1.3	3.6	73.5	53
Lupin	3	72	1.5	1.5	50	80
Maize	3	29	0.7	2.3	76.7	16
Peas	4.1	41	1.5	2.6	63.4	23
Rapeseed	10.9	33	3.8	7.1	65.1	10
Rice bran	17.2	16	3.1	14.1	82	2
Soy bean (heat treated)	5.5	54	2	3.5	63.6	49
Soybean meal	7.1	61	2.8	4.3	60.6	61
Sunflower meal	11.9	38	4.2	7.7	64.7	19
Wheat	3.4	48	0.9	2.5	73.5	46
Wheat middlings	10.8	36	2.8	8	74.1	26

* Values were determined in broilers fed diets containing 1.8 g calculated retainable P and 5.0 g Ca per kg diet from 10 days of age onward. P retention was measured in a three-day balance period (from days 21 to 24) in which dietary P intake and excreta P output were measured quantitatively; ** Phytate P degradability was calculated as: 100 × (AP/0.8–NPP)/PP [2].

6.2. Animal Feedstuffs

There are a number of animal-sourced feedstuffs (fish meals, meat meals, meat and bone meals) used in poultry diets mainly as protein supplements. Among these, meat meal is also a potential P source depending on its bone content. Meat meals are prepared by rendering offal from abattoirs. There can be a wide variation between batches depending on the raw materials used and rendering conditions [72]. If the ash content is high, this indicates that it contains a higher amount of bone and is referred to as meat and bone meal. If the ash content is low it is referred to as meat meal.

Meat and bone meal is a variable source of protein for poultry diets, but can be an excellent source of Ca and P and other minerals. Therefore, it is commonly used as a P supplement in poultry diets [7]. However, there are few reports that have measured AP from animal-derived feedstuffs. Limited RP values of animal-derived feedstuffs are shown in Table 4.

Table 4. Total P (TP) (g/kg), retainable P (RP) (% of TP) of animal-derived feedstuffs in three-week-old broilers [2].

Animal Feedstuffs	TP (g/kg)	RP (% TP) *
Bone meal	76	59
Fish meal	22	74
Meat meal	29	65
Meat and bone meal	60	66

* Values were determined in broilers fed diets containing 1.8 g calculated retainable P and 5.0 g Ca per kg diet from 10 days of age onward. P retention was measured in a three-day balance period (from days 21 to 24) in which dietary P intake and excreta P output were measured quantitatively.

6.3. Inorganic Phosphorus Sources

Inorganic P sources are commonly derived from natural rock phosphates. These natural phosphates are not readily utilised by poultry or other animals but must be chemically modified before inclusion in animal feed. Rock phosphates are changed into various orthophosphate forms: monocalcium phosphate, dicalcium phosphate, monodicalcium phosphate, defluorinated rock phosphate and monosodium phosphate [44]. Some commonly used mineral sources and their respective P composition and retention rate are listed in Table 5.

Table 5. Inorganic feed phosphates and their respective total P (TP) (g/kg) and retainable P (RP) (% of TP) [2].

Inorganic P Source	TP (g/kg)	RP (% TP) *	Reference
Calcium sodium phosphate	180	59	[2] *
Dicalcium phosphate	183	83	[74] [1,2]
Dicalcium phosphate (anhydrous)	197	55	[2] *
Dicalcium phosphate (hydrous)	181	77	[2] *
Monocalcium phosphate	226	84	[2] *
Monodicalcium phosphate (hydrous)	213	79	[2] *
Monodicalcium phosphate	203	77	[13] [1]
Monodicalcium phosphate	200	80	[13] [1]
Monodicalcium phosphate	216	81	[13] [1]
Monosodium phosphate	224	92	[2] *
Defluorinated phosphate	182	86	[74] [1,2]
Defluorinated phosphate	179	76	[74] [1,2]

* Values were determined in broilers fed diets containing 1.8 g calculated retainable P and 5.0 g Ca per kg diet from 10 days of age onward. P retention was measured in a three-day balance period (from days 21 to 24) in which dietary P intake and excreta P output were measured quantitatively; [1] Apparent utilisation of P from inorganic sources by broiler as determined under deficiency conditions; [1,2] Retainable P determined through broken line slope response.

Most of these studies determined apparent retention of P from each of the inorganic sources within the deficiency range of P. Leske and Coon [13] showed dramatic reductions in P retention from monocalcium phosphate as dietary P concentration approached the requirement level (98% retention at half of the P requirement to 59% retention at requirement level). Waldroup [73] reported that nearly 50% of excreted P is likely to be of inorganic origin.

Phosphorus must be in the phosphate form to be absorbed by birds. Improper processing and higher temperatures have a detrimental impact on the biological availability of calcium phosphates. As phosphates are heated, pyro- and meta- complexes are formed which greatly reduce the P bioavailability of inorganic sources. The availability of inorganic P sources is substantially affected by hydration, with hydrated phosphate sources generally more biologically available than anhydrous sources of the same type [75]. Feed phosphate with larger particle size has been shown to be more biologically available. Larger particle phosphate is retained longer in the gizzard allowing more complete digestion [76].

7. FACTORS AFFECTING PHOSPHORUS AVAILABILITY

Determination of the bioavailability of P in feedstuffs is a major challenge. There are many factors affecting P bioavailability in birds, including: experimental techniques; the chemical form of P; dietary concentrations of Ca, vitamin D3, energy, protein, fat, and vitamin K; the availability and interaction of other nutrients in the diet, feed processing, and particle size; bird physiological and health factors, such as feed consumption, growth rate, sex and age; management factors, including ambient temperature and lighting program. Some of these factors have been investigated in detail [14] and the major factors are briefly described below.

7.1. Phosphorus and Phytate

High levels of dietary inorganic P reduce PP availability [49,77] due to either the inhibitory effect of P on phytase activity [78], a negative feedback mechanism by excessive amounts of dietary P, or inhibition by products of phytate hydrolysis [49,58].

Phytate is the principal storage form of P in plant feedstuffs and is considered to be an anti-nutritive factor [79]. Phytate carries a strong negative charge and is capable of binding di- and trivalent cations such as calcium, cobalt, copper, iron, magnesium, manganese, nickel and zinc in very stable complexes [58,80], reducing the availability of P as well as these minerals to the animal [2,54,55,81].

Phytate has been shown to inhibit activities of some digestive enzymes such as pepsin, α-amylase and trypsin. Phytate may inhibit proteolysis by changing the protein configuration of digestive enzymes [82]. Inhibition may also result from the chelation of Ca ions which are essential co-factors for the activity of trypsin and α-amylase leading to reduced protein and starch digestion [50,83,84,85,86].

It has also been reported that phytate can reduce fat digestibility by forming insoluble Ca-phytate complexes with fatty acids in the gut lumen

[87]. In its chelated form, the phytate molecule is difficult to hydrolyse by phytase. The pH affects the solubility of phytate. Most phytate mineral complexes are soluble at low pHs (less than 3.5) with maximum insolubility occurring between 4 and 7 [88]. Champagne [89] found that Ca-phytate complexes precipitate at pH between 4 and 6, which is the approximate pH of the intestine where the Ca ions should be absorbed. Taylor [90] has suggested that the primary factor affecting PP utilisation is the Ca ion concentration in the small intestine where insoluble Ca-phytate complexes form. A precipitated phytate mineral complex would not be accessible for hydrolysis or absorption in the intestine.

7.2. Calcium

Calcium is an essential element for bone and egg shell formation, blood clotting, muscle contraction and transmission of nerve impulses. Calcium is also an important co-factor for many enzymes and hormones.

Increases in dietary P and Ca concentrations may affect apparent digestion of these nutrients. High levels of dietary Ca and low levels of P had detrimental effects on broiler performance [16,17,91,92]. Elevated dietary Ca levels increase pH in the gut and, as a result, P absorption [93] and retention [91] are decreased. High dietary Ca concentration can significantly increase gastrointestinal pH [94] and decrease not only phytate hydrolysis, but also pepsin activity in the proventriculus/gizzard and reduce apparent ileal protein digestibility [51]. High plasma P levels decrease Ca absorption from the gut [95]. The optimum dietary Ca-to-P ratio is of utmost importance for broiler production. With the Ca: P ratio, there will be a range of values that can meet metabolic demands for any production outcome. When dietary P concentration is high, birds can tolerate a wider range of Ca levels than when dietary P marginally meets the requirement (Li *et al.* unpublished data [96]).

Total Ca values are currently used for feed formulation since the bioavailability of Ca for feed ingredients have not been determined. Bioavailable Ca from plant raw materials is expected to be low. This is due to the high phytate content of these raw materials and the low contribution of Ca to the diet. Also, to some extent, phytate will negatively influence the availability of Ca from this dietary source by binding Ca in phytate-mineral complexes. Nevertheless, there is a need to determine the bioavailability of Ca in feed ingredients to further delineate the interaction of this nutrient with other nutrients and to improve the precision of feed formulation.

7.3. Vitamin D

Vitamin D_3 is essential for absorption of Ca and P utilization as depicted in Figure 1. Several reports have demonstrated that dietary addition of vitamin D_3 can significantly enhance the retention of P in birds [52]. However, excess vitamin $D_3$and its metabolites have not been shown to have a further beneficial effect on bird productivity when birds are already consuming adequate vitamin D_3. Like many other minerals, understanding of the molecular mechanisms of P absorption, especially the role of the hormonal form of vitamin D_3, is still limited.

7.4. Phytase

Numerous studies have demonstrated that dietary addition of phytase increases hydrolysis of PP and improves P utilisation in pigs and poultry [15,16,17,18,53]. Maximum utilisation of phytate by broilers fed diets with supplemental phytase was approximately 50% on average [97] following dietary P supplementation. There are many factors affecting the efficacy of microbial phytase such as phytase source and dose [98], physical factors of feedstuffs, feed particle size, Ca to P ratio [99], Ca bioavailability and inorganic P supplements, vitamin D status [100], bird age [101] and body weight [102], gut pH, digesta retention time, and Ca and P status. Phytase was found to be most efficacious when added to diets containing low AP and adequate Ca concentrations (Li *et al.* unpublished data [96]).

Dietary Ca level affects the efficacy of phytase in broilers [103], but not to the same extent for all phytases [104]. The same source of phytase can produce different responses in different flocks [105,106]. Moreover, the efficacy of feed phytase may be improved by the simultaneous dietary addition of other exogenous enzymes. All these variables make predicting phytase responses very difficult and lead to variation in P availability.

7.5. Birds

The ability of poultry to absorb dietary P is known to be influenced by age and physiological state. Young chicks with a rapidly developing skeletal system are critically sensitive to P deficiency [107] and tend to use P more efficiently than do mature birds [108]. Older birds have a greater ability to use PP, possibly due to more phytase activity in their intestinal tract [56]. A concern often raised by commercial nutritionists is the relevance of digestibility values generated with birds of one age group to birds of another age or other species.

Applegate *et al.* [106] conducted two experiments in which chicks were fed the same diet and found that the apparent PP hydrolysis varied nearly two-fold between the first and second experiment. Similar results have been reported by Nelson and Kirby [109] and Tamim *et al.* [103]. Bird variations make it difficult to predict treatment responses of different flocks. The variation between experiments and between flocks arises from many sources, some of which are summarised in Table 6.

Table 6. Summary of variation in processes associated with feeding of phosphorus to broilers [10].

Source of Variation	Coefficient of Variation (%)
Sampling variation	5 to 10
Analytical variation	5
Mixer variation	5 to 10
Bird utilisation	16
Ingredient variation (corn and soybean meal)	8 to 13

8. CONCLUSIONS

Re-examination of the phosphorus requirement of modern broilers has demonstrated that it is lower than the NRC [29] recommendations and the values currently used by the chicken meat industry. However, there are practical limitations in applying this information to industry as there is limited data on the bioavailability and variability of P in feed ingredients; there are few reports that have measured P bioavailability of feedstuffs for poultry. Furthermore, the different methods used have made comparisons between the studies and application to practice extremely difficult. It is only when commercial poultry nutritionists have this information that they will be in a position to reduce the safety margins for P when formulating diets. To tackle these challenges, cooperative research on a global scale is needed to standardise measurement procedures in order to produce a robust and reliable database which can be used by nutritionists to formulate diets to meet the bird's P requirement precisely; this approach is appropriate as only a couple broiler strains dominate international markets. As a starting point, the universal adoption of ileal P digestibility [60] as the preferred procedure to determine AP would be a major step forward. Moreover, to make sense of the data derived and to allow comparison of the results, it will be necessary to report ingredient composition of experimental diets, dietary concentrations of Ca and non-phytate P, dietary electrolyte balance and mineral content, all of which influence P absorption. Achievement of these goals will assist endeavours to sustain the global supply of phosphorus.

ACKNOWLEDGMENTS

The Australian Government Rural Industries Research and Development Corporation -Chicken Meat Program, Canberra, ACT, has supported the phosphorus research conducted by the authors.

AUTHOR CONTRIBUTIONS

All the authors contributed equally to the paper.

CONFLICTS OF INTEREST

The authors declare no conflicts of interest.

REFERENCES

1. Asimov, I. Asimov on Chemistry; Doubleday: Garden City, NY, USA, 1974.
2. Van der Klis, J.D.; Versteegh, H.A.J. Phosphorus nutrition of poultry. In Recent Developments in Poultry Nutrition 2; Garnsworthy, P.C., Wiseman, J., Eds.; Nottingham University Press: Nottingham, UK, 1999; pp. 309–320.
3. Shastak, Y.; Witzig, M.; Hartung, K.; Rodehutscord, M. Comparison of retention and prececal digestibility measurements in evaluating mineral phosphorus sources in broilers. Poult. Sci. 2012, 91, 2201–2209.
4. Cordel, D.; Drangert, J.; White, S. The story of phosphorus: Global food security and food for thought. Glob. Environ. Chang. 2009, 19, 292–305.
5. Neset, T.S.; Cordell, D. Global phosphorus scarcity: Identifying synergies for a sustainable future. J. Sci. Food Agric.2012, 92, 2–6.
6. Devereux, C.; Smart, M.; Kalt, F.P.; Takei, N. Animal feeds: Phosphate supplements. In CEH Marketing Research Report; HIS Chemical: Englewood, CO, USA, 1994.
7. Waldroup, P.W. Nutritional approaches to reducing phosphorus excretion by poultry. Poult. Sci. 1999, 78, 683–691.

8. Van der Klis, J.D.; Versteegh, H.A.J. Phosphorus nutrition in broilers. In Recent Advances in Animal Nutrition; Garnsworty, P.C., Wiseman, J., Haresign, W., Eds.; Nottingham University press: Nottingham, UK, 1996; pp. 71–83.
9. Cossa, J.; Oloffs, K.; Kluge, H.; Jeroch, H. Investigation into the TP and PP content in different varieties of grain maize. In Proceedings of the 11th European Symposium on Poultry Nutrition; World's Poultry Science Association: Faaberg, Denmark, 1997; pp. 444–446.
10. Applegate, T.J.; Angel, R. Feasibility versus Practicality of Phosphorus Reduction in Poultry: Progress and Future Needs. USDA National Agricultural Library: West Lafayette, IN, USA. Available online: http://www.cals.ncsu.edu/waste_mgt/natlcenter/sanantonio/ApplegateandAngel.pdf (accessed on 8 December 2014).
11. Schindler, D.W. Evolution of phosphorus limitation in lakes. Sciences 1977, 195, 260–262.
12. Sharpley, A.N.; Chapra, S.C.; Wedepohl, R.; Sims, J.T.; Daniel, T.C.; Reddy, K.R. Managing agricultural phosphorus for protection of surface waters: Issues and options. J. Environ. Qual. 1994, 23, 437–451.
13. Leske, K.; Coon, C.N. The development of feedstuff retainable phosphorus values for broilers. Poult. Sci. 2002, 81, 1681–1693.
14. Angel, R. Calcium and phosphorus requirements in poultry. In Proceedings of the 1st International Phytase Summit, Washington, DC, USA, 28–30 September 2010; pp. 65–71.
15. Li, X.; Zhang, D.; Huang, K.H.; Bryden, W.L. Available phosphorus requirement of broilers fed sorghum based diet from day 22 to 49. In Proceedings of the 19th European Symposium on Poultry Nutrition, Potsdam, Germany, 26–29 August 2013; p. 200.
16. Li, X.; Zhang, D.; Huang, K.H.; Bryden, W.L. Dietary calcium levels and non-phytate phosphorus requirement of broilers from days 1 to 21. Aust. Poult. Sci. Symp. 2013, 24, 50.
17. Yang, T.Y.; Setiyawan, E.; Li, X.; Zhang, D.; Bryden, W.L. Dietary calcium levels and non-phytate phosphorus requirement of broilers from days 22–49. In Proceedings of Recent Advances in Animal Nutrition in Australia 2013; University of New England: Armidale, New South Wales, Australia, 2013.
18. Zhang, D.; Li, X.; Huang, K.H.; Bryden, W.L. Available phosphorus requirement of broilers fed sorghum based diet from day 1 to 21. In Proceedings of the 19th European Symposium on Poultry Nutrition, Potsdam, Germany, 26–29 August 2013; p. 196.
19. Li, X.; Zhang, D.; Bryden, W.L. Bioavailability of phosphorus in feedstuffs for broilers. In Final Report for Rural Industries Research and Development Corporation; Rural Industries Research and Development Corporation: Kingston, Australia, 2015.
20. Klasing, K.C. Comparative Avian Nutrition; CAB International: Wallingford, UK, 1998.
21. Vitti, D.M.S.S., Kebreab, E., Eds.; Phosphorus and Calcium Utilization and Requirements in Farm Animals; CAB International: Wallingford, UK, 2010.
22. Yan, F.; Kersey, J.H.; Waldroup, P.W. Phosphorus requirements of broiler chicks three to six weeks of age as influenced by phytase supplementation. Poult. Sci. 2001, 80, 455–459.
23. Persia, M.E.; Saylor, W.W. Effects of broiler strain, dietary nonphytate phosphorus, and phytase supplementation on chick performance and tibia ash. J. Appl. Poult. Res. 2006, 15, 72–81.
24. Angel, R.; Applegate, T.J.; Christman, M. Effect of dietary non-phytate phosphorus (nPP) on performance and bone measurements in broilers fed on a four-phase feeding system. Poult. Sci. 2000, 79 (Suppl. 1), 21–22.
25. Angel, R.; Applegate, T.J.; Christman, M.; Mitchell, A.D. Effect of dietary non-phytate phosphorus (nPP) level on broiler performance and bone measurements in the starter and grower phase. Poult. Sci. 2000, 79 (Suppl. 1), 22.

26. Moran, E.T., Jr.; Todd, M.C. Continuous submarginal phosphorus with broilers and the effect of preslaughter transportation: Carcass defects, further-processing yields, and tibia-femur integrity. Poult. Sci. 1994, 73, 1448–1457.
27. Waldroup, P.W.; Kersey, J.H.; Saleh, E.A.; Fritts, C.A.; Yan, F.; Stilborn, H.L.; Crum, R.C., Jr.; Raboy, V. Non-phytate phosphorus requirement and phosphorus excretion of broilers fed diets composed of normal or high available phosphate corn with and without microbial phytase. Poult. Sci. 2000, 79, 1451–1459.
28. Yan, F.; Waldroup, P.W. Non-phytate phosphorus requirement and phosphorus excretion of broiler chicks fed diets composed normal or high available phosphate corn as influenced by phytase supplementation and vitamin D source. Int. J. Poult. Sci. 2006, 5, 219–228.
29. NRC. Nutrient Requirements of Poultry, 9th ed.; National Academy Press: Washington, DC, USA, 1994.
30. Yan, F.; Kersey, J.H.; Fritts, C.A.; Waldroup, P.W. Phosphorus requirements of broiler chicks six to nine weeks of age as influenced by phytase supplementation. Poult. Sci. 2003, 83, 294–300.
31. Dhandu, A.S.; Angel, R. Broiler non-phytin phosphorus requirement in the finisher and withdrawal phases of a commercial four-phase feeding system. Poult. Sci. 2003, 82, 1257–1265.
32. Gillis, M.B.; Norris, L.C.; Heuser, G.F. Studies on the biological value of inorganic phosphates. J. Nutr. 1954, 52, 115–125. [PubMed]
33. Grau, C.R.; Zweigart, P.A. Phosphatic clay as a phosphorus source for chicks. Poult. Sci. 1953, 32, 500–503.
34. Fritz, J.C.; Roberts, T.; Bochne, J.W.; Hove, E.L. Factors affecting the chick's requirement for phosphorus. Poult. Sci.1969, 48, 307–320.
35. Yoshida, M.; Hoshii, H. Improvement of biological assay to determine available phosphorus with growing chicks. J. Poult. Sci. 1977, 14, 33–43.
36. Hoshii, H.; Yoshida, M. Comparison of phosphorus availabilities estimated biologically based on toe ash content and carcass phosphorus retention. J. Poult. Sci. 1977, 14, 279–283.
37. Potchanacom, M.; Potter, L.M. Biological values of phosphorus from various sources for young turkeys. Poult. Sci.1987, 66, 505–513.
38. Potter, L.M. Bio-availability of phosphorus from various phosphates based on body weight and toe ash measurements. Poult. Sci. 1988, 67, 96–102.
39. Potter, L.M.; Potchanacom, M.; Ravindran, V.; Kornegay, E.T. Bio-availability of phosphorus in various phosphate sources using body weight and toe ash as response criteria. Poult. Sci. 1995, 74, 813–820.
40. Ketels, E.; De Groote, G. The relative bio-availability and the ileal digestibility of phosphorus in mineral and animal sources. In Proceedings of the 18th World's Poultry Congress, Nagoya, Japan, 4–9 September 1988; pp. 873–875.
41. De Groote, G.; Ketels, P.; Seynaeve, P. Relative bio-availability of phosphorus in meat meals as influenced by particle size. In Proceedings of the International Poultry Congress, Istanbul, Turkey, 6–10 September 1991; pp. 42–52.
42. De Groote, G. Biological availability of phosphorus in feed phosphates for broilers. In Proceedings of the 4th European Symposium on Poultry Nutrition, Tours, France, 17–20 October 1983; Larbier, M., Ed.; pp. 91–102.
43. Sauveur, B. Bio-availability to poultry of plant-origin phosphorus-Methodological criticisms and results. In Proceedings of the 4th European Symposium on Poultry Nutrition, Tours, France, 17–20 October 1983; Larbier, M., Ed.; pp. 103–113.
44. Payne, S.G. The Phosphorus Availability of Feed Phosphates in Broilers. MS Thesis, University of Stellenbosh, Stellenbosh, South Africa, 2005.
45. Edwards, H.M., Jr.; Gillis, M.B. A chromic oxide balance method for determining phosphate availability. Poult. Sci.1959, 38, 569–574.
46. Sibbald, I.R. Measurement of mineral bio-availability: Extension of true metabolisable energy methodology. Poult. Sci. 1982, 61, 485–487.

47. Simons, P.C.M.; Versteegh, H.A.J.; Jongbloed, A.W.; Kemme, P.A.; Slump, P. Improvement of phosphorus availability by microbial phytase in broilers and pigs. Br. J. Nutr. 1990, 64, 525–540.
48. NRC. Nutrient Requirements of Poultry, 7th ed.; National Academy Press: Washington, DC, USA, 1977.
49. Ankra-Badu, G.A.; Aggrey, S.E.; Pesti, G.M.; Bakalli, R.I.; Edwards, H.M., Jr. Modelling of parameters affecting phytate phosphorus bioavailability in growing birds. Poult. Sci. 2004, 83, 1083–1088.
50. Thompson, L.U.; Yoon, H. Starch digestibility as affected by poly phenols and phytic acid. J. Food Sci. 1984, 49, 1228–1229.
51. Walk, C.L.; Bedford, M.R.; McElroy, A.P. Influence of limestone and phytase on broiler performance, gastrointestinal pH, and apparent ileal nutrient digestibility. Poult. Sci. 2011, 91, 1371–1378.
52. Ravindran, V.; Bryden, W.L.; Kornegay, E.T. Phytates: Occurrence, bioavailability and implications in poultry nutrition. Poult. Avian Biol. Rev. 1995, 6, 125–143.
53. Selle, P.H.; Ravindran, V. Microbial phytase in poultry nutrition. Anim. Feed Sci. Technol. 2007, 135, 1–41.
54. Angel, R.; Saylor, W.W.; Mitchell, A.D.; Powers, W.; Applegate, T.J. Effect of dietary phosphorus, phytase, and 25-hydroxycholecalciferol on broiler chicken bone mineralization, litter phosphorus, and processing yields. Poult. Sci.2006, 85, 1200–1211.
55. Rodehutscord, M.; Dieckmann, A.; Witzig, M.; Shastak, Y. A note on sampling digesta from the ileum of broilers in phosphorus digestibility studies. Poult. Sci. 2012, 91, 965–971.
56. Axe, D. Phosphorus value in ingredient sources examined. Feedstuffs 1998, 70, 17–28.
57. Savage, J.E.; Yohe, J.M.; Pickett, E.E.; O'Dell, B.L. Zinc metabolism in the growing chick. Tissue concentration and effect of phytate on absorption. Poult. Sci. 1964, 43, 420–426.
58. Wise, A. Dietary factors determining the biological activities of phytase. Nutr. Abs. Rev. 1983, 53, 791–806.
59. Shastak, Y.; Rodehtscord, M. Determination and estimation of phosphorus availability in growing poultry and their historical development. World's Poult. Sci. J. 2013, 69, 569–586.
60. Rodehutscord, M.; WPSA (Working Group No. 2: Nutrition of the European Federation of Branches of World Poultry Science Association). Determination of phosphorus availability in poultry. World's Poult. Sci. J. 2013, 69, 687–698.
61. Allison, P. When Can You Safely Ignore Multicollinearity? 2012. Available online: http://statisticalhorizons.com/multicollinearity (accessed on 4 May 2015).
62. Bryden, W.L.; Li, X. Amino acid digestibility and poultry feed formulation: Expression, limitations and application. Rev. Bras. Zootec. 2010, 39 (Suppl.), 279–287.
63. Adeola, O.; Applegate, T.J. Phosphorus and calcium equivalency of phytase. In Proceedings of the 1st International Phytase Summit, Washington, DC, USA, 28–30 September 2010; pp. 45–64.
64. Adedokun, S.A.; Sands, J.S.; Adeola, O. Determining the equivalent phosphorus released by an Escherichia coli derived phytase in broiler chicks. Can. J. Anim. Sci. 2004, 84, 437–444.
65. Dilger, R.N.; Adeola, O. Estimation of true phosphorus digestibility and endogenous phosphorus loss in growing chicks fed conventional and low-phytate soybean meals. Poult. Sci. 2006, 85, 661–668.
66. Jendza, J.A.; Dilger, R.N.; Sands, J.S.; Adeola, O. Efficacy and equivalency of an Escherichia coli-derived phytase for replacing inorganic phosphorus in the diets of broiler chickens and young pigs. J. Anim. Sci. 2006, 84, 3364–3374.

67. Nyannor, E.K.D.; Adeola, O. Corn expressing an Escherichia coli-derived phytase gene: Comparative evaluation study in broiler chicks. Poult. Sci. 2008, 87, 2015–2022.
68. Nelson, T.S. The hydrolysis of phytate phosphorus by chicks and laying hens. Poult. Sci. 1976, 55, 2262–2264.
69. Barrier-Guillot, B.; Casado, P.; Maupetit, P.; Jondreville, C.; Gatel, F. Wheat phosphorus availability: 1-In vitro study; Factors affecting endogenous phytasic activity and phytic phosphorus content. J. Sci. Food Agric. 1996, 70, 62–68.
70. Li, X.; Zhang, D.; Bryden, W.L. University of Queensland: Gatton, Australia Unpublished Work. , 2015. University of Queensland: Gatton, Australia Unpublished Work. , 2015.
71. De Groote, G.; Huyghebaert, G. The bio-availability of phosphorus from feed phosphates for broilers as influenced by bio-assay method, dietary Ca-level and feed form. Anim. Feed Sci. Technol. 1997, 69, 329–340.
72. Skurray, G.R. The nutritional evaluation of meat meals for poultry. World's Poult. Sci. J. 1974, 30, 129–136.
73. Waldroup, P.W. Phosphorus, phytase, and the environment—A retrospective look. Proc. MD Nutr. Conf. Feed Manuf.2002, 49, 195–202.
74. Coon, C.N.; Seo, S.; Manangi, M.K. The determination of retainable phosphorus, relative biological availability, and relative biological value of phosphorus sources for broiler. Poult. Sci. 2007, 86, 857–868.
75. Gillis, M.B.; Edwards, H.M., Jr.; Young, R.J. Studies on the availability of calcium orthophosphates for chickens and turkeys. J. Nutr. 1962, 78, 155–162. [PubMed]
76. Lima, F.R.; Mendonca, C.X., Jr.; Alvarez, J.C.; Garzillo, J.M.F.; Ghion, E.; Leal, P.M. Biological evaluations of commercial dicalcium phosphates as sources of available phosphorus for broiler chicks. Poult. Sci. 1997, 76, 1707–1713.
77. Sanders, A.M.; Edwards, H.M., Jr.; Roland, G.N. Calcium and phosphorus requirements of the chick as determined response surface analysis. Br. J. Nutr. 1992, 67, 21.
78. Ravindran, V.; Cabahug, S.; Ravindran, G.; Selle, P.H.; Bryden, W.L. Response of broiler chickens to microbial phytase supplementation as influenced by dietary phytic acid and non-phytate phosphorus levels. II. Effects on apparent metabolisable energy, nutrient digestibility and nutrient retention. Br. Poult. Sci. 2000, 41, 193–200.
79. Bryden, W.L.; Selle, P.H.; Ravindran, V.; Acamovic, T. Phytate: An anti-nutrient factor in animal diets. In Poisonous Plants: Global Research and Solutions; Panter, K.E., Wierenga, T.L., Pfister, J.A., Eds.; CABI Publishing: Wallingford, UK, 2007; pp. 279–284.
80. Maenz, D.D.; Engle-Schaan, C.M.; Newkirk, R.W.; Classen, H.L. The effect of minerals and mineral chelators on the formation of phytase-resistant and phytase susceptible forms of phytic acid in solution and in a slurry of canola meal. Anim. Feed Sci. Technol. 1999, 81, 177–192.
81. Pallauf, J.; Rimbach, G.; Pippig, S.; Schindler, B.; Most, E. Effect of phytase supplementation to a phytate-rich diet based on wheat, barley and soya on the bioavailability of dietary phosphorus, calcium, magnesium, zinc and protein in piglets. Agribiol. Res. 1994, 47, 39–48.
82. Singh, M.; Krikorian, A.D. Inhibition of trypsin activity in vitro by phytase. J. Agric. Food Chem. 1982, 30, 799–800.
83. Deshpande, S.S.; Cheryan, M. Effects of phytate, divalent cations and their implications on alpha-amylase activity. J. Food Sci. 1984, 49, 516–519.
84. Liener, I.E. Anti-nutritional factors in legume seeds: State of art. In Recent Advances of Research on Anti-Nutritional Factors in Legume Seeds; Huismen, J., Van Der Poel, T.F.B., Liener, I.E., Eds.; Pudoc: Wageningen, The Netherlands, 1989; pp. 6–13.
85. Knuckles, B.E.; Betschart, A.A. Effect of phytate and other myo-inositol phosphate esters on alpha-amylase digestion of starch. J. Food Sci. 1987, 52, 719–721.

86. Caldwell, R.A. Effect of calcium and phytic acid on the activation of trypsinogen and the stability of trypsin. J. Agric. Food Chem. 1992, 40, 43–46.
87. Leeson, S.; Summers, J.D.; Caston, L. Response of brown egg strain layers to dietary calcium or phosphorus. Poult. Sci. 1993, 72, 1510–1514.
88. Selle, P.H.; Ravindran, V.; Caldwell, R.A.; Bryden, W.L. Phytate and phytase: Consequences for protein utilisation. Nutr. Res. Rev. 2000, 13, 255–278.
89. Champagne, E.T. Effects of pH on mineral-phytate, protein-mineral-phytate and mineral-fiber interactions. Possible consequences of atrophic gastritis on mineral bioavailability from high fiber foods. J. Am. Coll. Nutr. 1988, 38, 199–215.
90. Taylor, T.G. Dietary phosphorus and egg shell thickness in the domestic fowl. Br. Poult. Sci. 1965, 6, 79–87.
91. Keshavarz, K. The effect of dietary levels of calcium and phosphorus on performance and retention of these nutrients by laying hens. Poult. Sci. 1986, 65, 114–121.
92. Keshavarz, K. Phosphorus requirement of laying hens with and without phytase. In Proceedings of the Cornell Nutrition Conference for Feed Manufacturers, Syracuse, NY, USA, 19–21 October 1999; pp. 81–92.
93. Hurwitz, S.; Bar, A. Absorption of calcium and phosphorus along the intestinal tract of the laying fowl as influenced by dietary calcium and eggshell formation. J. Nutr. 1965, 86, 433–438. [PubMed]
94. Guinotte, F.; Gautron, J.; Nys, Y.; Soumarmon, A. Calcium solubilisation and retention in the gastrointestinal tract in chicks (Gallus domesticus) as a function of gastric acid secretion inhibition and of calcium carbonate particle size. Br. J. Nutr. 1995, 73, 125–139.
95. Keshavarz, K.; Austic, R.E. Effects of dietary minerals on acid–base balance and eggshell quality in chickens. J. Nutr.1990, 120, 1360–1369. [PubMed]
96. Li, X.; Zhang, D.; Bryden, W.L.; Huang, K.H. University of Queensland: Gatton, Australia Unpublished Work. , 2013. University of Queensland: Gatton, Australia Unpublished Work. , 2013.
97. Selle, P.H.; Ravindran, V.; Bryden, W.L.; Scott, T. Influence of dietary phytate and exogenous phytase on amino acid digestibility in poultry: A review. J. Poult. Sci. 2006, 43, 89–103.
98. Shirley, R.B.; Edwards, H.M., Jr. Graded levels of phytase past industry standards improves broiler performance. Poult. Sci. 2003, 82, 671–680.
99. Manangi, M.K.; Coon, C. Phytate phosphorus hydrolysis in broilers in response to dietary phytase, calcium, and phosphorus concentrations. Poult. Sci. 2008, 87, 1577–1586.
100. Mohammed, A.; Gibney, M.J.; Taylor, T.G. The effects of dietary levels of inorganic phosphorus, calcium and cholecalciferol on the digestion of phytate phosphorus by the chick. Br. Poult. Sci. 1991, 66, 251–259.
101. Kornegay, E.T. Nutritional, environmental and economic considerations for using phytase in pig and poultry diets. In Nutrient Management of Food Animals to Enhance and Protect the Environment; Kornegay, E.T., Ed.; Lewis Publishers: New York, NY, USA; CRC Press: Boca Raton, FL, USA, 1996; pp. 277–302.
102. Punna, S.; Roland, D.A., Sr. Influence of supplemental microbial phytase on first cycle laying hens fed phosphorus-deficient diets from day one of age. Poult. Sci. 1999, 78, 1407–1411.
103. Tamim, N.M.; Angel, R.; Christman, M. Influence of dietary calcium and phytase on phytate phosphorus hydrolysis in broiler chickens. Poult. Sci. 2004, 83, 1358–1367.
104. Augspurger, N.R.; Baker, D.H. Phytase improves dietary calcium utilisation in chicks, and oyster shell, carbonate, citrate, and citrate-malate forms of calcium are equally bioavailable. Nutr. Res. 2004, 24, 293–301.
105. Angel, R.; Dhandu, A.S.; Applegate, T.J. Phosphorus requirements of broilers and the impact of exogenous phytases. In Proceedings of the Arkansas Poultry Nutrition Conference; University of Arkansas: Fayetteville, AR, USA, 2002; pp. 1–22.

106. Applegate, T.J.; Angel, R.; Classen, H.L. Effect of dietary calcium, 25-hydroxycholecalciferol, or bird strain on small intestinal phytase activity in broiler chickens. Poult. Sci. 2003, 82, 1140–1148.
107. Nelson, T.S.; Peeler, H.T. The availability of phosphates from single and combined phosphates to chicks. Poult. Sci.1961, 40, 1321–1327.
108. Waldroup, P.W. Bioassays remain necessary to estimate phosphorus, calcium bioavailability. Feedstuffs 1996, 68, 13–23.
109. Nelson, T.S.; Kirby, L.K. The calcium binding properties of natural phytate in chick diets. Nutr. Rep. Int. 1987, 35, 949–956.

CHAPTER 5

Oral Bioavailability and Plasma Disposition of Pefloxacin in Healthy Broiler Chickens

María-Aránzazu Martínez*, Irma Ares, José-Luis Rodríguez, Marta Martínez, María-Rosa Martínez-Larrañaga, Gerardo Isea and Arturo Anadón

Department of Toxicology and Pharmacology, Faculty of Veterinary Medicine, Universidad Complutense de Madrid, Madrid, Spain

ABSTRACT

The pharmacokinetics of pefloxacin after single 10 mg/kg BW intravenous (IV) and oral doses were studied in healthy broiler chickens. For 24 h, serial blood samples were obtained after IV and oral administration. Concentrations of pefloxacin and its major metabolite *N*-demethyl pefloxacin (norfloxacin) were measured by use of high-performance liquid chromatography. The plasma concentrations–time data were found to fit a two-compartment open model. For pefloxacin, the elimination half-life ($t_{½β}$) was 8.44 ± 0.48 and 13.18 ± 0.82 h after IV and oral administration, respectively. After single oral dose, pefloxacin was rapidly absorbed with an absorption half-life ($t_{½a}$) and T_{MAX} of 0.87 ± 0.07 and 2.01 ± 0.12 h, respectively. Maximum plasma concentration (C_{MAX}) was 4.02 ± 0.31 µg/mL.

Oral bioavailability of pefloxacin was found to be 70 ± 2%. Pefloxacin was converted to *N*-demethyl pefloxacin (norfloxacin). This metabolite represented 5% of the parent drug plasma concentrations. The maximal plasma concentration (C_{MAX}) of *N*-demethyl pefloxacin (norfloxacin) was calculated as 0.19 ± 0.01 mg/mL. The $t_{½β}$ of *N*-demethyl pefloxacin after oral pefloxacin administration was 10.93 ± 0.80 h. The results indicate that an oral dose of 10 mg pefloxacin/kg BW, every 24 h, should be effective in treatment of the most systemic infections in poultry.

Keywords: pefloxacin, broiler chickens, IV and oral dose, kinetics, PK/PD modeling

INTRODUCTION

Pefloxacin [1-ethyl-6-fluoro-1-4-dihydro-4-oxo-7(4-methyl-1-piperazinyl) quinolone-3 carboxylic acid] is a fluorinated quinolone, which is structurally related to nalidixic acid. Pefloxacin has high antibacterial activity against Gram-negative bacteria including the most species of Enterobacteriaceae and *Neisseria, Campylobacter, Haemophilus* species, and Gram-positive bacteria such as *Staphylococcus*, and *Streptococcus*, between others (1). Antibacterial active structural analogs of nalidixic acid inhibit prokaryotic DNA gyrase *in vitro*. Pefloxacin and other 4-quinolones inhibit DNA gyrase activity and DNA replication. DNA gyrase inserts negative superhelical turns into bacterial DNA and belongs to a group of enzymes known as DNA topoisomerases that are responsible for controlling the spatial geometry of DNA *in vivo*. Inhibiting the action of DNA gyrase prevents the supercoiling and relaxation of DNA. The bactericidal effect of the 4-quinolones probably results from the inhibition of the resealing of open nicks in the DNA strand produced by DNA gyrase (1, 2).

A large number of structurally related analogs of nalidixic acid have been developed. In particular, the 4-quinolones such as ciprofloxacin, enoxacin, fleroxacin, lomefloxacin, and ofloxacin have broad spectrums of antibacterial activity, are rapidly bactericidal, penetrate most body fluids and tissues, and have demonstrated promising results in the treatment of a variety of infectious conditions (3–6). Pefloxacin is another 4-quinolone, which has similar *in vitro* antibacterial activity against most Gram-negative and Gram-positive bacteria. It is rapidly absorbed following oral administration, achieving steady state serum concentrations that are in excess of the MIC values for most pathogens. In addition, pefloxacin is extensively distributed throughout the body, producing high tissue concentrations. Because of its broad spectrum activity, it is likely that pefloxacin should have potential therapeutic application for many types of systemic infections. Pefloxacin has been shown to be generally well tolerated during short- and long-term oral administration. The pefloxacin metabolism is extensive (85–90%). The piperazinyl ring is the main site of metabolism. The ring is hydroxylated, N-oxidized, demethylated, formylated, and acetylated. Six metabolites of pefloxacin have been identified (Figure 1); the major metabolites are *N*-demethyl pefloxacin (norfloxacin) and pefloxacin *N*-oxide; the latter has low antibacterial activity (7). In poultry medicine, the potential usefulness of pefloxacin for treatment of common infections in chickens for fattening requires detailed information on pharmacokinetics properties to establish the orally administered dose necessary for maintaining bactericidal drug concentrations in the body. Although studies on the kinetic behavior of pefloxacin in many species including man (8–10), calves (11), rabbits (12), goats (13, 14), pigs (15), and sheep (16) are available; pharmacokinetic investigations on pefloxacin in poultry are limited (17, 18). Results of a preliminary study in our laboratory in chickens were presented as a poster at 9th International EAVPT Congress (19). The objective of this study was to determine the pharmacokinetics of pefloxacin after IV and oral administration in broiler chickens.

MATERIALS AND METHODS

Chemicals and Reagents

Pefloxacin mesylate dihydrate (P0106 Sigma) (≥97% purity) and its metabolite *N*-demethyl pefloxacin (norfloxacin) (Y0001301 Sigma-Aldrich) analytical standard (≥98% purity) were provided by Sigma-Aldrich, Spain. All other chemicals used were obtained from usual commercial sources and were of the highest grade available.

Animals

All experimental procedures involving animals were reviewed in accordance with ethics requirements and authorized by the official ethical committee of the Universidad Complutense de Madrid. Eighteen healthy Ross male chickens for fattening that were 40-days old and that each weighed 2.5 kg were used. The chickens, purchased at a poultry breeding farm (Nutreco, S.A., Sada Division, Cazalegas, Toledo, Spain) were placed individually in cages in the animal house of our university 1 week before the start of the study. Clinical signs of disease were not apparent. The animal house was maintained at 25 ± 2°C and 45–65% relative humidity as previously described (20). Antibiotic free commercial feed and water were available *ad libitum*.

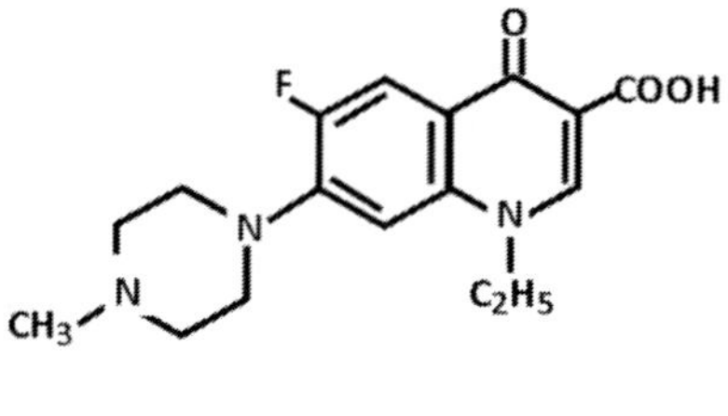

Pefloxacin

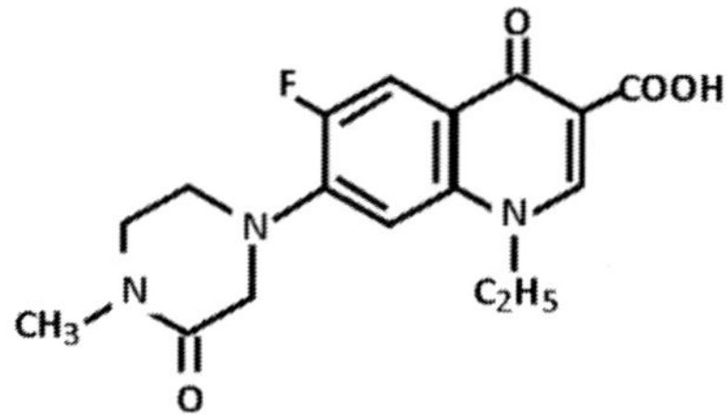

4-oxo-pefloxacin

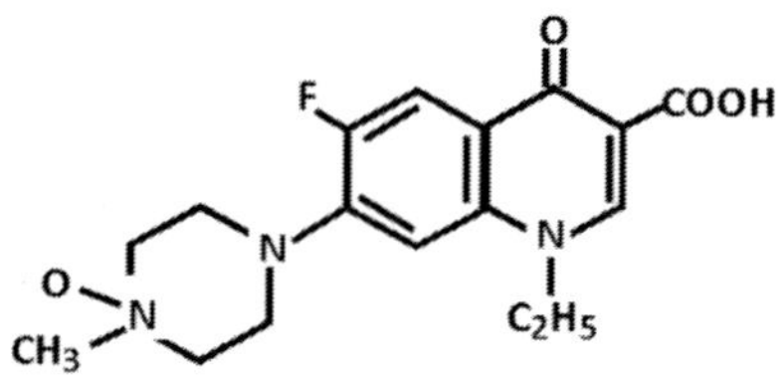

***N*-oxide pefloxacin**

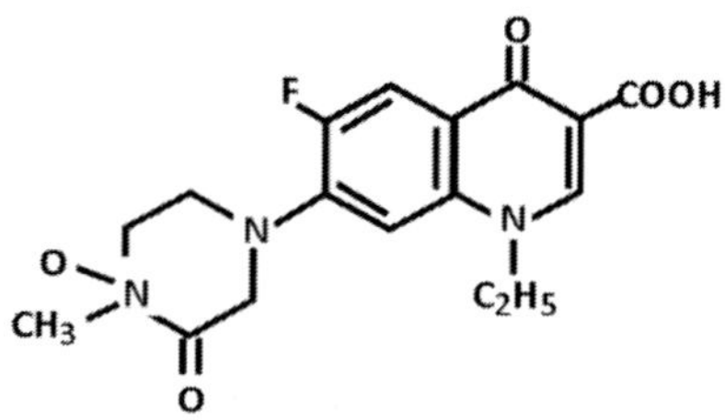

4-oxo *N*-oxide pefloxacin

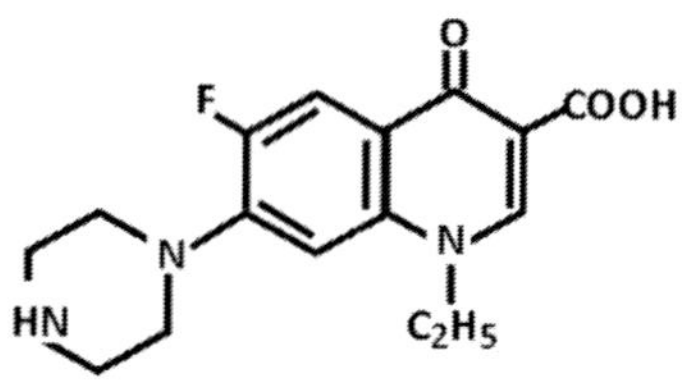

***N*-demethyl pefloxacin (norfloxacin)**

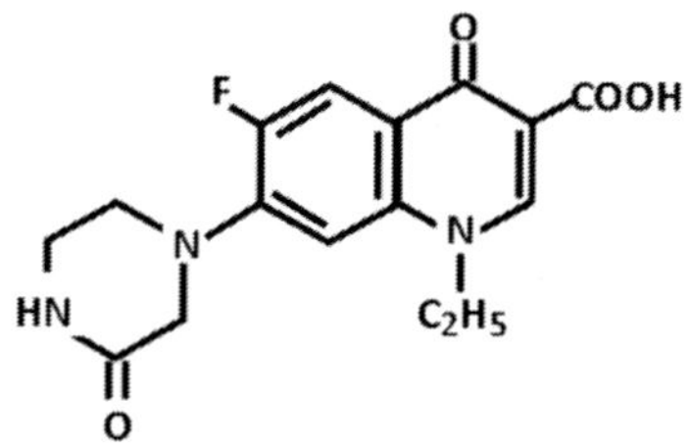

4-oxo-norfloxacin

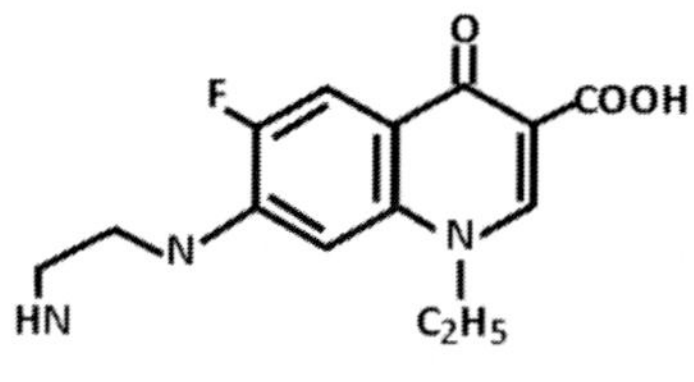

Desethylenenorfloxacin

Figure 1. Chemical structures of pefloxacin and metabolites.

Experimental Design

Eighteen birds were allotted to three groups. Chickens in groups A and B (eight broiler chickens/group) were given single IV or oral doses of pefloxacin (10 mg/kg BW). Chickens of group C (n = 2) did not receive any treatment and were used to determine the validation criteria of the analytical method. For IV administration, 500 mg of pefloxacin was dissolved in 10 mL water (sterilized 0.9% saline solution) to give a stock solution of 50 mg/mL (0.5 mL of stock solution is given IV for each animal of 2.5 kg BW, equivalent to 10 mg/kg BW). For oral gavage administration, 100 mg of pefloxacin was dissolved in 10 mL water (sterilized 0.9% saline solution) to give a stock solution of 10 mg/mL (2.5 mL of stock solution is given orally for each animal of 2.5 kg BW, equivalent to 10 mg/kg BW). Pefloxacin was administered IV into the right brachial vein of chickens in group A or orally directly into the crop using a thin plastic tube attached to a syringe. Food but not water was withheld for 12 h before oral dosing.

Blood samples (1 mL) were drawn from all chickens *via* cannula from the left brachial vein into heparinized syringes at 0.16, 0.33, 0.5, 1, 2, 4, 6, 8, 12, and 24 h after drug administration. Plasma was separated after centrifugation (1,500 *g* for 10 min) and was stored frozen at −45°C until assayed for pefloxacin and *N*-demethyl pefloxacin (norfloxacin) concentrations; experimental design previously described (21).

Analytical Method and Validation

Pefloxacin and *N*-demethyl pefloxacin (norfloxacin) concentrations in plasma were measured using an HPLC technique (22) with modifications.

Plasma Extraction

Plasma samples were separately extracted in methylene chloride as described (23, 24). The plasma sample was added to 8 mL of methylene chloride and 0.5 mL of 0.5 M sodium phosphate buffer pH 7.5. The tube was mechanically shaken and centrifuged at 2,500 *g* for 10 min. The organic phase (lower layer) was transferred into other disposable tube. This extraction was repeated three times, and all organic phases were pooled. Sodium hydroxide (0.5 M, 0.5 mL) was then added to the plasma extract, and the tube was shaken at 1,500 *g* for 10 min. The aqueous phase (upper layer) was collected and frozen (−45°C) until HPLC assay. A 20 μL aliquot was injected into the HPLC column for assay of pefloxacin and *N*-demethyl pefloxacin (norfloxacin).

HPLC Analysis

Plasma concentrations of pefloxacin and *N*-demethyl pefloxacin (norfloxacin) were determined using a Shimadzu liquid chromatographic system equipped with a system controller CBM-20A/CBM-20 Alite, two solvent delivery modules LC-20AD, a spectrofluorimetric detector RF-$10A_{XL}$, and LC solution software. All samples were analyzed using a 5 μm particle size Nucleosil C18 column (12.5 cm × 0.4 cm) preceded by a C18 guard column. The mobile phase (pH 4.8) was acetonitrile (150 mL), sodium acetate trihydrate (2 g), citric acid monohydrate (2 g), trimethylamine (1 mL), and water (850 mL), and a flow rate of 2 mL/min was used. The excitation and emission wavelengths of the detector were 330 and 440 nm, respectively. Peak areas in the sample chromatograms were quantitated by external standard technique using solutions of pefloxacin and norfloxacin reference standards.

The analytical method was fully validated according to EU requirements for the compounds pefloxacin and *N*-demethyl pefloxacin (norfloxacin) norfloxacin [linearity, recovery rate, accuracy, precision, trueness, quantification limit (LOQ), detection limit (LOD) and specificity] (25). Drug concentrations were determined from peak areas and the use of calibration curves obtained by running plasma samples from broiler chickens not administered pefloxacin (i.e., chickens of group C) that were fortified with pefloxacin as well as with *N*-demethyl pefloxacin (norfloxacin). For plasma samples as determined by use of the linear least squares regression procedure, a linear relationship existed in the calibration curve of pefloxacin and *N*-demethyl pefloxacin over the range of 0.01–20 μg/mL, which always yielded a correlation coefficient exceeding 0.9998. Overall mean recovery of pefloxacin and *N*-demethyl pefloxacin (norfloxacin) from plasma was greater than 96%. Within-day and day-to-day precision were <5.5%. The LOQ was 0.02 μg/mL for pefloxacin and 0.03 μg/mL for *N*-demethyl pefloxacin (norfloxacin) in the plasma. Interference of endogenous compounds was verified on blank plasma from untreated chickens, which provided the specificity of the method. This method differs from those reported by Pant et al. (17) and Dimitrova et al. (18) who used, respectively, chloroform-isopentanol, acetonitrile or only protein precipitation with perchloric acid, instead of methylene chloride for the extraction of pefloxacin and *N*-demethyl pefloxacin (norfloxacin) from plasma.

Data Analysis

Plasma concentration *versus* time data was sequentially fitted to 1-, 2- and multiple-compartment models, using the computer program Phoenix (Version 7.0; Pharsight Corporation, Mountain View, CA, USA). The model was determined for best fit on the basis of a smaller value for the Akaike information criterion (26). The two-compartment model was the best fit for all broiler chickens. This model was used to establish pharmacokinetic parameters as described for other drugs (20, 21, 27). Plasma curves of pefloxacin after a single IV and oral administration and those of *N*-demethyl pefloxacin (norfloxacin) (the main metabolite in plasma) after a single IV and oral administration of pefloxacin were obtained for each chicken and were fitted to the following exponential equations (20, 21, 27):

$$C = A_1\, e^{-\alpha t} + A_2\, e^{-\beta t}\,(\mathrm{IV})$$

$$C = A_1\, e^{-\alpha t} + A_2\, e^{-\beta t} - A_3\, e^{-K_a t}\,(\mathrm{oral})$$

where C is the plasma concentration of drug; A_1, A_2, and A_3 are mathematical coefficients (i.e., A_1 and $A_2$are the plasma concentrations extrapolated to time 0 of the first and second elimination phases of drug and A_3 for the absorption phase); α is the hybrid rate constant for the distribution phase; β is the hybrid rate constant for the elimination terminal phase (i.e., α and β are the slopes of the first and second elimination phases of the drug disposition); and K_a the first-order absorption rate constant and t is the time. Absorption half-life ($t_{½a}$), half-life of α phase ($t_{½\alpha}$), half-life of β phase ($t_{½\beta}$), distribution rate constants for transfer of the drug from the central to the peripheral compartment (K_{12}) and from the peripheral to the central compartment (K_{21}), and the elimination rate constant (K_{10}) were calculated using standard equations as described (28, 29). After IV and oral administration, the area under the concentration–time curves (AUC) was calculated as follows:

$$\mathrm{AUC} = (A_1 / \alpha) + (A_2 / \beta);\ \text{or}$$

$$\mathrm{AUC} = (A_1 / \alpha) + (A_2 / \beta) - (A_3 / K_a)$$

Total plasma clearance (CL) was calculated, using the following formula:

$$\mathrm{CL} = (\mathrm{dose/kg})/\mathrm{AUC};\ \text{or}\ \mathrm{CL} = (\mathrm{dose/kg})\,(F/\mathrm{AUC})$$

Oral bioavailability (F) was determined as follows:

$$F = (\mathrm{AUC}_{\mathrm{oral}}) / (\mathrm{AUC}_{\mathrm{IV}})$$

Oral bioavailability (F) was calculated from the ratio between the value of $\mathrm{AUC}_{\mathrm{oral}}$ for each chicken and the mean value of $\mathrm{AUC}_{\mathrm{IV}}$ for the eight chickens used in the IV administration study. Complete absorption was determined on the basis of AUC_{IV}, which represents the mean AUC for the eight broiler chickens to which pefloxacin was administered. Because of the small individual variation in AUC_{IV} and the fact that the same eight chickens were not available for oral and IV studies, the mean AUC_{IV} rather than AUC_{IV} for each chicken was used to estimate bioavailability after oral administration of pefloxacin.

Mean residence time (MRT) was calculated as follows:

$$MRT = (A_1 / \alpha^2 + A_2 / \beta^2) \times (1 / AUC)$$

Apparent volume of distribution [$V_{d(area)}$] was determined as follows:

$$V_{d(area)} = (dose/kg)/AUC \cdot \beta; \text{ or } V_{d(area)} = (dose/kg)(F)/AUC \cdot \beta$$

Volume of distribution at steady state (V_{ss}) was determined as follows:

$$V_{ss} = MRT \times CL$$

Maximum drug plasma concentration (C_{MAX}) after oral administration and the time at which C_{MAX} was achieved (T_{MAX}) was determined directly from the concentration *versus* time curve.

Mean pharmacokinetic variables were obtained by averaging the variables calculated for drug disposition after each pefloxacin administration in each broiler chicken.

Differences in pharmacokinetic data between dosing routes were analyzed for statistical significance by the Mann–Whitney *U* test. Differences of $P < 0.05$ were considered significant. All data were tabulated as mean ± SD.

RESULTS

Plasma Pefloxacin Disposition after Single IV and Oral Administration

Mean plasma concentrations (in micrograms per milliliter, ±SD) of pefloxacin and its metabolite *N*-demetil pefloxacin (norfloxacin) obtained after single oral and IV dose of pefloxacin are presented in Figure 2. The plasma concentration–time profile of pefloxacin and *N*-demetil pefloxacin after oral and IV administration of pefloxacin for each chicken were similar to the overall means. Values of the parameters that described absorption and disposition kinetics of pefloxacin in broiler chickens are presented in Table 1. The kinetic parameters of the metabolite *N*-demetil pefloxacin (norfloxacin) after oral administration of pefloxacin are summarized in Table 2.

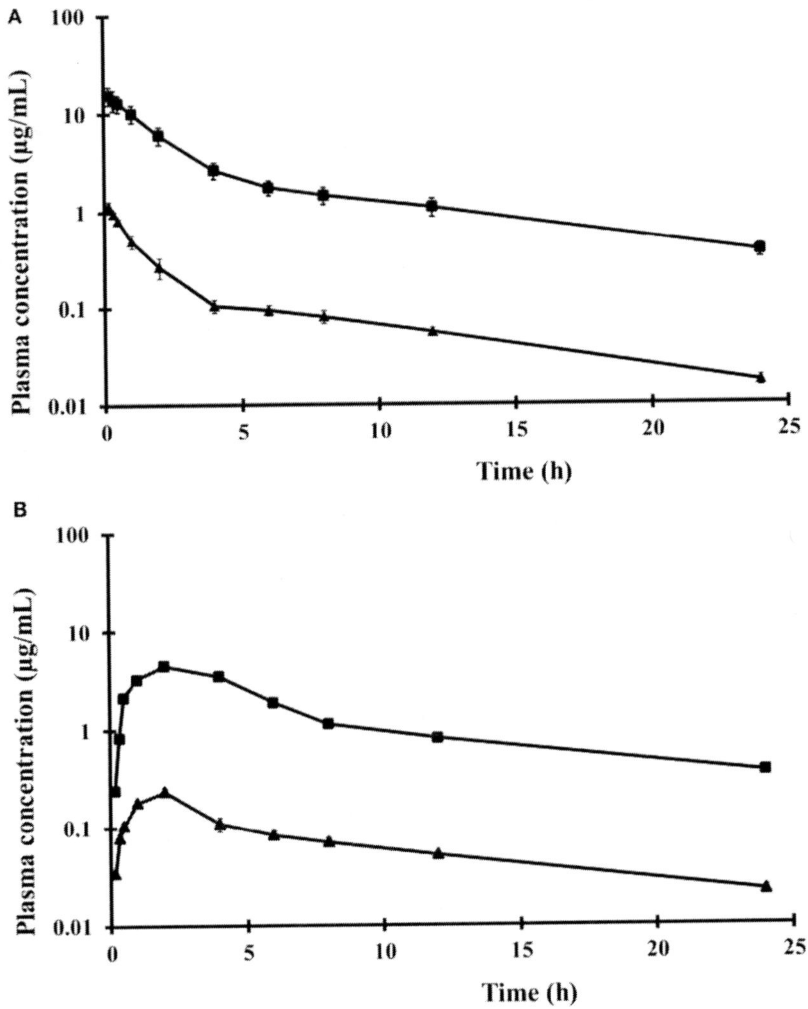

Figure 2. PLASMA CONCENTRATIONS OF PEFLOXACIN (■) AND *N*-DEMETHYL PEFLOXACIN (NORFLOXACIN) (▲) AFTER SINGLE IV ADMINISTRATION (A) AND AFTER SINGLE ORAL ADMINISTRATION (B) OF 10 MG PEFLOXACIN/KG BW. Data are expressed as mean ± SD values for eight broiler chickens. Symbols without bars indicate that the SD is within the symbols.

Table 1. Pefloxacin kinetic parameters for broiler chickens after a single iv and oral administration of pefloxacin (10 mg/kg bw).

Parameter	IV	Oral
A_1 (μg/mL)	14.77 ± 3.50	16.33 ± 5.13
A_2 (μg/mL)	2.69 ± 0.63	1.27 ± 0.16
A_3 (μg/mL)	–	18.40 ± 5.33
α (h^{-1})	0.70 ± 0.06	0.43 ± 0.07
β (h^{-1})	0.082 ± 0.005	0.054 ± 0.004
K_a (h^{-1})	–	0.80 ± 0.07
$t_{½\alpha}$ (h)	1.00 ± 0.09	1.68 ± 0.36
$t_{½\beta}$ (h)	8.44 ± 0.48	13.18 ± 0.82***
$t_{½a}$ (h)	–	0.87 ± 0.07
$V_{d(area)}$ (L/kg)	2.36 ± 0.46	3.55 ± 0.21
V_{ss} (L/kg)	1.54 ± 0.33	–
K_{12} (h^{-1})	0.28 ± 0.03	0.16 ± 0.04
K_{21} (h^{-1})	0.18 ± 0.02	0.10 ± 0.01
K_{10} (h^{-1})	0.32 ± 0.04	0.21 ± 0.03
AUC (mg/h/L)	53.56 ± 10.09	37.71 ± 0.96
F (%)	–	70 ± 2
MRT (h)	7.94 ± 0.50	13.57 ± 0.72
CL (L/h/kg)	0.19 ± 0.04	0.19 ± 0.004
C_{MAX} (μg/mL)	–	4.02 ± 0.31
T_{MAX} (h)	–	2.01 ± 0.12

*Values are the mean ± SD, n = 8. Significantly different ***P < 0.001.*

A_1, A_2, and A_3, mathematical coefficients; α, hybrid rate constant for distribution phase; β, hybrid rate constant for terminal elimination phase. K_a, first-order absorption rate constant; $t_{½a}$, absorption half-life; $t_{½\alpha}$, half-life at α phase; $t_{½\beta}$, half-life at β phase; $V_{d(area)}$, apparent volume of distribution; V_{ss}, volume of distribution at steady state; K_{12}, distribution rate constant for transferring the drug from the central to the peripheral compartment; K_{21}, distribution rate for transfer from peripheral to central compartment; K_{10}, elimination rate constant; AUC, area under the concentration–time curve; MRT, mean residence time; CL, total plasma clearance; C_{MAX}, maximal concentration in plasma after oral administration; T_{MAX}, time needed to reach C_{MAX}.

–, Not applicable.

Table 2. *N*-demethyl pefloxacin (norfloxacin) kinetic parameters for broiler chickens after a single iv and oral administration of the parent drug, pefloxacin (10 mg/kg bw).

Parameter	IV	Oral
$T_{½α}$ (h)	0.60 ± 0.20	1.25 ± 0.13
$t_{½β}$ (h)	8.02 ± 0.47	10.93 ± 0.80
K_{12} (h^{-1})	0.61 ± 0.19	0.24 ± 0.05
K_{21} (h^{-1})	0.21 ± 0.03	0.21 ± 0.02
K_{10} (h^{-1})	0.51 ± 0.07	0.17 ± 0.01
AUC (mg/h/L)	2.70 ± 0.33	1.93 ± 0.18
MRT (h)	7.54 ± 0.47	13.77 ± 0.86
C_{MAX} (μg/mL)	–	0.19 ± 0.01
T_{MAX} (h)	–	1.51 ± 0.12

Values are the mean ± SD, n = 8.
*See **Table 1** for abbreviations.*

After IV administration of pefloxacin, a rapid distribution phase ($t_{½α}$ = 1.00 ± 0.09 h) and a slower elimination phase ($t_{½β}$ = 8.44 ± 0.48 h) were observed (Table 1). The apparent volume of distribution [$V_{d(area)}$] and at steady state (V_{ss}) and clearance (CL) values were 2.36 ± 0.46 L/kg, 1.54 ± 0.33 L/kg, and 0.19 ± 0.04 L/h/kg, respectively (Table 1). Pefloxacin was rapidly and widely absorbed after oral administration (10 mg/kg BW). Pefloxacin concentrations in plasma 10 and 20 min were 0.24 ± 0.02 and 0.84 ± 0.02 μg/mL, and plasma drug concentrations were 1.15 ± 0.03 and 0.81 ± 0.02 μg/mL for 8 and 12 h, respectively. The half-life of oral absorption ($t_{½a}$) was 0.87 ± 0.07 h. Bioavailability (*F*) of pefloxacin after oral administration was 70 ± 2%. Maximal plasma concentration of pefloxacin (C_{MAX}: 4.02 ± 0.31 μg/mL) was detected at 2.01 ± 0.12 h after oral administration.

A fraction of pefloxacin was biotransformed to *N*-demethyl pefloxacin (norfloxacin) after oral and IV administration of pefloxacin. This metabolite represented 5% of the parent drug plasma concentrations, as calculated by the ratio between the mean AUC for *N*-demetil pefloxacin and mean AUC for pefloxacin after oral and IV administration of pefloxacin. Plasma concentration of *N*-demethyl pefloxacin (0.19 ± 0.01 μg/mL) peaked at 1.51 ± 0.12 h after oral administration of pefloxacin. The $t_{½β}$ of *N*-demethyl pefloxacin after oral pefloxacin administration was 10.93 ± 0.80 h (Table 2).

Efficacy Predictors

Table 3 shows the estimated values for AUC_{24}/MIC and C_{MAX}/MIC for the MIC90 (upper value) and MIC50 (lower value) (1, 30–33). The applied C_{MAX} and AUC_{24} values for single oral administration were 4.02 μg/mL and 37.71 mg/h/L, respectively.

Table 3. Efficacy predictors (c_{max}/mic and auc_{24}/mic) estimated for pefloxacin against susceptible bacteria in broiler chickens after a single oral dose of 10 mg/kg bw.

Bacteria	C_{MAX}/MIC	AUC_{24}/MIC (h)
Escherichia coli—MIC 0.25–0.06 (μg/mL)	16.08–67	150.84–628.5
Salmonella spp.—MIC 0.25–0.03 (μg/mL)	16.08–134	150.84–1,257
Pseudomonas aeruginosa—MIC 2–0.06 (μg/mL)	2.01–67	18.85–628.5
Haemophilus influenza—MIC 0.06–0.03 (μg/mL)	67–134	628.5–1,257
Shigella spp.—MIC 0.2–0.06 (μg/mL)	20.1–67	188.55–628.5
Staphylococcus aureus—MIC 0.50–0.25 (μg/mL)	8.04–16.08	75.42–150.84
Mycoplasma gallisepticum—MIC 0.1 (μg/mL)	40.2	377.1

For calculations, the MIC applied values were MIC_{90} upper and MIC_{50} lower (26–30).

DISCUSSION

The present paper is the first to describe the kinetic parameters for pefloxacin and its metabolite *N*-demethyl pefloxacin (norfloxacin) in chickens for fattening. The other major metabolite *N*-oxide pefloxacin was not investigated because it is practically inactive (7). In this study, the kinetics of pefloxacin and *N*-demethyl pefloxacin (norfloxacin) after a single IV and oral administration of pefloxacin (10 mg/kg BW) were determined. Disposition of pefloxacin and *N*-demethyl pefloxacin (norfloxacin) after IV and oral administration of pefloxacin in chickens was best described by use of a two-compartment model, in accordance with most of the results reported for humans (8, 10), calves (11), goats (13, 14), pigs (15), and sheep (16), but not for ducks and chickens (17, 18).

The kinetic variables obtained after IV administration showed for pefloxacin a rapid distribution ($t_{½\alpha}$ = 1.00 ± 0.09 h) and a slow elimination phase ($t_{½\beta}$ = 8.44 ± 0.48 h). Pefloxacin is well distributed to the tissues [$V_{d(area)}$ = 2.36 ± L/kg and V_{ss} = 1.54 ± 0.33 L/kg, respectively]. The elimination half-life of pefloxacin after IV administration ($t_{½β}$ = 8.44 ± 0.48 h) was much longer than those previously reported in other studies: 2.21 h in calves (10), 1.12–1.6 h in goats (12, 13), and 2.84–3.25 in ducks (18) but comparable to that reported in sheep ($t_{½β}$ = 6.88 h) (16).

Pefloxacin was rapidly ($t_{½a}$ = 0.87 ± 0.07 h) and extensively absorbed after oral administration with a comparable maximal plasma concentration (C_{MAX} = 4.02 ± 0.31 μg/mL) but a shorter T_{MAX} (2.01 ± 0.12 h) to those values previously reported in the literature for broiler chickens (C_{MAX} = 3.78 ± 0.23 μg/mL at a T_{MAX} = 3.33 ± 0.21 h) (17). However, the C_{MAX} and T_{MAX} detected in our study for pefloxacin were both much higher than those reported for ducks (C_{MAX} = 1.42 ± 0.20 μg/mL at a T_{MAX} = 1.35 ± 0.23 h) (18). Oral bioavailability of pefloxacin was 70 ± 2% in our study, which was comparable to that reported in ducks (60–68%) (18). In the present study, plasma concentrations (0.84 ± 0.02 μg/mL) were achieved in a relatively short time (20 min) and maintained up to 12 h (1.15 and 0.81 μg/mL), equivalent values previously observed (18). In our study, $t_{½β}$ (13.18 ± 0.82 h) after oral dose of pefloxacin was statistically higher than the $t_{½β}$ (8.44 ± 0.48 h) after IV administration. This difference may be the result of continued absorption of pefloxacin probably by an enterohepatic circulation with reabsorption of the drug from the gastrointestinal tract.

Pefloxacin was poorly metabolized in chickens by *N*-demethylation. In the present study, in broiler chickens the metabolite *N*-demethyl pefloxacin (norfloxacin) only represented 5% of the parent drug plasma concentrations, showing that this metabolite with a lower potency of activity than parent compound (7) should not be taken into account to calculate antimicrobial dosing regimen of pefloxacin for broiler chickens. The maximal plasma concentration (C_{MAX}) of *N*-demethyl pefloxacin (norfloxacin) was calculated as 0.19 ± 0.01 mg/mL, value much lower than that previously reported in broiler chickens (C_{MAX}: 0.80 ± 0.07 mg/mL) (17). The small value of C_{MAX} also indicates that in chickens the antibiotic action can mostly be attributed to the parental drug pefloxacin rather than its metabolite of *N*-demethyl pefloxacin (norfloxacin). These results of the present study are comparable to those previously reported in dog, monkey, and humans after oral doses of pefloxacin (8). The antimicrobial activity in dog, monkey, and human plasma was largely due to the parent drug which, respectively, accounted for 64, 94, and 84% of the total activity (8).

In the present study, the rate of elimination of *N*-demethyl pefloxacin (norfloxacin) ($t_{½β}$: 10.93 ± 0.80 h) after oral pefloxacin administration was more rapid than that of pefloxacin ($t_{½β}$: 13.18 ± 0.82 h) ($P < 0.01$) but not after IV pefloxacin administration ($t_{½β}$ of *N*-demethyl pefloxacin: 8.02 ± 0.47 h and $t_{½β}$ of pefloxacin: 8.44 ± 0.48). In a previous study in chickens, the elimination half-life of *N*-demethyl pefloxacin (norfloxacin) after oral dose of 10 mg pefloxacin/kg BW was 5.66 h (17). This difference is probably the result of the different kinetic profile (two-compartment open model) applied in our study for the metabolite norfloxacin.

The integration of pharmacokinetic–pharmacodynamic (PK–PD) data represents the most approach to determining dosing regimens of antimicrobial drugs for subsequent evaluation in disease models and clinical trials (34, 35). As fluoroquinolones, against most if not all susceptible pathogens, kill bacteria by a concentration-dependent killing action, the PK–PD variables widely used to predict effective doses are the ratio of the AUC at 24 h to the MIC (AUC_{24h}/MIC) and the ratio of the maximal drug concentration to the MIC (C_{MAX}/MIC) (36). The AUC_{24h}/MIC ratio is the most important variable in predicting such effects, with the rate of clinical cure being greater than 80% when this ratio is higher that 100–125 h (37–39). The second predictor of efficacy for concentration-dependent antibiotic is the ratio C_{MAX}/MIC, considering that values above 8–10 would lead to better clinical results, as well as, to prevent resistant bacterial mutants surviving treatment (38, 40, 41). Nevertheless, it must be emphasized that these values provide general guidance, and lower or higher numerical values may apply against organisms of all classes (36). In the present study, considering the AUC_{24h}/MIC and C_{MAX}/MIC ratios obtained (Table 3), it can be concluded that pefloxacin administered orally at a dose rate of 10 mg/kg BW in broiler chickens might be effective against bacteria with MIC values ≤0.50 μg/mL but not sufficient for treating infections of bacteria with MIC over 0.8 μg/mL. Nevertheless, further investigation on pefloxacin plasma disposition in broiler chickens after multiple oral doses (10 mg pefloxacin/kg BW, daily for 5 days) should be carried out to recommend that oral dose rates of 10 mg/kg p.c. once daily produce blood concentrations that should be effective against susceptible bacteria in broiler chickens.

ETHICS STATEMENT

The study was undertaken in accordance with the ethics requirements and authorized by the official ethical committee for animal testing of our university.

AUTHOR CONTRIBUTIONS

The authors have contributed in the paper to fix the objectives based on the related works reported in the literature, apply methodology, and calculate the results and discuss them.

CONFLICT OF INTEREST STATEMENT

The authors declare that the research was conducted in the absence of any commercial or financial relationships that could be construed as a potential conflict of interest.

FUNDING

This work was supported by projects (ALI-BIRD-CM Program) Ref. S2013/ABI-2728 from Comunidad de Madrid.

REFERENCES

1. Gonzalez JP, Henwood JM. Pefloxacin. A review of its antibacterial activity, pharmacokinetic properties and therapeutic use. *Drugs* (1989) 37(5):628–68.
2. Shen LL, Perne AG. Mechanism of inhibition of DNA gyrase by analogues of nalidixic acid: the target of the drugs in DNA. *Proc Natl Acad Sci U S A* (1985) 82(2):307–11.
3. Campoli-Richards DM, Monk JP, Price A, Benfield P, Todd PA, Ward A. Ciprofloxacin: a review of its antibacterial activity, pharmacokinetic properties and therapeutic use. *Drugs* (1988) 35(4):373–447.
4. Chin NX, Novelli A, Neu HC. *In vitro* activity of lomefloxacin (SC-47111; NY-198), a difluoroquinolone 3-carboxylic acid, compared with those of other quinolones. *Antimicrob Agents Chemother* (1988) 32(5):656–62.
5. Henwood JM, Monk JP. Enoxacin: a review of its antibacterial activity, pharmacokinetic properties and therapeutic use. *Drugs* (1988) 36(1):32–66.
6. Monk J-P, Campoli-Richards DM. Ofloxacin: a review of its antibacterial activity, pharmacokinetic properties and therapeutic use. *Drugs* (1987) 33(4):346–91.
7. Neuman M. Comparative pharmacokinetic parameters of new systemic fluoroquinolones: a review. *Chemioterapia* (1987) 6(2):105–12.
8. Montay G, Goueffon Y, Roquet F. Absorption, distribution, metabolic fate and elimination of pefloxacin mesylate in mice, rats, dogs, monkeys and humans. *Antimicrob Agents Chemother* (1984) 25(2):463–72.
9. Frydman AM, Le Roux Y, Lefebvre MA, Djebbar F, Fourtillan JB, Gaillot J. Pharmacokinetics of pefloxacin after repeated intravenous and oral administration (400 mg bid) in young healthy volunteers. *J Antimicrob Chemother* (1986) 17(Suppl B):65–79.
10. Barre J, Houin G, Tillement JP. Dose-dependent pharmacokinetic study of pefloxacin, a new antibacterial agent, in humans. *J Pharm Sci* (1984) 73(10):1379–82.
11. Srivastava AK, Dumka VK, Deol SS. Disposition kinetics and urinary excretion of pefloxacin after intravenous injection in crossbred calves. *Vet Res Commun* (2000) 24(3):189–96.
12. Marrakchi-Benjaafar S, Cochereau I, D'Hermies F, Pocidalo JJ. Tolerability, kinetics, and efficacy of subconjunctival pefloxacin in pigmented rabbits. *Antimicrob Agents Chemother* (1995) 39(4):834–8.
13. Abd El-Aty AM, Goudah A. Some pharmacokinetic parameters of pefloxacin in lactating goats. *Vet Res Commun* (2002) 26(7):553–61. doi:10.1023/A:1020243615928

14. Malik JK, Rao GS, Ramesh S, Muruganandan S, Tripathi HC, Shukla DC. Pharmacokinetics of pefloxacin in goats after intravenous or oral administration. *Vet Res Commun* (2002) 26(2):141–9.
15. Dimitrova D, Katsarov V, Tsoneva D. Pharmacokinetics of pefloxacin in pigs after single intramuscular application. *Agric Sci Technol* (2012) 4:215–9. d
16. Moutafchieva R, Djouvinov D. Pharmacokinetics of pefloxacin in sheep. *J Vet Pharmacol Ther* (1997) 20(5):405–7.
17. Pant S, Rao GS, Sastry KVH, Tripathi HC, Agmohant J, Malik JK. Pharmacokinetics and tissue residues of pefloxacin and its metabolite norfloxacin in broiler chickens. *Br Poult Sci* (2005) 46(5):615–20.
18. Dimitrova G, Moutafchieva R, Kanelov I, Dinev T, Yanev S, Pandova B, et al. Pharmacokinetics of pefloxacin and its metabolite norfloxacin in male and female ducks. *J Vet Pharmacol Ther* (2008) 31(2):167–70.
19. Isea G, Martinez MA, Martinez-Larrañaga MR, Diaz MJ, Anadón A. Pharmacokinetic characteristics of pefloxacin and its metabolite *N*-demethyl pefloxacin in chickens. 9th International EAVPT Congress. Lisboa (Portugal), July 13-18. *J Vet Pharmacol Ther* (2003) 26(Suppl 1):113–4.
20. Anadón A, Martínez MA, Martínez M, De La Cruz CO, Díaz MJ, Martínez-Larrañaga MR. Oral bioavailability, tissue distribution and depletion of flumequine in the food producing animal, chickens for fattening. *Food Chem Toxicol* (2008) 46:662–70.
21. Anadón A, Suarez FH, Martínez MA, Castellano V, Martinez M, Ares I, et al. Plasma disposition and tissue depletion of difloxacin and its metabolite sarafloxacin in the food producing animals, chicken for fattening. *Food Chem Toxicol* (2011) 49:441–9.
22. Montay G, Tassel JP. Improved high-performance liquid chromatographic determination of pefloxacin and its metabolite norfloxacin in human plasma and tissue. *J Chromatogr* (1985) 339(1):214–8.
23. Forchetti C. High performance liquid chromatographic procedure for the quantification of norfloxacin un urine, serum and tissues. *J Chromatogr* (1984) 309:277–82.
24. Anadón A, Martínez-Larrañaga MR, Díaz MJ, Fernandez R, Martínez MA, Fernandez MC. Pharmacokinetics and tissue residues of norfloxacin and its N-desethyl- and oxo-metabolites in healthy pigs. *J Vet Pharmacol Ther* (1995) 18(3):220–5.
25. The European Agency for the Evaluation of Medicinal Products (EMEA). *Guideline on Validation of Analytical Procedures: Methodology (VICH GL2)*. (1998) London, UK: EMEA. p. 1–10. CVMP/VICH/591/98-FINAL.
26. Yamaoka K, Nakagawa T, Uno T. Application of Akaike's information criterion (AIC) in the evaluation of linear pharmacokinetic equations. *J Pharmacokinet Biopharm* (1978) 6(2):165–75. doi:10.1007/BF01117450
27. Anadón A, Martínez MA, Martínez M, Rios A, Caballero V, Ares I, et al. Plasma and tissue depletion of florfenicol and florfenicol-amine in chickens. *J Agric Food Chem* (2008) 56:11049–56.
28. Wagner JG, editor. *Fundamentals of Clinical Pharmacokinetics*. Hamilton, IL: Drug Intelligence Publications (1975).
29. Wagner JG. Linear pharmacokinetic equations allowing direct calculation of many needed pharmacokinetic parameters from the coefficients and exponents of poly-exponential equations which have been titled to the data. *J Pharmacokinet Biopharm* (1975) 4:443–67.
30. Appelbaum PC, Spangler SK, Sollenberger L. Susceptibility of non-fermentative gram-negative bacteria to ciprofloxacin, norfloxacin, amifloxacin, pefloxacin and cefpirome. *J Antimicrob Chemother* (1986) 18(6):675–9.
31. Clarke AM, Zemcov SJ, Campbell ME. *In-vitro* activity of pefloxacin compared to enoxacin, norfloxacin, gentamicin and new β-lactams. *J Antimicrob Chemother* (1985) 15(1):39–44.
32. Felmingham D, O'Hare MD, Robbins MJ, Wall RA, Williams AH, Cremer AW, et al. Comparative *in vitro* studies with 4-quinolone antimicrobials. *Drugs Exp Clin Res* (1984) 11(5):317–29.

33. Maple P, Brumfitt W, Hamilton-Miller JM. A review of the antimicrobial activity of the fluoroquinolones. *J Chemother* (1990) 2(5):280–94.
34. Ambrose PG, Bhavnani SM, Rubino CM, Louie A, Gumbo T, Forrest A, et al. Pharmacokinetics-pharmacodynamics of antimicrobial therapy: it's not just for mice anymore. *Clin Infect Dis* (2007) 44(1):79–86.
35. Toutain PL, del Castillo JR, Bousquet-Melou A. The pharmacokinetic-pharmacodynamic approach to a rational dosage regimen for antibiotics. *Res Vet Sci* (2002) 73:105–14.
36. Lees L. Pharmacokinetics, pharmacodynamics and therapeutics of pradofloxacin in the dog and cat. *J Vet Pharmacol Ther* (2013) 3:209–21.
37. Forrest A, Nix DE, Ballow CH, Goss TF, Birmingham MC, Schentag JJ. Pharmacodynamics of IV ciprofloxacin in seriously ill patients. *Antimicrob Agents Chemother* (1993) 37:1073–81.
38. Madaras-Kelly KJ, Ostergaard BE, Baeker Hovde L, Rotschafer JC. Twenty-four-hour area under the concentration-time curve/MIC ratio as a generic predictor of fluoroquinolone antimicrobial effect by using three strains of *Pseudomonas aeruginosa* and an *in vitro* pharmacodynamic model. *Antimicrob Agents Chemother* (1996) 40:627–32.
39. Lode H, Borner K, Koeppe P. Pharmacodynamics of fluoroquinolones. *Clin Infect Dis* (1998) 27:33–9.
40. Dudley MN. Pharmacodynamics and pharmacokinetics of antibiotics with special reference to the fluoroquinolones. *Am J Med* (1991) 91(Suppl 6A):45–50.
41. Walker RD. The use of fluoroquinolones for companion animal antimicrobial therapy. *Aust Vet J* (2000) 78:84–90.

CHAPTER 6

Evidence Based Phytopharmacological Potential of Herbal Extracts in Post-Ingestion Management of Mycotoxins in Animal Models

Juma KK[1*], Fulakeza RMJ[2], Ngeranwa JN[1], Ngugi MP[1] and Mburu ND[1]

[1]*Department of Biochemistry and Biotechnology, Kenyatta University, P.O. Box 43844-00100, Nairobi, Kenya*

[2]*Department of Basic Medical Sciences, University of Malawi, College of Medicine, P/Bag 360, Chichiri, Blantyre 3, Malawi, Kenya*

ABSTRACT

Aflatoxins and ochratoxins are common mycotoxic food poisons and acute doses are responsible for causing liver and kidney damage and hepatocarcinoma. The strategy for managing mycotoxins has mainly been pre-ingestive. However, few solutions have been proposed for post-ingestion management and likewise there is no antidote for managing the condition. Studies have demonstrated that Saccharomyces cerevisiae and herbal extracts have potential of limiting the damage caused by reactive oxygen species and toxic effects associated with mycotoxins. Further evidence also shows that herbal extracts provides protection to a variety of cancer cell lines associated with mycotoxins. This study therefore examines the probable potential and options of using medicinal herbal extracts, on their own or in combinations with Saccharomyces cerevisiae, in the management of mycotoxicosis. A combination of the two has potential to offer a multifunctional approach in the management of post ingested mycotoxins such as aflatoxins and ochratoxins.

Keywords: Aflatoxins; Ochratoxins; **MYCOTOXICITY**; Herbal extracts; *Saccharomyces cerevisiae*; Hepatoprotection; Nephroprotection

INTRODUCTION

Aflatoxins are naturally occurring mycotoxins produced by the fungi *Aspergillus flavus* and *Aspergillus parasitus.* They were first isolated from peanut meal [1]. These mycotoxins are one of the greatest health hazards and they are major contaminants of food consumed by humans and animals. There are over twenty types of aflatoxins with aflatoxins B1 being the most dangerous. In animals, it has been associated with anorexia, reduced feed intake and, poor digestion [2]. It also has an impact on the immune system leading to immunosuppression and low weight gain [3]. Some cases of aflatoxin poisoning have been associated with reduced reproductive potencies [4]. Aflatoxin B1 has also been shown to be teratogenic, mutagenic and tremorgenic [4]. It is also associated with destructive effects of the vital organs of the body such as the kidneys, central nervous system and more importantly the liver.

An assessment of the effect of exposure of laboratory animals to different doses of aflatoxin B1 demonstrates that short exposure to large doses of aflatoxin B1 leads to acute toxicity characterized by fever, abdominal pains, lethargy, vomiting, edema, fatal loss of function of the liver and rough hair coat. Histological examination of the tissues in acute poisoning leads to the leucocytic infiltrations and vascular congestion of tissues. The liver fibroblast and mononuclear cells are also affected by the reactions of aflatoxin B1 in the liver [5]. In males aflatoxicosis is known to affect spermatogenesis and androgen biosynthesis in the testis. Epididymis, seminiferous tubules and vas deferens have also been cited to be targeted by aflatoxin B1 [5-9]. Deleterious effects on sexual maturation, hormone levels, gestation and growth of fetus are also reported [10,11]. Aflatoxins can cause death in 72 hours when administered in chronic doses [4-8].

The most toxic and prevalent ochratoxin is ochratoxin type A. The relevance of ochratoxins type B and C have not been determined yet and are considered less important. Ochratoxin A is common in cereals, coffee, fruits and red wine and is a potent human carcinogen. It bio accumulates in the bodies of animals. Meat and meat products contamination has been well demonstrated in cases of ochratoxins [12]. When ingested in diet, it is known to induce acute nephrotoxicity in mammals [12].

TYPES AND STRUCTURES OF MYCOTOXINS

Aflatoxins: There are four major type of aflatoxins designated as B1, B2, G1 and G2. They are designated so because of their RI values on thin layer chromatography plates. The structure of the aflatoxins is referred to as difurocoumarolactones which are also referred to as difurocoumarin derivatives. The structures consist of a bifuran ring that is fused to a coumarone nucleus with a pentanone ring in aflatoxin B and M while aflatoxin G is a six membered ring [12].

OCHRATOXINS: Like the aflatoxins ochratoxins are produced mainly by the fungus Aspergillus ochraceus. However, they are also produced by Aspergillus niger which has significant industrial applications. Some penicillum species have also been documented to be associated with production of ochratoxins (*P. verrucosum* and *P. carbonarius*) [12].

Ochratoxics, like aflatoxins, are secondary metabolites. Biosythetically, their structures are those of a pentaketide derived from dihydrocoumarins family of compounds and are linked to a β-phenylalanine [13]. Metabolites of ochratoxin A include ochratoxin B which is a dechloro analog and ochratoxin C which is an ethyl ester. Others forms of ochratoxins that have been identified are ochratoxins α which is a chlorinated molecule and its analogue ochratoxin β. Recent findings have demonstrated variants of ochratoxins A. One is a dechlorinated ochratoxin A [14] and another quinone/hydroquinone derived metabolites [15] with serious toxic properties. The composition of the various ochratoxin A derived metabolites has been well demonstrated by Khoury and Atoui, [16].

Pharmacodynamics and Biotransformation of Mycotoxins

The absorption of aflatoxins occurs through the gastrointestinal tracts with ingestion of small doses. The toxin is then transported to the body tissues and organs including the lipophilic tissues. Aflatoxin B1 accumulates in the liver with a very high concentration [17]. Aflatoxin B1 is metabolized through microsomal enzymes by hydroxylation, hydration, demethylation, and epoxidation reactions in liver cells. The biotransformation pathway for aflatoxin B1 has been demonstrated by Patterson and Allcroft [18]. Hydroxylation of aflatoxin B1 at carbon 4 or 22 yield's the production of aflatoxin M1 and aflatoxin Q1. Hydration reactions at carbon 2-3 double bonds lead to the formation of aflatoxin B2a. Aflatoxin B1-epoxide is formed by epoxidation at carbon 2, 3 double bonds. On the other hand, O-demethylation leads to the formation of aflatoxin P1 [18].

Aflatoxin B1 causes damage to the liver and adversely affects the key metabolic pathways of carbohydrates, proteins and lipids [19]. The impairment in the metabolism of proteins has been associated with the reduction in resistance of poultry to infections from Salmonella spp, coccidian and *Candida albicans* [20]. This has also led to the anticoagulation of blood as a result of the impairment of factor II and VII used for the biosynthesis of prothrombin and in the activation of the clotting process [21]. Aflatoxin is a cytotoxin and it interferes with the RNA and and ultimately proteins synthesis. Aflatoxin B1 is excreted in urine, feaces and in milk of the lactating mothers. It occurs as unchanged or other metabolites [22]. The Milk aflatoxin M1 is a metabolite of aflatoxin B1 [23].

Mycotoxins and Health

Aflatoxin B1 has been classified as a class one carcinogen and its concentration in the body of animals and humans have been correlated to the health effects of liver cancer and cirrhosis [24]. Moreover, it has also been associated with immune suppression, reproductive disorders, hepatic damage, hepatocarcinoma and nutritional disorders such as stunted growth. The nature of the response of the body to the toxin correlates to the levels of exposure to aflatoxins [25]. The activity of aflatoxins is associated with the pro-oxidant potential associated with aflatoxin B1. The reactions of aflatoxin B1 releases reactive oxygen species (ROS) such as the superoxide anion (O-), hydrogen peroxide (H_2O_2) and hydroxyl radical (-OH) promoted by liver enzymes [26]. The generation of ROS induces oxidative stress to the liver leading to damage of the liver cell membrane and cellular components. In contrast, ochratoxins are known to cause damage to the kidney [12]. There is a possibility that ochratoxins promote hepatotoxicity following evidence of hepatomegaly when used alone on laboratory animals [27,28].

EFFECTS ON BODY AND ORGAN WEIGHTS

Ingestion of aflatoxins and ochratoxin individually or in combination has been demonstrated to affect the gross body weight and performance in animal study models [28-33]. Ingestion of aflatoxin contaminated food alters the percentage organ weight. In most of the cases, hepatomegaly has been reported while there was hypertrophy of the spleen; thymus and bursa of fabricius in cases of aflatoxicosis [34,35]. Similar findings have also been demonstrated in ochratoxicosis [28,36,37]. Moreover, in mycotoxicosis involving a combination of aflatoxin and ochratoxin, a similar effect was observed in broiler chicken [29,32].

EFFECTS ON BIOCHEMICAL LIVER PROFILES

Damage to liver cells is indicated by the elevation of liver enzymes levels of aspartate amino transferase (AST), alanine amino transferase (ALT), alkaline phosphatase (ALP), lactate dehydrogenase (LDH) and gamma glutamyl transferase (γ-GT) among other key liver enzymes [38,39]. This follows the distortion of the integrity of the permeable membrane leading to the leakage of intracellular enzymes out of the cells detectable in blood. Aflatoxins have been shown to significantly increase the levels of ALT, AST, uric acid, and creatine levels in the blood serum in mice fed on aflatoxin [40].

Aflatoxicosis and ochratoxicosis reduces serum total protein and albumin in animal models [33]. In addition, when aflatoxin is used alone, it has a similar effect [41]. Similarly, ochratoxins reduce the serum proteins in broilers [28,36,42,43]. The synergistic effect of combining aflatoxins and ochratoxins in chicks also lower the serum proteins and albumins [36,44]. It has been suggested that the reduction of the serum proteins and albumin could be as a result of the pathological damage induced on the tissues of the animal models by mycotoxins and ochratoxins [33].

Alkaline phosphatases have been reported to increase in dietary aflatoxicosis and ochratoxicosis alone and in combinations. This has been suggested to be as a result of impaired liver metabolism [33]. Similarly, serum creatine and uric acid levels were significantly elevated in broilers [33]. This has also been established in other study findings [35,36,42-44]. This elevation in creatinine and uric acid levels has been associated with the effect of aflatoxin and ochratoxin to the renal dysfunction.

EFFECTS ON HEMATOLOGICAL PROFILES

Both aflatoxins and ochratoxins significantly reduce the levels of hemoglobin in laboratory experimental animals [33,43-46]. However, this may not have a significant effect on other hematological parameters [33]. Many other studies have been able to identify a reduction in the total erythrocyte count and plasma cell volume following aflatoxin poisoning [47,48]. Similar findings have also been recorded in ochratoxin poisoning [44,46,49]. But, a decrease in total leucocyte count may occur as a result of aflatoxicosis [49] and ochratoxicosis [46]. Leucocytosis may also occur in cases of ochratoxin A as reported by Stove et al. [28]. The role of mycotoxins and ochratoxins in the induction of bone marrow toxicity is yet to be determined [33]. It has also been suggested that the reduction in hemoglobin levels may be as a result of the reduced capability of protein synthesis in both cases of aflatoxin and ochratoxin toxicosis [33]. Probably, similar effects take place in humans, and have a significant impact on the health of humans. A search for alternative intervention for mycotoxicosis is therefore significant.

EFFECTS ON LIPID PROFILE

Aflatoxin poisoning also affects the lipid profiles of the liver. The total cholesterol, triglycerides and LDL cholesterol are significantly elevated while HDL cholesterol values are significantly reduced [38,50,51]. However, in separate studies, it was shown that the aflatoxins [45,52] cause a significant reduction in the levels of serum cholesterol and triglyceride. This has also been suggested to be the same case in ochratoxin [28,36,42,43] and in cases of combinations of mycotoxins [35] Studies have proposed the impairment of lipid metabolism as the main cause to the changes in the serum cholesterol and triglycerides. A similar effect in humans would have a serious health effect and requires the search for post ingestion intervention strategies.

EFFECTS ON ANTIOXIDANT ACTIVITY

The effect of aflatoxins on the antioxidant activity through the observation of the oxidative stress markers has also been demonstrated. It lowers the total antioxidant capacity, sodium/potassium (Na/K) – ATPase Activity), glutathione (GSH) levels are also depleted. On the contrary, malonyl dialdehyde levels are significantly increased in cases of aflatoxicosis [38,53-55]. The use of the antioxidants has been proposed to be one of the major mechanisms through which the effect of the ROS can be reduced and hence reducing the damage to the liver cells. Liver damage has been demonstrated to be associated with DNA damage in aflatoxin poisoning during increased protein oxidation [56]. These are the same effects reported on the health of humans following mycotoxicosis and hence there is need for an alternative interventions besides the regimens used currently.

EFFECTS ON IMMUNE SYSTEM

Aflatoxin alone and in combinations causes a reduction in the humoral and cell mediated immune responses [43,57,58]. The immune response is also reduced when ochratoxin is used alone [27,28,48,59,60]. In cases of combined aflatoxin and ochratoxin, Campbell et al. [61] have indicated a reduced immune response in chicks. The reduction of the immune response is associated with the reduction in the proteins and globulins. It is also caused by impairment of the antigen processes during phagocytosis [37,41]. Moreover, it induces lymphotoxic activities in cases of ochratoxin A as an effect on lymphocytes [62].

EFFECTS ON HISTOLOGICAL FINDINGS

Hepatocytes swell and become vacuolated suggesting the toxic effects of aflatoxins. Regions of necrosis also become diffusely spread out in the liver parenchyma and are infiltrated with hetrophils and lymphocytic aggregates [33]. Tissue damage occurs in the spleen, kidney and other tissues of the body in cases of aflatoxicosis [41,47]. Similar findings have also been suggested in ochratoxicosis [27,33,36,63].

Current Management Strategies of Mycotoxicosis in Humans and Animals

So far, no antidotes have been developed for the **MANAGEMENT** of mycotoxin poisoning and more specifically for aflatoxin B1 which is the most dangerous [64]. L-methionine administered at 200 mg/kg body weight has been the most preferred remedy for management of aflatoxin B1. They are administered intraperitoneally (IP) after every 8 hours until the clinical symptoms diminish. Methionine is also a regimen used in management of hepatotoxicity caused by acetaminophen poisoning [65,66]. Similarly, sodium thiosulfate is administered IP at 50 mg/kg to manage aflatoxicosis. They are administered IP at 8 hours interval and are proven to offer therapeutic potential in aflatoxin management [64]. The main strategies proposed for managing contamination before ingestions includes processes such as thermal inactivation, irradiation, drying, solvent extraction, adsorption from a solution, inactivation of the microbes and fermentation. The action levels for aflatoxins have also been documented [67]. Limited post ingestion solutions are available for management of aflatoxicosis and ochratoxicosis (Figure 1).

Chemoprotection and enterosorption techniques have also been applied in management of aflatoxin poisoning [64]. Esterified glucomanoses and extracts from yeast have been very useful in management of aflatoxins. It has been suggested that they promote detoxification and also reduce the formation of aflatoxin epoxide; a potent toxin which cause damage to DNA and RNA. This reduces the effect of aflatoxin in causing hepatocellular damage and carcinoma. Chlorophyl has also been suggested to lower the biologically effective dose of aflatoxins [64].

Enterosorptive agents are suggested to bind to aflatoxins and make them biologically unavailable to humans and animals [68]. Adsorbents such as hydrated sodium calcium aliminosilicate have been indicated to protect against aflatoxicosis. Zeolites have also been suggested to be effective reactive oxygen species scavengers [69]. However, the use of enterosorptive and adsorbents is not effective since they bind to aflatoxins along with other nutrients essential to the body [68]. This may not be useful especially among pregnant women.

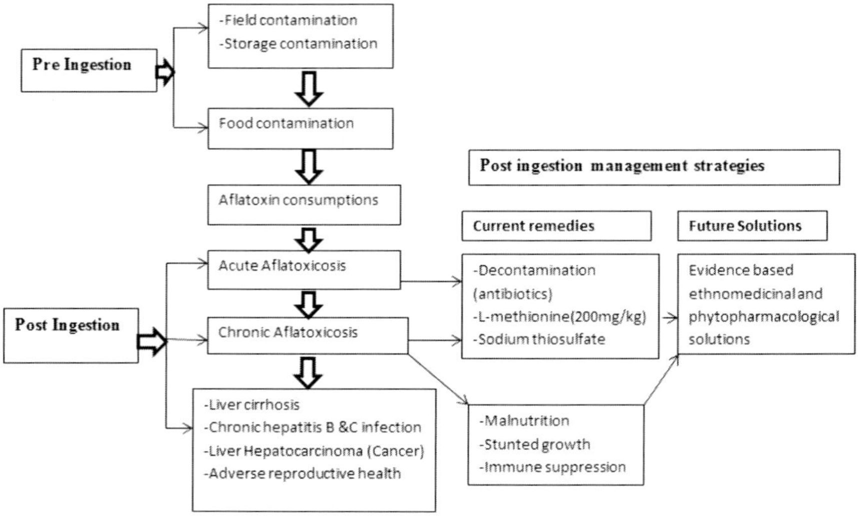

Figure 1: Pre and post ingestive causes, effects and management strategies in aflatoxin poisoning.

Alternative and Complimentary Management of Mycotoxicosis

HERBAL EXTRACTS

Urtica dioica is the main plant with anti-oxidative properties that has been tested on animal models. In these studies, treatment of animals with *Urtica dioica* significantly reduced superoxide dismutase levels, and elevated the reduced levels of malonyldialdehyde and lowered glutathione levels. *Urtica dioica* (2 ml/kg) has demonstrated that it has potential in regeneration of the liver following partial hepatectomy in rat model when administered orally for seven days [70]. It was able to reduce the highly elevated levels of malonyl dialdehyde after partial hepatectomy. Similarly, the levels of superoxide dismutase and glutathione were elevated significantly following the administration of the *Urtica dioica* in the tissues of the rat models. Hence, *Urtica dioica* has potential in the regeneration of the liver after liver damage caused by oxidative stress, proliferation and apoptosis after partial hepatectomy [70]. An oral dose of *Urtica dioica* (450 mg/kg) has also proven to be hepatoprotective and immunoprotective in mice exposed to 250 and 500 mg/kg acetaminophen poisoning administered IP [39].

Nigella sativa L. (2 ml/kg) and *Urtica dioica* (0.2 ml/kg) has been shown to reverse the toxic effect of carbon tetrachloride on lipid peroxidation, antioxidant and liver enzymes in hepatotoxicity in rats as animal models administered IP for 60 days. *Urtica dioica* and Nigella sativa when used alone and in a combination has been shown to decrease lipid peroxidation, liver enzymes and increase the antioxidant defense system activity in rats with induced damage [71]. It has been suggested that the high antioxidants; vitamin C, polyphenols and flavonoids present in *Urtica dioica* (2ml/kg) significantly inhibit aflatoxins B1 (25 μg/kg) induced oxidative damage through the scavenging of the reactive oxygen species (ROS) by stimulating the antioxidants defense systems in the body administered orally for 90 days [72].

Several studies on management of aflatoxicosis have been performed using various herbal extracts. In vitro hepatoprotective effect of both the ethanolic and aqueous extract of papaya fruits on rats administered at a dose of 250 mg/kg for 4 days have been demonstrated in aflatoxicosis by Hassan et al. [38]. It was shown that aqueous extracts had higher phenols and DPPH scavenging abilities compared to the ethanolic extracts. Oral aflatoxin (2 mg/kg body weight) alone causes significant increase in the biochemical parameters: ALT, AST and ALP. There was better improvement in lipid profiles for both aqueous extracts compared to ethanolic extracts administered orally on female Sprague-Dawley rats administered at a dose of 250 mg/kg body weight. Similarly, the oxidative stress markers; total antioxidant capacity, MDA and Na/K ATPase activity. Malonyl dialdehyde activity was significantly decreased in aqueous treated groups compared to the ethanolic extract treated groups. Hence, oral administrations of 250 mg/kg of papaya fruit extracts have significant hepatoprotective abilities [38]. Besides the hepatoprotective abilities, papaya has significant medicinal properties for many other medical conditions. Papaya contains alpha-tocopherol, lycopene, benzylisothiocynate, proteolytic enzymes such a papain and chymopapain [73-76]. The phytochemical constituent of papaya contains various forms of alkaloids; carpain and carpasemine [77], triterpenes, organic acids [78,79]. Moreover, they also have cystatin, ascorbic acid, cyanogenic glucosides and glucosinolates [80] phenols [38] and flavonoids [80]. It is suggested that this group of chemicals provide protection to cells against cellular damage caused by exposure to high levels of free radicals [81-83], while it also aids in digestion [83-85]. Moreover, Osaka et al. [86] have reported that papaya is a good source of vitamins A, C, E, K and folate. It is also rich in fiber; in addition it is fat-free, cholesterol-free and is low in sodium, which further suggests its potential beneficial effects.

Encouraging findings have been established for *Ixora coccinea* Linn. (Rubiaceae), *Rhinacanthus nasuta*. Linn (Acanthaceae) and *Spilanthes ciliate* (Asteraceae) at a dose of 100, 200, 300 and 400 mg/kg administered orally for two days before a dose of 1.5 mg/kg of aflatoxins was administered IP and sacrificed by decapitation 72 hours later [87-90]. *Camelia sinensis* (green tea leaves) has shown cancer chemo protective activities as a result of the inhibitory effects on diverse cellular activities related to the development of cancer in aflatoxin treated laboratory models [91]. In addition, sage leaves (*Salvia officinalis*) have also been indicated to have suppressive effects in tumor growth promotion on a variety of cell lines: Amelanotic melanoma cell line C32, renal cell adenocarcinoma ACHN, hormone-dependent prostate carcinoma LNCaP and finally the human breast cancer cell line MCF-7 in rats using a dose range of 5-400 μg/ml for a period of 48 hours [92]. This suggests strong anticancer properties. A similar effect can be replicated in the hepatocarcinoma caused by mycotoxins. Fenugreek seeds (*Trigonella foenum gracum*) from the family of Legomenaca have been demonstrated to have anticancer properties and antioxidant effects on MCF-7 breast cancer cell lines through the induction of apoptosis [93]. *Carum carvi* also referred to as caraway seeds belonging to Umbellifera have also demonstrated the antioxidant profile [94]. Galangal rhizomes (*Alpinia officinarum*) from the botanical family of Zingiberaceae have chemoprevention effect on tumor [95]. The same has been reported for ginger (*Zingiber officinale*) [96], Frankincense resins (*Gum olibanum*) [97], Myrrh (*Commiphora myrrha*) [98], Cinnamon barks (*Cimmomum zeylanicum*) [99]. Herbal extracts from these plants demonstrate possible significant hepatoprotection potential in cancer induced by aflatoxin B1 [100]. Similarly, combination of these herbs may also improve the biochemical profiles following post ingestion of mycotoxins.

Combined herbal extracts prepared as supplements have shown to be effective in the improvement of body weight in experimental models [33]. Performances of the animal models in studies have also demonstrated improvement [101]. Polyherbal supplementation in aflatoxin poisoning confers partial protection against significant change in organ weight [33]. The use of herbal extracts also confers partial protection in cases of either aflatoxin [58] or **OCHRATOXINS** [28].

The effects of polyherbal extracts on tissues following aflatoxin poisoning showed pre-protection against significant change in hematological parameters [33]. The protection against significant elevation in liver enzymes: AST, ALT, ALP, Albumin and serum proteins have also been demonstrated in various studies in mycotoxicosis [33]. Herbal extracts have therefore been suggested to possess restorative role in protein synthesis. The use of curcumin and *Curcuma longa* has also demonstrated improvement in the normal serum proteins following cases of aflatoxicosis in animal laboratory models [58,102]. Polyherbal extracts are also protective against significant damage to the kidney. They have been shown to reduce the significant elevation in the values of creatinine and uric acid [33]. This therefore suggests the protective effects of the two herbs in mycotoxicosis. This is also the case with the protection against the significant changes in lipid profiles; cholesterol and triglycerides suggesting their protective role in the mycotoxin poisoning [33]. Protection against immunotoxicity by herbal supplementations has also been suggested by various studies. It is suggested that it protects the immune organs from histotoxic effects caused by the mycotoxins [33]. Using *Curcuma longa* and 5% extract of artichoke has demonstrated improvement in the immune response in aflatoxicosis and ochratoxin poisoning [58]. No mechanism of action responsible for immunoprotection has been suggested. Histological improvements in liver tissues and other organs have also been improved by curcumin and *Curcuma longa*, 5% aqueous extract of artichoke [58,102]. It has been suggested that herbal extracts cause toxin neutralization [33,103]. No precise protective mechanism of action has been determined for the protective action of the herbal extracts. Other studies using carbon tetrachloride and acetaminophen have shown that protective effects of polyphenols and flavonoids target the antioxidation processes in the body. The mechanism of action of most herbal extracts has been reviewed by Sabeena and Ajay, [104].

Saccharomyces cerevisiae

Studies have evaluated the potential of a dose of 4 × 1016 CFU of ***Saccharomyces*** *cerevisiae* administered for 7 days and tested for its ability to reduce aflatoxins induced toxicity in animal models has been demonstrated. It was shown that aflatoxin at 0.7 mg/kg can reduce the weight of experimental animals [40]. However, animal models fed on *Saccharomyces cerevisiae* for 7 days showed improvement in the gross weight gain following aflatoxin poisoning [40]. The levels of superoxide dismutase, glutathione, total antioxidant capacity, and NA^+/K^+ATPase are also significantly reduced in both the liver and kidney in aflatoxin treated groups [38,40]. There was also a significant increase in the non-enzymatic antioxidants superoxide dismutase and glutathione levels in *Saccharomyces cerevisiae* treated models and a decrease in the levels of malonyldialdehyde compared to the controls. Histopathological studies also showed a significant damage to the liver tissues in aflatoxin B1 treated mouse models. It can be concluded that *Saccharomyces cerevisiae* is a safe and successful agent in counteracting the aflatoxin toxicity and is also hepatoprotective against aflatoxin induced damage. It is therefore a beneficial agent in the modulation of the activity of aflatoxins [40]. In another study of *Saccharomyces cerevisiae* and lactic acid bacteria (*Lactobacillus rhamnosus*) the combination was able to significantly reduce serum ALT, AST, gamma glutamyl tranferase, creatinine and blood urea nitrogen in comparison to the control groups. Glutathione was significantly increased in the studies more than the control. In addition, they were diminished products of aflatoxin in both the in vitro and in vivo studies [105]. The beneficial effects of *Saccharomyces cerevisiae* in management of aflatoxin B1 toxicity has also been evaluated in broilers [106] and Quails [107]. Bueno et al. [108] has also demonstrated that *Saccharomyces cerevisiae* is an important and efficient micro-organism in aflatoxin B1 quenching. Moreover, it is suggested that *Saccharomyces cerevisiae* cell wall have glucan, they reduce the frequency of micro nuclei induced by cyclophosphamide. Fermented yeast products from yeast have also been suggested to offer protection against damage of body tissues, elevation of liver and kidney enzymes [40]. *Saccharomyces cerevisiae* are known to produce glutathione. Glutathione has many beneficial effects such as in the protection against UV [109], heavy metals [110]. Glutathione has a significant role in reactive oxygen species damage to cells in organisms and protects against tissue damage [70]. Yeasts have now been adopted for the massive commercial production of glutathione [111].

POTENTIAL OF COMBINATION OF HERBAL EXTRACTS AND *Saccharomyces cerevisiae*

Herbs have been demonstrated to work through the anti-oxidative mechanisms serving as free radical scavengers promoted by polyphenols, flavonoids, ascorbic acid, terpenoids for the superoxide's generated by the ROS pathways that cause damage to vital organs in the body [112,113]. The polyphenols also inhibit the actions of Cytochrome P-450 preventing the metabolism of aflatoxins at post ingestion [113,114]. Mineral elements also boost the protective abilities against damage by toxins [114]. They also inhibit the growth of cancer therefore offering potential for protection against development of hepatocarcinoma caused by aflatoxin poisoning [92-99]. *Saccharomyces cerevisiae* quenches the toxins in the GIT, while it elevates the levels of glutathione in the body conferring hepatoprotection [40]. A combination of the two mechanisms will therefore be able to complements each other and offer multifunctional approach to management of mycoxicosis in humans.

Animal Models in the Assessment of Mycotoxicosis

A comparative study on aflatoxin B1 metabolism in mice and rats has been done by Steyn et al. [115]. This study was able to show that albino mice have almost the same levels of susceptibility to the acute oral effects of aflatoxin B1 [116]. However, mice have been found to be resistant to the carcinogenic actions of aflatoxin B1 compared to rat models. In terms of metabolism, it is has been shown that the mouse has a higher rate of metabolism of the aflatoxin B1 compared to the rats. This has been associated to the high activity of the cytochrome P450 in mice than in the rats hence inducing the resistance to the carcinogenic effects of aflatoxin B1 in the mouse model [115]. The mechanism in rat models has been associated with the conversion of the aflatoxin B1 to aflatoxin M1 in vivo [117]. The same mechanism has been proved to work in vitro [118]. Hence, acute effect of aflatoxins can be best demonstrated in mice while chronic effects are better when using rat as animal models. No comparative inter and intra specie studies of the suitability of birds and primates to rodents have been demonstrated in assessing the effects of aflatoxins and ochratoxins.

Relevance of Alternative Interventions of Mycotoxicosis to Humans

It is estimated that 55 billion people all over the globe suffer from uncontrolled aflatoxin poisoning worldwide [119]. The number of people with hepatocarcinoma is also growing. Furthermore, there have been no attempts to establish the number of liver cancer associated with mycotoxins exposure. There are also no drugs for management of mycotoxicosis in both humans and animals [64]. Four findings have been suggested by the Aflatoxin Workgroup, convened by the Centers for Disease Control and Prevention and World Health Organization. First, it was suggested that the burden of the diseases and impact on human health need to be quantified. Secondly; an inventory of the ongoing intervention strategies was to be created. Moreover, an evaluation on efficacy of the interventions was to be determined. Finally, results should be disseminated [120]. This review addresses the impact of mycotoxins on animal and human health and alternative post ingestion management strategies conducted in animal models. It has been shown that the alternative intervention have potential for management of mycotoxins and specifically aflatoxins and ochratoxins. Further clinical trials are needed using higher animals such as apes for evaluation of their effectiveness in management of aflatoxicosis before they can be adopted for use by humans and animals. No comparative studies have been done on the efficacy of either oral or other route of administration of herbal extracts to be considered for human application. However, oral administration of *Saccharomyces cerevisiae* is considered beneficial due to its ability to adsorb aflatoxins while at the same time exerting its effect in protecting the body organs against tissue damage through the anti-oxidative mechanisms.

CONCLUSION

The review has addressed aspects of mycotoxins on health and why there is need to conduct these studies in order to determine mechanisms that will help reduce the impact of mycotoxins on health. This report also cites relevant studies that show independent efficacy of use of herbal extracts and *Saccharomyces cerevisiae* in post ingestion management of mycotoxins in animal models. A study investigating the efficacy of the combined therapy of the two regimens is therefore justified. It is therefore expected that the overall effect of the herbal extracts and *Saccharomyces cerevisiae* alone and when combined will significantly reduce the effect of mycotoxins in humans as demonstrated in animal models. Moreover, it is anticipated that combinations of herbal extract and *Saccharomyces cerevisiae* will target different pathways for management of mycotoxins allowing for multi-functional management of mycotoxins. However, clinical trials are needed to ascertain the adoption of the regimens for use in humans.

REFERENCES

1. SCHOENTAL R (1961) Liver changes and primary liver tumours in rats given toxic guinea pig diet (M.R.C. Diet 18). Br J Cancer 15: 812-815.
2. Akande KE, Abubakar MM, Adegbola TA, Bogoro SE (2006) Nutritional and Health Implications of Mycotoxins in Animal Feeds: A Review. Pak J Nutr 5: 398-403.
3. Meissonnier GM, Marin DE, Galtier P, Bertin G, Taranu I, et al. (2006) Modulation of the immune response by a group of fungal food contaminant, the aflatoxins. In: Mengheri E, Roselli M, Britti MS, Finamore A. (Eds.), Nutr. Immun Research Sign post, Kerala 147-166.
4. Abu El-Saad AS, Mahmoud HM (2009) Phytic Acid Exposure Alters AflatoxinB1-induced Reproductive and Oxidative Toxicity in Albino Rats (Rattus norvegicus). Evid Based Complement Alternat Med 6: 331-341.
5. Avinash W, Lakkawar K, Chattopadhyay S, Tripurari SJ (2004) Experimental aflatoxin B1 toxicosis in young rabbits-a clinical and pathoanatomical study. Slov Vet Res 41: 73-81.
6. Richburg JH (2000) The relevance of spontaneous- and chemically-induced alterations in testicular germ cell apoptosis to toxicology. Toxicol Lett 112-113: 79-86.
7. Akbarsha MA, Averal HI, Girija R, Anandhi S, Faridha Banu A (2000) Male reproductive toxicity of vincristine: ultrastructural changes in the epididymal epithelial apical cell. Cytobios 102: 85-93.
8. Krausz C, Forti G (2000) Clinical aspects of male infertility. In: K. McElreavey, the Genetic Basis of Male Infertility. Springer, Heidelberg 121.
9. Shuaib FM, Ehiri J, Abdullahi A, Williams JH, Jolly PE (2010) Reproductive health effects of aflatoxins: a review of the literature. Reprod Toxicol 29: 262-270.
10. Turner PC, Collinson AC, Cheung YB, Gong Y, Hall AJ, et al. (2007) Aflatoxin exposure in utero causes growth faltering in Gambian infants. Int J Epidemiol 36: 1119-1125.

11. Gupta RC (2011) Aflatoxins, ochratoxins and citrinin in reproductive and developmental toxicology. In Gupta, RC ed. Academic Press/Elsevier, Amsterdam 753-763.
12. Pfohl-Leszkowicz A1, Manderville RA (2007) Ochratoxin A: An overview on toxicity and carcinogenicity in animals and humans. Mol Nutr Food Res 51: 61-99.
13. Miller JD, Trenholm HL (1994). Mycotoxins in Grain: Compounds Other than Aflatoxin. Eagan Press; St. Paul, MN, USA.
14. Faucet-Marquis V, Pont F, Størmer FC, Rizk T, Castegnaro M, Pfohl Leszkowicz A (2006) Evidence of a new dechlorinated ochratoxin A derivative formed in opossum kidney cell cultures after pretreatment by modulators of glutathione pathways: Correlation with DNA-adduct formation. Mol Nutr Food Res 50: 530-542.
15. Tozlovanu M, Faucet-Marquis V, Pfohl-Leszkowicz A, Manderville RA (2006) Genotoxicity of the hydroquinone metabolite of ochratoxin A: structure-activity relationships for covalent DNA adduction. Chem Res Toxicol 19: 1241-1247.
16. el Khoury A, Atoui A (2010) Ochratoxin a: general overview and actual molecular status. Toxins (Basel) 2: 461-493.
17. Mintzlaff HJ, Lotzsch R, Tauchmann F, Meyer W, Leistner L (1974) Aflatoxin residues in the liver of broiler chicken given aflatoxin-containing feed. Fleischwirtschaft 54: 774-778.
18. Patterson DS, Allcroft R (1970) Metabolism of aflatoxin in susceptible and resistant animal species. Food Cosmet Toxicol 8: 43-53.
19. Smith RH (1965) The Inhibition of amino acid activation in liver and E. coli preparations by aflatoxin in vivo. Biochem J 95: 438-448.
20. Smith JW, Prince WR, Hamilton PB (1969) Relationship of aflatoxicosis to Salmonella gallinarum infections of chickens. Appl Microbiol 18: 946-947.
21. Bababunmi EA, Bassir O (1969) The effect of aflatoxin on blood clotting in the rat. Br J Pharmacol 37: 497-500.
22. Allcroft R, Roberts BA, Lloyd MK (1968) Excretion of aflatoxin in a lactating cow. Food Cosmet Toxicol 6: 619-625.
23. Holzapfel CW, Steyn PS, Purchase IF (1966) Isolation and structure of aflatoxins M1 and M2. Tetrahedron Lett 25: 2799-2803.
24. Newberne PM, Butler WH (1969) Acute and chronic effects of aflatoxin on the liver of domestic and laboratory animals: a review. Cancer Res 29: 236-250.
25. Wu F1, Khlangwiset P (2010) Health economic impacts and cost-effectiveness of aflatoxin-reduction strategies in Africa: case studies in biocontrol and post-harvest interventions. Food Addit Contam Part A Chem Anal Control Expo Risk Assess 27: 496-509.
26. Preston RJ, Williams GM (2005) DNA-reactive carcinogens: mode of action and human cancer hazard. Crit Rev Toxicol 35: 673-683.
27. KozaczyÅ„ski W (1994) Experimental ochratoxicosis A in chickens. Histopathological and histochemical study. Arch Vet Pol 34: 205-219.
28. Stoev SD, Anguelov G, Ivanov I, Pavlov D (2000) Influence of ochratoxin A and an extract of artichoke on the vaccinal immunity and health in broiler chicks. Exp Toxicol Pathol 52: 43-55.
29. Huff WE, Doerr JA (1981) Synergism between aflatoxin and ochratoxin A in broiler chickens. Poult Sci 60: 550-555.
30. Giambrone JJ, Diener UL, Davis ND, Panangala VS, Hoerr FJ (1985) Effects of aflatoxin on young turkeys and broiler chickens. Poult Sci 64: 1678-1684.
31. El-Karim SA, Arbid MS, Soufy AH, Bastamy M, Effat MM (1991) Influence of metabolite ochratoxin a on chicken immune response. Egyp J Comp Pathol Clin Pathol 64:159-172.
32. Raju MVLN, Devegowda G (2000) Influence of esterified glucomannan on performance and organ morphology, serum biochemistry and haematology in broilers exposed to individual and combined mycotoxicosis (aflatoxin, ochratoxin and T-2 toxin). Brit Poult Sci 41:640-650.

33. Kalorey DR, Kurkure NV, Ramgaonkar JS, Sakhare PS. Shubhangi W, Nigot NK (2005) Effect of Polyherbal Feed Supplement "Growell" during Induced Aflatoxicosis, Ochratoxicosis and Combined Mycotoxicoses in Broilers. Asian - Austr J Anim Sci 2005. 18: 375-383.
34. Reddy NP, Rao PV, Reddy VR, Yadagiri B (1984) Effect of selected levels of dietary aflatoxin on the performance of broiler chicken. Ind J Anim Sci 54:68-73.
35. Huff WE, Kubena LF, Harvey RB, Phillips TD (1992) Efficacy of hydrated sodium calcium aluminosilicate to reduce the individual and combined toxicity of aflatoxin and ochratoxin A. Poult Sci 71: 64-69.
36. Sreemannarayana O, Marquardt RR, Frohlich AA, Abramson D, Phillips GD (1989) Organ weights, liver constituents, and serum components in growing chicks fed ochratoxin A. Arch Environ Contam Toxicol 18: 404-410.
37. Singh GS, Chauhan HV, Jha GJ, Singh KK (1990) Immunosuppression due to chronic ochratoxicosis in broiler chicks. J Comp Pathol 103: 399-410.
38. Hassan NS, Abdel-Wahhab KG, Khadrawy YA, El-Nekeety AA, Mannaa FA, et al. (2013) Evaluation of radical scavenging properties and the protective role of papaya fruits extract against oxidative stress in rats fed aflatoxin-contaminated diet. Communicata Scientiae 4:43-57.
39. Juma KK, Maina SG, Muriithi JN, Mwangi BM, Mworia KJ, Mwonjoria MJ, Ngeranwa JN, Mburu ND (2015) Protective Effects of Urtica dioica and cimetidine® on liver function following acetaminophen Induced Hepatotoxicity in Mice. J Develop Drugs 4: 130.
40. Darwish HR, Omara EA, Abdel-Aziz KB, Farag IM, Nada SA, Tawfek NS (2011) Saccharomyces cerevisiae modulates Aflatoxin-induced toxicity in male Albino mice. Rep Opin 3: 32-43.
41. Kalorey DR. (1993) Effect of aflatoxin on humoral immune system of chicks. Ph.D. Thesis approved by Panjabarao Krishi Vidyapeeth, Akola, India.
42. Manning RO, Wyatt RD (1984) Toxicity of Aspergillus ochraceus contaminated wheat and different chemical forms of ochratoxin A in broiler chicks. Poult Sci 63: 458-465.
43. Ramadevi NR, Gopal Naidu R. Ravikumar P (2000) An assessment of the protective effect of bentonite on ochratoxicosis in broiler with reference to certain hematological profile. Ind Vet J 77:303-306.
44. Doerr JA, Huff RB (1980) Interactive effects of aflatoxin and ochratoxin A on some blood constituents in broiler chickens. Int J Poult Sci 59:1600-1602
45. Mani K, Nrhari D, Kumaraj R, Ramamoorthy N (1993) Influence of dietary aflatoxin B1 on certain haematological and biochemical characters of broiler chicken. Ind Vet J 70:801-804.
46. Mohiuddin SM, Warasi SMA, Reddy MV (1993) Haematological and biochemical changes in experimental ochratoxicosis in broiler chicken. Ind Vet J 70:613-617.
47. Balachandran C, Ramkrishnan R. (1986). Influence of dietary aflatoxin on growth rate, feed consumption and haematology in broiler chicken. Proceed Nat Sem on "Avian Diseases". Hyderabad, India.
48. Singh A, Satija KC, Mahipal SK (1992) Haematological and biochemical studies on broiler chicks fed aflatoxin B1 and after its withdrawal. Ind J Poult Sci 27:153-156.
49. Aved IAM, Dafella R, Yogi AF and Adam SEI (1991) Effect of ochratoxin A on Lohaman type Chicks. Vet Hum Toxicol 33:357-560.
50. El-Nekeety AA, Mohamed SR, Hathout AS, Hassan NS, Aly SE, et al. (2011) Antioxidant properties of Thymus vulgaris oil against aflatoxin-induce oxidative stress in male rats. Toxicon 57: 984-991.
51. Hathout AS, Mohamed SR, El-Nekeety AA, Hassan NS, Aly SE, et al. (2011) Ability of Lactobacillus casei and Lactobacillus reuteri to protect against oxidative stress in rats fed aflatoxins-contaminated diet. Toxicon 58: 179-186.
52. Vassan P, Ravi R, Purshothaman MR (1998) Effect of feeding graded levels of aflatoxin (AFB1) on performance of broilers chicks. Ind J Poult. Sci 33:214-216.

53. Abdel-Wahhab MA, Aly SE (2003) Antioxidants and radical scavenging properties of vegetable extracts in rats fed aflatoxin-contaminated diet. J Agric Food Chem 51: 2409-2414.
54. Abdel-Wahhab MA, Saeed A, Hufner A (2005) NMR and radical scavenging activities of Patulet in from Urtica urens L. against aflatoxin B1. Pharma Biol 43: 515-525.
55. Abdel-Wahhab MA, Hassan NS, El-Kady AA, Khadrawy YA, El-Nekeety AA, et al. (2010) Red ginseng extract protects against aflatoxin B1 and fumonisins-induced hepatic pre-cancerous lesions in rats. Food Chem Toxicol 48: 733-742.
56. Abdel-Aziem SH, Hassan AM, Abdel-Wahhab MA (2011) Dietary supplementation with whey protein and ginseng extract counteracts oxidative stress and DNA damage in rats fed an aflatoxin-contaminated diet. Mutat Res 723: 65-71.
57. Ilgaz A (1987) Immunosuppressive effect of aflatoxicosis in chicks vaccinated against New castle disease. Rev Med Vet Mycol 22(Abst.):1405.
58. Kurkure NV, Pawar SP, Kognole SM, Bhandarkar AG, Ganorkar AG, Kalorey DR (2000) Ameliorative effect of turmeric (Curcuma longa) in induced aflatoxicosis in cockrels. Ind J Vet Pathol 24:26-28.
59. El-Karim SA, Arbid M S, Soufy AH, Bastamy M, Effat MM (1991) Influence of metabolite ochratoxin A on chicken immune response. Egyptain J Com Pathol Clin Pathol 4:159-172.
60. Santin E, Paulillo AC, Maiorka PC, Alessi AC, Krabbe EL, et al. (2002) The effects of ochratoxin/aluminosilicate interaction on the tissues and humoral immune response of broilers. Avian Pathol 31: 73-79.
61. Campbell ML Jr, May JD, Huff WE, Doerr JA (1983) Evaluation of immunity of young broiler chickens during simultaneous aflatoxicosis and ochratoxicosis. Poult Sci 62: 2138-2144.
62. Lea T, Steien K, Størmer FC (1989) Mechanism of ochratoxin A-induced immunosuppression. Mycopathologia 107: 153-159.
63. Dwivedi P, Burns RB (1984) Pathology of ochratoxicosis A in young broiler chicks. Res Vet Sci 36: 92-103.
64. Gupta RC (2012) Mycotoxins exposure: Symptoms, diagnosis, and pathophysiology of mycotoxins exposure. 2012 conference preceedings 1274: 74-86.
65. Dargan PI, Jones AL (2003) Management of paracetamol poisoning. Trends Pharmacol Sci 24: 154-157.
66. Brok J, Bickley N, Gluud C (2006) Interventions for Paracetamol (acetaminophen) overdose. Cochrane database of systemic reviews 2: CD003328.
67. Heathcote JG, Hibbert JR (1978) Aflatoxins: Chemical and biological aspects. Elsevier, New York: 173-186.
68. Williams JH, Phillips TD, Jolly PE, Stiles JK, Jolly CM, et al. (2004) Human aflatoxicosis in developing countries: a review of toxicology, exposure, potential health consequences, and interventions. Am J Clin Nutr 80: 1106-1122.
69. Pellegrino P, Mallet B, Delliaux S, Jammes Y, Guieu R, et al. (2011) Zeolites are effective ROS-scavengers in vitro. Biochem Biophys Res Commun 410: 478-483.
70. Oguz S, Kanter M, Erboga M, Toydemir T, Sayhan MB, et al. (2015) Effects of Urtica dioica on oxidative stress, proliferation and apoptosis after partial hepatectomy in rats. Toxicol Ind Health 31: 475-484.
71. Kanter M, Meral I, Dede S, Gunduz H, Cemek M, Ozbek H, Uygan I (2003) Effects of Nigella sativa L. and Urtica dioica L. on lipid peroxidation, antioxidant enzyme systems and some liver enzymes in CCl4-treated rats. J Vet Med Series A - physiol Pathol Clin Med 50: 264-268.
72. Yener Z, Celik I, Ilhan F, Bal R (2009) Effects of Urtica dioica L. seed on lipid peroxidation, antioxidants and liver pathology in aflatoxin-induced tissue injury in rats. Food Chem Toxicol 47: 418-424.
73. Ching LS, Mohamed S (2001) Alpha-tocopherol content in 62 edible tropical plants. J Agric Food Chem 49: 3101-3105.
74. van Breemen RB, Pajkovic N (2008) Multitargeted therapy of cancer by lycopene. Cancer Lett 269: 339-351.

75. Basu A, Haldar S (2008) Dietary isothiocyanate mediated apoptosis of human cancer cells is associated with Bcl-xL phosphorylation. Int J Oncol 33: 657-663.
76. Seigler DS, Pauli GF, Nahrstedt A, Leen R (2002) Cyanogenic allosides and glucosides from Passiflora edulis and Carica papaya. Phytochemistry 60: 873-882.
77. Iyer D, Sharma BK, Patil UK (2011) Effect of ether- and water-soluble fractions of Carica papaya ethanol extract in experimentally induced hyperlipidemia in rats. Pharm Biol 49: 1306-1310.
78. Cowan MM (1999) Plant products as antimicrobial agents. Clin Microbiol Rev 12: 564-582.
79. Osuna-Torres L, Tapia-Pérez ME, AguilarContreras A (2005) Plantas medicinales dela medicina traditional Mexicana Para tartar affections gastrointestinales: Estudio etnobotánico fitoquimicoy farmacológico. Universidad de Barcelona, Barcelona, Spain.
80. Miean KH, Mohamed S (2001) Flavonoid (myricetin, quercetin, kaempferol, luteolin, and apigenin) content of edible tropical plants. J Agric Food Chem 49: 3106-3112.
81. Ames BN, Shigenaga MK, Hagen TM (1993) Oxidants, antioxidants, and the degenerative diseases of aging. Proc Natl Acad Sci U S A 90: 7915-7922.
82. Dillard CJ, German JB (2000) Phytochemicals: nutraceuticals and human health. J Sci Food Agric 80: 1744-1756.
83. Prior RL, Cao G. (2000). Antioxidant phytochemicals in fruits and vegetables: diet and health implications. Hort Sci 35: 588-592.
84. Weisburger JH, Reddy BS, Rose DP, Cohen LA, Kendall ME, et al. (1993) Protective mechanisms of dietary fibers in nutritional carcinogenesis. Basic Life Sci 61: 45-63.
85. AACC (2001) American association of cereal chemists. The definition of dietary fiber. Cer Foods World 46: 112-126.
86. Otsuki N, Dang NH, Kumagai E, Kondo A, Iwata S, et al. (2010) Aqueous extract of Carica papaya leaves exhibits anti-tumor activity and immunomodulatory effects. J Ethnopharmacol 127: 760-767.
87. Latha PG, Suja SR, Acham A, Rajasekharan S, Panikkar KR. (2003) Hepatoprotective effects of Ixora coccinea flower extract on rats. J Trop Med Plants 4: 33-38.
88. Suja SR, Latha PG, Rajasekharan S, Pushpangadan P (2003) Antihepatotoxic activity of Spilanthes ciliate on paracetamol induced liver damage in rats. J Pharmaceut Biol 41:536-540.
89. Suja SR, Latha PG, Pushpangadan P, Rajasekharan S (2003) Assessment of hepatoprotective and free radical scavenging effects of Rhinacanthus nasuta (Linn.) Kurz in Wistar rats. J Natural Remed 4:66-72.
90. Suja SR, Latha PG, Pushpangadan P, Rajasekharan S (2003) Evaluation of hepatoprotective effects of Rhinacanthus nasuta root extracts. J of Trop Med Plants 4:151-157.
91. Révész K, Tüttô A, Konta L (2007) Effect of green tea flavonols on the function of the endoplasmic reticulum. Orv Hetil 148: 1903-1907.
92. Loizzo MR, Tundis R, Menichini F, Saab AM, Statti GA, et al. (2007) Cytotoxic activity of essential oils from labiatae and lauraceae families against in vitro human tumor models. Anticancer Res 27: 3293-3299.
93. Sebastian KS, Thampan RV (2007) Differential effects of soybean and fenugreek extracts on the growth of MCF-7 cells. Chem Biol Interact 170: 135-143.
94. Kamaleeswari M, Nalini N (2006) Dose-response efficacy of caraway (Carum carvi L.) on tissue lipid peroxidation and antioxidant profile in rat colon carcinogenesis. J Pharm Pharmacol 58: 1121-1130.
95. Heo MY, Sohn SJ, Au WW (2001) Anti-genotoxicity of galangin as a cancer chemopreventive agent candidate. Mutat Res 488: 135-150.
96. Vijaya Padma V, Arul Diana Christie S, Ramkuma KM (2007) Induction of apoptosis by ginger in HEp-2 cell line is mediated by reactive oxygen species. Basic Clin Pharmacol Toxicol 100: 302-307.

97. Bhushan S, Kumar A, Malik F, Andotra SS, Sethi VK, et al. (2007) A triterpenediol from Boswellia serrata induces apoptosis through both the intrinsic and extrinsic apoptotic pathways in human leukemia HL-60 cells. Apoptosis 12: 1911-1926.
98. Shoemaker M, Hamilton B, Dairkee SH, Cohen I, Campbell MJ (2005) In vitro anticancer activity of twelve Chinese medicinal herbs. Phytother Res 19: 649-651.
99. Lee KG, Shibamoto T (2002) Determination of antioxidant potential of volatile extracts isolated from various herbs and spices. J Agric Food Chem 50: 4947-4952.
100. Yousef, JM (2009) The protective effects of selected mixed herbal extracts on weight, serum and liver tissues in rats before and after exposure to aflatoxin B1. World J Med Sci 4: 01-08.
101. Godbole SM, Kalorey DR, Ingle VC, Kurkure NV, Ganorkar AG, Harne SD (2001) Effect of growell during induced aflatoxicosis in chicks: Growth and haematological studies. Ind J Compa Microbiol, Immunol Infect Dis 22:183-185.
102. Soni KB, Rajan A, Kuttan R (1992) Reversal of aflatoxin induced liver damage by turmeric and curcumin. Cancer Lett 66: 115-121.
103. Shubhangi W (2001) Studies on effect of aqueous herbal extract on growth of A. ochraceus, ochratoxin production and its neutralization. MV.Sc. Thesis approved by Dr. Panjabarao Krishi Vidyapeeth, Akola, India.
104. Sabeena HS, Ajay GN (2014) Current status of natural products for the treatment of diseases-A review. Int J phytopharm 4: 37-43.
105. Nada S A, Amra HA, Deabes MMY, Omara EA (2010) Saccharomyces cerevisiae and Probiotic Bacteria Potentially Inhibit Aflatoxins Production In vitro and In vivo Studies. Int J Toxicol 8.
106. Çelyk K, Denly M, Ertürk M, Öztürkcan O, Doran F (2001) Evaluation of dry yeast (Saccharomyces cerevisiae) compounds in the feed to reduce aflatoxin B1 (AFB1) residues and toxicity to Japanese quails. J Appl Anim Res 20: 245-250.
107. Celyk K, Denly M, Savas T (2003) Reduction of toxin effects of Aflatoxin B1 by using baker's yeast (Saccharomyces cerevisiae) in growing broilers checks diet. Revista brasileira de zootecnia 32.
108. Bueno DJ, Casale CH, Pizzolitto RP, Salvano MA, Oliver G (2007) Physical adsorption of aflatoxin B1 by lactic acid bacteria and Saccharomyces cerevisiae: a theoretical model. J Food Prot 70: 2148-2154.
109. Sollod CC, Jenns AE, Daub ME (1992) Cell surface redox potential as a mechanism of defense against photosensitizers in fungi. Appl Environ Microbiol 58: 444-449.
110. Perego P, Vande Weghe J, Ow DW, Howell SB (1997) Role of determinants of cadmium sensitivity in the tolerance of Schizosaccharomyces pombe to cisplatin. Mol Pharmacol 51: 12-18.
111. Wei G, Li Y, Du G, Chen J (2003) Application of a two-stage temperature control strategy for enhanced glutathione production in the batch fermentation by Candida utilis. Biotechnol Lett 25: 887-890.
112. Gilani AH, Janbaz KH (1995) Preventive and curative effects of Artemisia absinthium on acetaminophen and CCl4-induced hepatotoxicity. Gen Pharmacol 26: 309-315.
113. Bray BJ, Perry NB, Menkes DB, Rosengren RJ (2002) St. John's wort extract induces CYP3A and CYP2E1 in the Swiss Webster mouse. Toxicol Sci 66: 27-33.
114. Iyanda AA, Anetor JI, Oparinde DP, Adeniyi FA (2010) Effects of methionine containing paracetamol formulation on serum vitamins and trace elements in male rats. Niger J Physiol Sci 25: 129-134.
115. Steyn M, Pitout MJ, Purchase IF (1971) A comparative study on aflatoxin B 1 metabolism in mice and rats. Br J Cancer 25: 291-297.
116. BUTLER WH (1964) ACUTE TOXICITY OF AFLATOXIN B-1 IN RATS. Br J Cancer 18: 756-762.
117. Patterson DS, Roberts BA (1970) The formation of aflatoxins B2a and G2a and their degradation products during the in vitro detoxification of aflatoxin by livers of certain avian and mammalian species. Food Cosmet Toxicol 8: 527-538.

118. Schabort JC, Steyn M (1969) Substrate and phenobarbital inducible aflatoxin-4-hydroxylation and aflatoxin metabolism by rat liver microsomes. Biochem Pharmacol 18: 2241-2252.

119. Strosnider H, Azziz-Baumgartner E, Banziger M, Bhat RV, Breiman R, et al. (2006) Workgroup report: public health strategies for reducing aflatoxin exposure in developing countries. Environ Health Perspect 114: 1898-1903.

120. WHO 2008. The Global Burden of Disease: 2004 Update. Geneva: World Health.

CHAPTER 7

Household Food Items Toxic to Dogs and Cats

Cristina Cortinovis and Francesca Caloni*

Department of Health, Animal Science and Food Safety, Universitá degli Studi di Milano, Milan, Italy

ABSTRACT

Several foods that are perfectly suitable for human consumption can be toxic to dogs and cats. Food-associated poisoning cases involving the accidental ingestion of chocolate and chocolate-based products, *Allium* spp. (onion, garlic, leek, and chives), macadamia nuts, *Vitis vinifera* fruits (grapes, raisins, sultanas, and currants), products sweetened with xylitol, alcoholic beverages, and unbaked bread dough have been reported worldwide in the last decade. The poisoning episodes are generally due to lack of public knowledge of the serious health threat to dogs and cats that can be posed by these products. The present review aims to outline the current knowledge of common food items frequently involved in the poisoning of small animals, particularly dogs, and provides an overview of poisoning episodes reported in the literature.

Keywords: chocolate, ethanol, grape, macadamia nuts, onion, poisoning, pets, xylitol

INTRODUCTION

Several foods, while safe for humans, may pose a serious threat to the health of dogs and cats. Food-associated poisoning cases involving the accidental ingestion of chocolate, onions, macadamia nuts, *Vitis vinifera* fruits (grapes, raisins, sultanas, and currants), xylitol, and ethanol have been recorded worldwide in the last decade (1–3). Foods accounted for 14.8% of hazardous exposure cases reported to the Kansas State Veterinary Diagnostic Laboratory. Chocolate- and cocoa-based products were most commonly involved, followed by products sweetened with xylitol, onions and garlic, grapes and raisins, macadamia nuts, and ethanol (3). In general, the poisoning episodes resulted from a lack of public knowledge of the health hazard to small animals that may be posed by these products. Dogs and cats may be fed harmful foodstuffs by owners unaware of the danger or given the wide occurrence of these products in the home, pets may easily have accidental access to them. Dogs are undiscriminating in their eating habits and will readily ingest potentially harmful foodstuffs, thus being far more commonly affected than cats (4). While some foodstuffs, such as chocolate, have long been known to cause poisoning in dogs and cats, others such as grapes had previously been considered unlikely to cause problems and have emerged as a potential concern only in the last few years (5). As a consequence, cases of significant exposure had been wrongly diagnosed for many years (5). This review aims to outline the current knowledge of common household foods frequently involved in the poisoning of small animals, particularly dogs, and provides an overview of poisoning episodes reported in the literature.

Allium spp.

Onion (*Allium cepa*), garlic (*Allium sativum*), leek (*Allium porrum*), and chives (*Allium schoenoprasum*) are all members of the genus *Allium* (Amaryllidaceae family). These bulbous plants are strongly aromatic, producing a characteristic odor when crushed, and are commonly used (fresh, cooked, or dehydrated) as ingredients in many dishes. The components responsible for their toxicity are organosulfoxides. Chewing the plant converts organosulfoxides to a complex mixture of sulfur compounds. The primary toxicological mechanism of *Allium*-derived sulfur compounds is oxidative hemolysis characterized by the development of methemoglobinemia and Heinz body formation in the erythrocytes (6, 7). Cooking, drying, and processing do not eliminate the toxic effect of *Allium* spp. (7, 8). Dogs and cats are highly susceptible to *Allium* toxicosis and the ingestion of 5 g/kg of onions by cats and 15–30 g/kg by dogs is enough to cause clinically important hematologic changes (8). In the case of dogs, hereditary high erythrocyte-reduced glutathione and potassium concentrations observed in certain breeds (e.g., Akita, Shiba, and Jindo) lead to greater susceptibility to onion-induced oxidative damage (9). Clinical signs of *Allium* toxicosis may appear 1 day or several days after consumption depending on the amounts ingested. Common clinical signs initially include vomiting, diarrhea, abdominal pain, loss of appetite, and depression. Due to the developing anemia, pale mucous membranes, weakness, rapid respiratory and heart rates, jaundice, and dark urine (reddish or brown) indicating hemoglobinuria are subsequently observed (7). Several cases of dog and cat poisoning by *Allium* spp. have been reported in the literature. Poisoning has been reported to occur after the ingestion of Catalan spring onion commonly known as "calcot" (10), baked garlic (11), onion soufflè (12), butter-cooked onions (13), and Chinese steamed dumplings containing Chinese chives (*Allium tuberosum*) and garlic (14). A case in which a dog was intentionally fed a large quantity of raw onions by the owner has also been reported (15). From 1994 to 2008, 69 cases of canine poisoning and 4 cases of feline poisoning by *Allium* spp. ingestion (16) were recorded by the Veterinary Poisons Information Service (VPIS). In the case of dogs, vomiting, diarrhea, abdominal pain and, less frequently, anemia, hematuria, and convulsions were reported. Two cases of death occurred and two dogs were euthanized. In the case of cats, gastrointestinal signs, lethargy, and polydipsia occurred in one case, while anemia and icterus were observed in the second case. One cat remained asymptomatic without treatment, and the other cat died from hemorrhage into the pleural and abdominal cavities (16). Recently, hypertension associated with garlic-induced hemolytic anemia has been reported in the case of a dog (11). No specific antidote is available for *Allium* toxicosis. Inducing vomiting should be considered in asymptomatic dogs and cats, provided there are no complicating factors and not more than 2 h have elapsed since ingestion (8). The administration of activated charcoal is indicated after vomiting has stopped. Once clinical signs have manifested themselves, treatment should consist of supportive care. Severely anemic animals may require a blood transfusion (8).

ETHANOL

Ethanol or ethyl alcohol is a two-carbon alcohol found in a variety of products, such as alcoholic beverages, paint and varnish, medication, perfume, mouthwash, certain types of thermometers, and certain forms of antifreeze. It is also used as a disinfectant, a fuel substitute, and if administered intravenously, as a competitive substrate in the treatment of dogs and cats poisoned by ethylene glycol. Ethanol toxicosis in small animals generally occurs as a result of accidental ingestion of alcoholic beverages (17–20). Ethanol intoxication has also been reported in the case of dogs, following ingestion of rotten apples (21), sloe berries used to make sloe gin (22), and uncooked bread and pizza dough (23–25). The latter contains the yeast *Saccharomyces cerevisiae*, which metabolizes carbohydrate substrates to ethanol and carbon dioxide (26). Once ingested, ethanol is rapidly absorbed from the gastrointestinal tract and crosses the blood–brain barrier (27). The mechanism of action of ethanol is not completely clear. Ethanol is suspected of inhibiting *N*-methyl-D-aspartate glutamate receptors in brain cells and the related production of cyclic guanosine monophosphate (27). Clinical signs usually develop within an hour of ingestion and include central nervous system (CNS) depression, ataxia, lethargy, sedation, hypothermia, and metabolic acidosis (27). Animals may become comatose and develop severe respiratory depression. A distended and painful abdomen due to excessive gas production may be noted in animals that ingest uncooked bread dough (25). Emesis should be induced with extreme caution and only in cases of very recent ingestion by animals that prove to be asymptomatic (27). Recently, hemodialysis has been shown to be beneficial for the rapid removal of ethanol in patients with severe toxicosis (20). Yohimbine, an alpha 2-adrenergic antagonist which readily crosses the blood–brain barrier, has been recommended as an arousal agent in the treatment of ethanol intoxication (25). In previous case reports, most patients recovered when monitored closely and given supportive care (22, 24, 28). Fatal ethanol intoxication was reported in the case of a dog, following the massive ingestion of rotten apples. The dog developed vomiting, ataxia, tremors, and dehydration, and died 48 h later. However, liver damage secondary to the chronic ingestion of rotten apples (presumed to reflect chronic ethanol toxicity) was suspected as a factor in the death of this dog (21).

GRAPES AND THEIR DRIED PRODUCTS (RAISINS, SULTANAS, AND CURRANTS)

Grapes, the fruits of *Vitis vinifera*, and their dried products (raisins, sultanas, and currants) have been reported to cause renal failure in dogs. The fruits may be ingested raw or cooked as ingredients of fruit cake, mince pies, malt loaf, snack bars, scones, and other baked goods (28). The toxic syndrome has also been observed with consumption of marc (the residue of grapes after pressing) (29). The toxic principle(s) and the exact mechanism of grape-induced nephrotoxicity are still unknown. The latter appears to involve a nephrotoxic agent or an idiosyncratic reaction, leading to hypovolemic shock and renal ischemia (30). There is considerable variation in the susceptibility of dogs to grapes and their dried products. In a recent study that reviewed 180 reports recorded by the VPIS between August 1994 and September 2007 on the ingestion of *Vitis* fruits by dogs, some animals were reported to remain asymptomatic after ingesting up to 1 kg of raisins while others died following the ingestion of just a handful (28). Published case reports have identified renal failure in dogs following the ingestion of estimated doses of raisins as low as 2.8 mg/kg (31) and as little as four to five grapes in a dog weighing 8.2 kg (32) (Table 1). Therefore, ingestion of any quantity of these fruits should be considered as a potential clinical problem. Vomiting within 24 h of ingestion is the typical clinical sign observed. Diarrhea, anorexia, lethargy, and abdominal pain have also been reported (28, 31). Partially digested grapes and grape products may be present in the vomit, fecal material, or both (31–33). This is followed by signs of renal insufficiency or failure (oliguria, anuria, polydipsia, proteinuria, and elevated serum concentrations of creatinine and urea) within a short period (28, 31, 32). In cases of dogs with oliguria or anuria, the prognosis is generally poor (30, 34). The time taken to administer treatment may also play a significant role in the outcome (28). Given the large variability in the tolerance exhibited by dogs, the ingestion of any quantity of grapes or grape products by dogs should be handled aggressively (28). Following recent ingestion, prompt decontamination using emetics and repeated doses of activated charcoal is highly recommended (28, 30). All dogs should receive aggressive intravenous fluid therapy for a minimum of 48–72 h, and their renal function should be monitored for at least 72 h (28, 30).

HOPS

Hops, the inflorescences of the female plant of the species *Humulus lupulus*, are used for beer brewing to add a bitter taste and hoppy aroma to beer and to stabilize the beer foam. As home brewing becomes increasingly popular, companion animals may be at risk of exposure. The ingestion of both fresh and spent hops has been associated with the development of malignant hyperthermia in dogs (41). Any breed of dog may be affected, but breeds predisposed to malignant hyperthermia (e.g., Greyhounds, Labrador Retrievers, Saint Bernards, Pointers, Dobermans, Border Collies, English Springer Spaniels, and northern breeds) appear to be particularly susceptible (42). Hops contain a variety of compounds, including resins, essential oils (hycrocarbons and oxygenated compounds), phenolic compounds (tannins), and nitrogenous compounds (43). These compounds or their metabolites may uncouple oxidative phosphorylation resulting in malignant hyperthermia (6, 42). Clinical signs may be seen within hours of consuming hops and include marked hyperthermia, anxiety, tachycardia, tachypnea, panting, vomiting, abdominal pain, and seizures. Dark brown urine suggesting muscle necrosis may be observed (6, 42). Mortality can be high despite the aggressive therapy for hyperthermia (6). Duncan et al. (44) reported five cases involving dogs (four of which were Greyhounds) that exhibited marked hyperthermia, restlessness, panting, vomiting, signs of abdominal pain, and seizures, after the ingestion of hops. Four of the five dogs died despite intensive care treatment (44). Dogs suspected of recent ingestion of hops should be treated aggressively to decontaminate the gastrointestinal tract. Cooling measures, such as ice baths and cold IV fluids, should be used to lower the high body temperature. If available, dantrolene sodium, a skeletal muscle relaxant that has been used in humans to reverse malignant hyperthermia, may be administered (6, 41, 42).

Table 1. Range of doses of grapes and their dried products (raisins, sultanas, and currants) reported to cause renal failure in dogs.

Breed	Age (years)	Sex	Product	Dose	Time to onset of clinical signs (h)	Outcome	Reference
Labrador Retriever (13 dogs) Golden Retriever (4 dogs) Others (26 dogs)	Range: 0.6–13, median: 4	Male (27 dogs) Female (16 dogs)	Raisins (28 dogs) Grapes (13 dogs) Raisins and grapes (2 dogs)	Raisins[a]: range: 2.8–36.4 g/kg, median: 19.6 g/kg Grapes[b]: range: 19.6–148.4 g/kg, median: 40.6 g/kg	<24 (35 dogs) 24–48 (8 dogs)	Recovery (23 dogs) Died (5 dogs) Euthanized (15 dogs)	(31)
Labrador Retriever	3	Male	Raisins	3 g/kg	–[c]	Died or euthanized	(35)
Dachshund	5	Male	Grapes	4–5 grapes in an 8.2-kg dog	–	Recovery	(32)
German Shepherd Dog cross	4	Male	Grapes	4 g/kg	–	Died or euthanized	(35)
Norwegian Elkhound	5	Male	Raisins	4.8 g/kg	–	Died	(33)
Labrador Retriever cross	3.5	Male	Grapes	6 g/kg	–	Died or euthanized	(35)
American Pit Bull Terrier	8.5	Male	Raisins	6 g/kg	–	Died or euthanized	(35)
Labrador Retriever	3.5	Male	Currants	7.8 g/kg	12	Recovery	(36)
–	–	–	Grapes (5 dogs) Raisins (5 dogs)	11.6–31.2 g/kg[d]	Few hours	Recovery (5 dogs) Died (2 dogs) Euthanized (3 dogs)	(33)
Labrador Retriever	6	Female	Raisins	15.7 g/kg	8	Died	(32)
Australian Shepherd Dog	7	Male	Raisins	18 g/kg	–	Died or euthanized	(35)
Maltese	2	Female	Grape skins	18.8 g/kg	12	Recovery	(37)
Galgo Espaniol	1	Female	Grapes and marc	20 g/kg	6–7	Euthanized	(38)
Maltese	3	Male	Grapes	20 g/kg	5	Died	(39)
Border Collie	3.5	Male	Raisins	20.6 g/kg	12	Recovery	(32)
Basenji	10	Female	Raisins	24 g/kg	–	Died or euthanized	(35)
Golden Retriever	2	Male	Raisins	30 g/kg	–	Died or euthanized	(35)
Labrador Retriever	7	Male	Raisins	30 g/kg	–	Died or euthanized	(35)
Maltese	2	Female	Grapes	65.2 g/kg	8	Euthanized	(40)

[a] *Data are limited to 24 dogs.*
[b] *Data are limited to four dogs.*
[c] *Unknown or not provided.*
[d] *Data are limited to four dogs.*

MACADAMIA NUTS

Macadamia nuts are produced by trees of the genus *Macadamia* (Proteaceae family). Only two species, *Macadamia integrifolia* and *Macadamia tetraphylla*, are plants of commercial importance as sources of food. All *Macadamia* spp. accumulate cyanogenic glycoside (proteacin and durrin) in their seeds but in very low concentrations in the case of commercial seeds (45). Macadamia nuts are very popular as snacks for human consumption, both as plain nuts or when used in cakes, cookies, or candy (46). They are considered to be a valuable food for their low cholesterol and sodium content and as an excellent source of manganese and thiamine (45). Macadamia nut toxicosis has only been reported in dogs to date (1, 47–49). The mechanism of action of their toxicity in the case of dogs is currently unknown and the dose required to induce toxicity has not been established precisely (46). The ingestion of as little as 0.7 g/kg of nuts has been associated with the development of clinical signs in dogs (48). In another case, a series of toxic doses ranging from 2.2 to 62.4 g/kg was reported (49). Clinical signs generally develop within 12 h of ingestion and may include weakness (particularly hind limb weakness), depression, vomiting, ataxia, tremors, hyperthermia, abdominal pain, lameness, stiffness, recumbency, and pale mucous membranes (47–49). Although macadamia nut poisoning is relatively infrequent, 83 cases were reported in Queensland, a major area for *Macadamia* spp. cultivation in Australia, over a period of 5 years (1). No deaths have been reported to date, and animals are expected to fully recover within 24–48 h with minimal veterinary intervention (46). The induction of emesis should be considered in cases of recent ingestion of macadamia nuts by dogs that prove to be asymptomatic (46).

METHYLXANTHINES (CAFFEINE, THEOBROMINE, AND THEOPHYLLINE)

Methylxanthines, including caffeine, theobromine, and theophylline, are plant-derived alkaloids that are commonly found in a variety of foods, beverages, human medication, and other products in the home. Caffeine is found in coffee (from the fruit of *Coffea arabica*), tea (from the leaves of *Thea sinensis*), guarana (from the seeds of *Paullinia cupana*), and as an additive in many soft drinks. Theobromine occurs in cacao seeds (*Theobroma cacao*) and in products manufactured from these seeds, such as chocolate. Theophylline is found in tea along with caffeine. Moreover, caffeine is used in human medication to increase mental alertness, and theophylline is widely used as a bronchodilator in anti-asthma drugs (50). Methylxanthines antagonize cellular adenosine receptors and inhibit cellular phosphodiesterases, causing an increase in cyclic adenosine monophosphate (cAMP) (50). Moreover, methylxanthines enhance the release of catecholamines and increase cellular calcium entry while inhibiting intracellular sequestration of calcium by the sarcoplasmic reticulum, leading to increased muscular contractility (51). These combined actions result in the stimulation of both the CNS and cardiac muscle, the relaxation of smooth muscle, most notably bronchial muscle, and diuresis (50). Methylxanthine poisoning cases have been reported after the ingestion of herbal supplements containing guarana (52), garden mulch made of cacao bean shells (53, 54), caffeine tablets (55, 56), and caffeine-containing bait (57). However, most poisoning cases occur as a result of chocolate ingestion (4, 58–62). Though the intoxication of cats may also occur, dogs are most commonly affected because of their indiscriminate eating habits (4). The presence of chocolate was noted in the 10 most common cases of toxicosis involving dogs reported to the VPIS and to the American Society for the Prevention of Cruelty to Animals (ASPCA), Animal Poisons Control Center (APCC) in the past few years (60, 63). Poisoning episodes frequently occur around holidays when there is a higher occurrence of chocolate products in the home (4, 60). In addition to theobromine, chocolate contains caffeine but in much lower concentrations. Theobromine and caffeine concentrations vary according to the type of chocolate (4). Unsweetened baking chocolate and cocoa powder usually contain more than 14 mg of theobromine per gram. Semisweet dark chocolate and milk chocolate often contain around 5 and 2 mg of theobromine per gram, respectively, while white chocolate is considered to be an insignificant source of theobromine (50). According to the ASPCA APCC data, mild clinical signs appear in dogs after ingesting 20 mg/kg of theobromine and caffeine, while severe clinical signs are observed at 40–50 mg/kg and seizures occur at 60 mg/kg (4). Dogs with CYP1A2 deficiency polymorphism 1117C > T may be more at risk of poisoning due to reduced metabolism (64, 65). In dogs, caffeine is absorbed rapidly after ingestion while theobromine is absorbed 10 times slower, reaching peak plasma levels at approximately 10 h (50). Initial clinical signs are generally observed within 2–4 h after ingestion and include restlessness, polydipsia, urinary incontinence, vomiting, and perhaps diarrhea. Dogs can be in an excited state and show marked hyperthermia and tachycardia. As intoxication progresses, hypertension, cardiac arrhythmias, premature ventricular contractions, muscular rigidity, ataxia, seizures, and coma may be observed. Death may occur from cardiac arrhythmia or respiratory failure (50). Decontamination *via* emesis or gastric lavage, administration of multiple doses of activated charcoal, and meticulous supportive care should be the mainstay of treatment (4, 60). Prognosis is usually good, if effective decontamination is obtained within 2–4 h of ingestion (50, 60).

XYLITOL

Xylitol is a five-carbon sugar alcohol primarily used as an artificial sweetener in many products, including sugar-free gum, candy, bread, cookies, and other baked goods. It can also be purchased as a granulated powder for cooking and baking. Because of its antibacterial activity and palatability, xylitol is also included in a variety of medical and dental care products. An additional concern is that the use of xylitol is not just confined to products intended for human use. Xylitol is also an ingredient in drinking water additives developed to help maintain dog and cat dental health (66). The increased marketing and use of xylitol as a sweetener in recent years has led to increased risk of pet exposure to this agent (66, 67). Dogs are the species at the risk of developing severe, life-threatening clinical signs. In dogs, xylitol is a potent stimulator of insulin release, leading to dramatic decrease in blood glucose levels (66, 67). Doses, as low as 0.03 g/kg, have resulted in hypoglycemia in this species (68). Moreover, xylitol ingestion has been associated with liver failure in dogs. The mechanisms responsible for hepatic injury are not clear yet. It is thought to be related to either adenosine triphosphate (ATP) depletion secondary to xylitol metabolism, leading to hepatic necrosis or the generation of hepatocyte-damaging reactive oxygen species or both (69). According to the ASPCA APCC data, the lowest dose associated with xylitol-induced liver failure in dogs is 0.5 g/kg (70). Clinical signs of xylitol toxicity in dogs may be related to hypoglycemia or hepatopathy or both. Vomiting is usually the initial clinical sign. Clinical signs of hypoglycemia, including lethargy, ataxia, collapse, and seizures, may develop within 30–60 min after ingestion or may be delayed up to 12 h after ingestion (66, 70). In dogs developing hepatopathy, lethargy, icterus, vomiting, and coagulopathic signs, such as petechiae, ecchymoses, and gastrointestinal hemorrhages, may be observed (67, 69). Several cases of xylitol ingestion have been reported in the last decade (69, 71–75). Table 2 summarizes the cases of xylitol-induced acute hepatic failure reported in the literature. Recently, a retrospective study of the 192 cases of xylitol ingestion by dogs, reported to three American university teaching hospitals from December 2007 to February 2012, has been performed (68). The median estimated dose ingested was 0.32 g/kg (range 0.03–3.64 g/kg). Clinical signs developed in 41 (21%) dogs and primarily included vomiting and lethargy, followed by diarrhea, ataxia, seizures, restlessness, and anorexia (68). Thirty dogs (15.6%) became hypoglycemic and no significant difference was observed between hypoglycemic and euglycemic dogs with regards to the estimated dose of xylitol ingested. No dogs developed clinical signs or showed biochemistry values consistent with liver failure. All dogs survived and were discharged, suggesting an excellent prognosis for dogs that receive prompt veterinary care and evade liver failure (68). Supportive care and monitoring are the mainstay of xylitol toxicity treatment. The induction of emesis should only be attempted early and in asymptomatic animals. Activated charcoal is not recommended because of its poor ability to bind to xylitol (68, 70). Blood glucose levels and liver function should be monitored. If hypoglycemia develops, intravenous dextrose should be administered (69, 70). Xylitol ingestion should always be considered by veterinarians as a differential diagnosis for any unexplained hypoglycemic presentation with or without accompanying liver dysfunction (76).

Table 2. Range of xylitol doses reported to cause acute hepatic failure in dogs.

Breed	Age (years)	Sex	Product	Dose (g/kg)	Time to onset of clinical signs (h)	Outcome	Reference
Standard Poodle	3	Male	5 or 6 cookies	1.4–2	24	Died	(69)
English Springer Spaniel	2.5	Male	Homemade bread containing xylitol	3.7	1–2	Recovered	(72)
Labrador Retriever	8	Female	Xylitol powder	4.1	2	Recovered	(69)
Cocker Spaniel	4	Male	45 pieces of gum	4.7	72	Died	(73)
Dalmatian	6	Female	8 muffins sweetened with xylitol	5	9	Recovered	(69)
Scottish Terrier	5	Female	30 pieces of gum	7	24	Euthanized	(69)
Labrador Retriever mixed breed	6	Female	Xylitol powder	13.9	24	Died	(69)
Welsh Springer Spaniel	4	Male	4 large chocolate-frosted muffins sweetened with xylitol	15.9	1–2	Euthanized	(69)
Miniature Dachshund	7	Female	100 pieces of gum	16	72	Recovered	(69)
Chihuahua	9	Male	Granulated xylitol	45	1–2	Recovered	(75)

CONCLUSION

The present review highlights the issue of exposure of small animals, particularly dogs, to potentially harmful foodstuffs commonly present in the home. A noticeable trend in exposure has emerged due to the increasing popularity of xylitol as a sweetener in several products. Obtaining an accurate history of exposure, early recognition of clinical signs, and rapid establishment of appropriate therapy can greatly improve the prognosis of food-related poisoning cases. Large gaps still exist in public knowledge of the hazard that certain foodstuffs may pose to the health of dogs and cats. Preventing exposure is the key to reducing the incidence of these poisoning episodes. Therefore, it is important to increase the knowledge of pet owners with regard to foodstuffs that must not be fed to dogs and cats and should be stored outside their reach.

AUTHOR CONTRIBUTIONS

CC and FC gave substantial contributions to the conception and design of the work; drafted the work and revised it critically for important intellectual content; gave final approval of the version to be published; and agreed to be accountable for all aspects of the work in ensuring that questions related to the accuracy or integrity of any part of the work are appropriately investigated and resolved.

CONFLICT OF INTEREST STATEMENT

The authors declare that the research was conducted in the absence of any commercial or financial relationships that could be construed as a potential conflict of interest.

The reviewer PZ and handling Editor declared their shared affiliation, and the handling Editor states that the process nevertheless met the standards of a fair and objective review.

REFERENCES

1. Mckenzie RA. Poisoning of companion animals by garden and house plants in Queensland: a veterinary practice survey. *Aust Vet J* (2007) **85**:467–8.
2. Caloni F, Berny P, Croubels S, Sachana M, Guitart R. Epidemiology of animal poisonings in Europe. 2nd ed. In: Gupta RC, editor. *Veterinary Toxicology: Basic and Clinical Principles*. San Diego, CA: Elsevier Inc. (2012). p. 88–97.
3. Mahdi A, Van der Merwe D. Dog and cat exposures to hazardous substances reported to the Kansas State Veterinary Diagnostic Laboratory: 2009-2012. *J Med Toxicol* (2013) **9**:207–11.
4. Gwaltney-Brant S. Chocolate intoxication. *Vet Med* (2001) **96**:108–11.
5. Campbell A. Grapes, raisins and sultanas, and other foods toxic to dogs. *UK Vet* (2007) **12**:1–3.
6. Knight AP. *A Guide to Poisonous House and Garden Plants*. Jackson, ID: Teton NewMedia (2007).
7. Salgado BS, Monteiro LN, Rocha NS. *Allium* species poisoning in dogs and cats. *J Venom Anim Toxins Incl Trop Dis* (2011) **17**:4–11.
8. Cope RB. *Allium* species poisoning in dogs and cats. *Vet Med* (2005) **100**:562–6.
9. Yamoto O, Maede Y. Susceptibility to onion-induced hemolysis in dogs with hereditary high erythrocyte reduced glutathione and potassium concentrations. *Am J Vet Res* (1992) **53**:134–7.
10. Guitart R, Mateu C, Lopez i Agullo A, Alberola J. Heinz body anaemia in two dogs after Catalan spring onion ("calcot") ingestion: case reports. *Veterinarni Medicina* (2008) **53**:392–5.
11. Kang MH, Park HM. Hypertension after ingestion of baked garlic (*Allium sativum*) in a dog. *J Vet Med Sci* (2010) **72**:515–8.

12. Spice RN. Hemolytic anemia associated with ingestion of onions in a dog. *Can Vet J* (1976) **17**:181–3.
13. Freed E. A pungent poisoning: onion toxicosis in a cat. *Vet Med* (2015) **110**(8).
14. Yamato O, Kasai E, Katsura T, Takahashi S, Shiota T, Tajima M, et al. Heinz body hemolytic anemia with eccentrocytosis from ingestion of Chinese chive (*Allium tuberosum*) and garlic (*Allium sativum*) in a dog. *J Am Anim Hosp Assoc* (2005) **41**:68–73.
15. Smith CH, Ellison RS. Concurrent onion poisoning and haematuria in a dog. *N Z Vet J* (1986) **34**:77–8.
16. Sturgeon K, Campbell A. A comparison of *Allium* species poisoning in cats and dogs. *Clin Toxicol* (2008) **46**:385.
17. Ratclife RC, Zuber RM. Acute ethyl alcohol poisoning in dogs. *Aust Vet J* (1977) **53**:48–9.
18. Houston DM, Head LL. Acute alcohol intoxication in a dog. *Can Vet J* (1993) **34**:41–2.
19. van Wuijckhuise L, Cremers GG. Alcohol poisoning in dogs. *Tijdschr Diergeneeskd* (2003) **128**:284–5.
20. Keno LA, Langston CE. Treatment of accidental ethanol intoxication with hemodialysis in a dog. *J Vet Emerg Crit Care* (2011) **21**:363–8.
21. Kammerer M, Sachot E, Blanchot D. Ethanol toxicosis from the ingestion of rotten apples by a dog. *Vet Hum Toxicol* (2001) **43**:349–50.
22. Lumeij JT. Ethanol poisoning as a differential diagnosis in a hunter's dog with tetraplegia. *Tijdschr Diergeneeskd* (2009) **134**:932–3.
23. Thrall MA, Freemyer FG, Hamar DW, Jones RL. Ethanol toxicosis secondary to sourdough ingestion in a dog. *J Am Vet Med Assoc* (1984) **184**:1513–4.
24. Suter RJ. Presumed ethanol intoxication in sheep dogs fed uncooked pizza dough. *Aust Vet J* (1992) **69**:20.
25. Means C. Bread dough toxicosis in dogs. *J Vet Emerg Crit Care* (2003) **13**:39–41.
26. Thrall MA, Hamar DW. Alcohols and glycols. 2nd ed. In: Gupta RC, editor. *Veterinary Toxicology: Basic and Clinical Principles*. San Diego, CA: Elsevier Inc. (2012). p. 735–44.
27. Richardson JA. Ethanol. 3rd ed. In: Petersen ME, Talcott PA, editors. *Small Animal Toxicology*. St. Louis, MO: Saunders (2013). p. 547–9.
28. Sutton NM, Bates N, Campbell A. Factors influencing outcome of *Vitis vinifera* (grapes, raisins, currants and sultanas) intoxication in dogs. *Vet Rec* (2009) **164**:430–1.
29. Lovell K, Harvey MA. Maggie the marc eater (grape residue nephro toxicosis). *Contr Therapy Series* (2006) **244**:1711–2.
30. Mostrom MS. Grapes and raisins. 3rd ed. In: Petersen ME, Talcott PA, editors. *Small Animal Toxicology*. St. Louis, MO: Saunders (2013). p. 569–72.
31. Eubig PA, Brady MS, Gwaltney-Brant SM, Khan SA, Mazzaferro EM, Morrow CM. Acute renal failure in dogs after the ingestion of grapes or raisins: a retrospective evaluation of 43 dogs (1992-2002). *J Vet Intern Med* (2005) **19**:663–74.
32. Mazzaferro EM, Eubig PA, Hackett TB, Legare M, Miller C, Wingfield WE, et al. Acute renal failure associated with raisin or grape ingestion in 4 dogs. *J Vet Emerg Crit Care* (2004) **14**:203–12.
33. Gwaltney-Brant S, Holding JK, Donaldson CW, Eubig PA, Khan SA. Renal failure associated with ingestion of grapes or raisins in dogs. *J Am Vet Med Assoc* (2001) **218**:1555–6.
34. Porterpan B. Raisins and grapes: potentially lethal treats for dogs. *Vet Med* (2005) **100**:346–50.
35. Morrow CM, Valli VE, Volmer PA, Eubig PA. Canine renal pathology associated with grape or raisin ingestion: 10 cases. *J Vet Diagn Invest* (2005) **17**:223–31.
36. Stanley SW, Langston CE. Hemodialysis in a dog with acute renal failure from currant toxicity. *Can Vet J* (2008) **49**:63–6.
37. Oh HW, Jun HK, Choi HJ, Lee YW, Song KH. Treatment for acute renal failure occurred by ingestion of grape skins in a dog. *Korean J Vet Res* (2008) **48**:215–8.

38. Usselmann D. Acute renal failure in a dog after ingestion of grapes and grape products. *Praktische Tierarzt* (2007) **88**:790–5.
39. Itoh T, Nishi A, Ikeda A, Kushima E, Kushima K, Uchida K, et al. Acute renal failure after ingestion of grapes in a dog. *J Japan Vet Med Assoc* (2010) **63**:875–7.
40. Kang J-H, Chang D, Ahn B, Na K-J, Yang M-P. Acute renal proximal tubular dysfunction in two dogs associated with grapes ingestion. *Lab Anim Res* (2006) **22**:409–12.
41. Gwaltney-Brant SM. Miscellaneous indoor toxicants. 3rd ed. In: Petersen ME, Talcott PA, editors. *Small Animal Toxicology*. St. Louis, MO: Saunders (2013). p. 291–308.
42. Lee JA. Hops. In: Osweiler GD, Hovda LR, Brutlag AG, Lee JA, editors. *Blackwell's Five-Minute Veterinary Consult Clinical Companion: Small Animal Toxicology*. Ames, IA: Wiley-Blackwell (2010). p. 436–40.
43. Stevens R. The chemistry of hop constituents. *Chem Rev* (1967) **67**:19–71.
44. Duncan KL, Hare WR, Buck WB. Malignant hyperthermia-like reaction secondary to ingestion of hops in five dogs. *J Am Vet Med Assoc* (1997) **210**:51–4.
45. Balhorn DJ. Cyanogenic glycosides in nuts and seeds. 1st ed. In: Preedy VR, Watson RR, Patel VB, editors. *Nuts and Seeds in Health and Disease Prevention*. San Diego, CA: Elsevier Inc. (2011). p. 129–36.
46. Gwaltney-Brant SM. Macadamia nuts. 3rd ed. In: Petersen ME, Talcott PA, editors. *Small Animal Toxicology*. St. Louis, MO: Saunders (2013). p. 625–7.
47. Hansen SR, Buck WB, Meerdink G, Khan SA. Weakness, tremors, and depression associated with macadamia nuts in dogs. *Vet Hum Toxicol* (2000) **42**:18–21.
48. McKenzie RA, Purvis-Smith GR, Allan SJ, Czerwonka-Ledez BJ, Hick LM, Dunn MS, et al. Macadamia nut poisoning of dogs. *Aust Vet Pract* (2000) **30**:6–10.
49. Hansen SR. Macadamia nut toxicosis in dogs. *Vet Med* (2002) **97**:274–6.
50. Dolder LK. Methylxanthines: caffeine, theobromine, theophylline. 3rd ed. In: Petersen ME, Talcott PA, editors. *Small Animal Toxicology*. St. Louis, MO: Saunders (2013). p. 647–52.
51. Craft EM, Powell LL. Chocolate and caffeine. In: Osweiler GD, Hovda LR, Brutlag AG, Lee JA, editors. *Blackwell's Five-Minute Veterinary Consult Clinical Companion: Small Animal Toxicology*. Ames, IA: Wiley-Blackwell (2010). p. 421–8.
52. Ooms TG, Khan SA, Means C. Suspected caffeine and ephedrine toxicosis resulting from ingestion of an herbal supplement containing guarana and ma huang in dogs: 47 cases (1997-1999). *J Am Vet Med Assoc* (2001) **21**:225–9.
53. Drolet R, Arendt TD, Stowe CM. Cacao bean shell poisoning in a dog. *J Am Vet Med Assoc* (1984) **185**:902.
54. Hansen S, Trammel H, Dunayer E, Gwaltney S, Farbman D, Khan S. Cocoa bean mulch as a cause of methylxanthine toxicosis in dogs. *Clin Toxicol* (2003) **41**:720.
55. Michael SJ. Caffeine poisoning in a dog. *J Am Vet Med Assoc* (1961) **139**:559.
56. Foor J, Stowe CM. Acute fatal caffeine toxicosis in a dog. *J Am Vet Med Assoc* (1975) **167**:379.
57. Tawde SN, Puschner B, Albin T, Stump S, Poppenga RH. Death by caffeine: presumptive malicious poisoning of a dog by incorporation in ground meat. *J Med Toxicol* (2012) **8**:436–40.
58. Siméon C, Charrueau H, Blanchard G. Chocolate poisoning in a dog. *Point Veterinaire* (2002) **33**:56–8.
59. Ghazaleh N, Aldavood SJ, Boluki Z, Akbarein H, Nekouie Jahromi OA. A case-series on chocolate poisoning in four Terrier dogs in Tehran. *Iranian J Vet Med* (2008) **22**:Article13.
60. Sturgeon K, Sutton NM. Theobromine toxicity in dogs – is it exaggerated? *Clin Toxicol* (2008) **46**:384.
61. Agudelo CF, Filipejova Z, Schanilec P. Chocolate ingestion-induced non-cardiogenic pulmonary oedema in a puppy: a case report. *Veterinarni Medicina* (2013) **58**:109–12.
62. Reddy BS, Reddy LSSV, Sivajothi S. Chocolate poisoning in a dog. *Int J Vet Health Sci Res* (2013) **1**:401.

63. Meadows I, Gwaltney-Brant S. The 10 most common toxicoses in dogs. *Vet Med* (2006) **101**:142–7.
64. Collica S. *Der Polymorphismus 1117C>T im Cytochrom P450 CYP1A2 beeinträchtigt die Metabolisierung von Theobromin beim Beagle Hund [Dissertation]*. Giessen: Justus-Liebig-University Giessen (2012).
65. Bates N, Rawson-Harris P, Edwards N. Common questions in veterinary toxicology. *J Small Anim Pract* (2015) **56**:298–306.
66. Murphy LA, Coleman AE. Xylitol toxicosis in dogs. *Vet Clin North Am Small Anim Pract* (2012) **42**:307–12.
67. Peterson ME. Xylitol. *Top Companion Anim Med* (2013) **28**:18–20.
68. DuHadway MR, Sharp CR, Meyers KE, Koenigshof AM. Retrospective evaluation of xylitol ingestion in dogs: 192 cases (2007-2012). *J Vet Emerg Crit Care* (2015) **25**:646–54.
69. Dunayer EK, Gwaltney-Brant SM. Acute hepatic failure and coagulopathy associated with xylitol ingestion in eight dogs. *J Am Vet Med Assoc* (2006) **229**:1113–7.
70. Dunayer EK. New findings on the effects of xylitol ingestion in dogs. *Vet Med* (2006) **12**:791–7.
71. Dunayer EK. Hypoglycemia following canine ingestion of xylitol-containing gum. *Vet Hum Toxicol* (2004) **46**:87–8.
72. Todd JM, Powell LL. Xylitol intoxication associated with fulminant hepatic failure in a dog. *J Vet Emerg Crit Care* (2007) **17**:286–9.
73. Lim C-Y, Yoo J-H, Kim C-G, Park C, Park H-M. Acute hepatic failure induced by xylitol toxicosis in two dogs. *J Vet Clin* (2008) **25**:510–3.
74. Fawcett A, Phillips A, Malik R. Hypoglycaemia and acute hepatic failure associated with accidental xylitol ingestion in a dog. *Aust Vet Pract* (2010) **40**:142–7.
75. Schmid RD, Hovda LR. Acute hepatic failure in a dog after xylitol ingestion. *J Med Toxicol* (2015).
76. Campbell A, Bates N. Xylitol toxicity in dogs. *Vet Rec* (2010) **167**:108.

CHAPTER 8

Signal Grass (*Brachiaria decumbens*) Toxicity in Grazing Ruminants

Susan G. Low

Department of Environment and Agriculture, Bentley Campus, Curtin University, Bentley WA 6102, Australia

ABSTRACT

Signal grass (*Brachiaria decumbens*) is a highly productive tropical grass that is widespread through South America, Australia, Indonesia, Vanuatu and Malaysia due to its adaptation to a wide range of soil types and environments. Animal production from these *B. decumbens* pastures is highly variable due to sporadic outbreaks of photosensitisation associated with low growth rates of young animals, anorexia and wasting. The identification of *B. decumbens* toxicity through clinical signs may grossly underestimate the impact and severity of the disease. Affected animals without clinical signs have elevated serum liver enzyme concentrations resulting from blockage of the bile ducts by birefringent crystals, identified as calcium salts of steroidal saponins found in leaves and stems. The concentrations of the steroidal saponins vary through the year and within the plant. Young, green leaves contain 5–10 times the saponin concentration of mature leaves indicating that *B. decumbens* pastures are likely to be more toxic during sprouting and early growth. Previous exposure, selective grazing, and avoiding toxic leaves may partly explain apparent resistance of some animals to *B. decumbens* toxicity. Further research is needed to define growing conditions that produce elevated saponin levels and to investigate the impact of *B. decumbens* on rumen function.

Keywords: *Brachiaria decumbens*; signal grass; toxicity; photosensitisation; steroidal saponins; ruminants

1. INTRODUCTION

Brachiaria spp., generally referred to as signal or palisade grasses based on the similarity of the flower head structure to a railway signal, originated in East Africa (Uganda, Kenya, Tanzani Rwanda, Burundi and Zaire) as a pasture resource for grazing by cattle and wildlife, growing at a range of altitudes from 500 to 2300 m a.s.l. These species have since been replaced with other forage species [1] that are more suitable to the changing climate. There has been confusion with Brachiaria species names being interchangeable over the past 50 years with *B. eminii*, *B. decumbens*, and *B. ruziziensis* all being used for the same plant material [2,3]. The most widely used cultivar as a grazing pasture globally is *B. decumbens* cv. Basilisk which was bred from seed introduced into Australia in 1930 from Uganda [4] and approved for registration in 1973. Signal grass (*Brachiaria decumbens*) has been introduced into many tropical countries including tropical and sub-tropical areas of Australia [5,6,7,8], Vanuatu [9], Colombia [10,11], Indonesia, Fiji [12], Trinidad [13] and Venezuela [14] with Brazil currently having more than 50 million hectares of *Brachiaria* spp. pastures [4,15,16]. In Malaysia, *Brachiaria* species have been planted on more than 80% of improved farming pastures with *B. decumbens* as the most favoured species [17]. Livestock production from these pasture systems has been variable even when both pasture quality and quantity were high. This review will focus on constraints to the use of *B. decumbens* cv. Basilisk as a pasture for grazing ruminants.

2. AGRONOMY OF *BRACHIARIA DECUMBENS* (SIGNAL GRASS)

B. decumbens is a low-growing decumbent perennial grass with flowering stems up to 100 cm high originating from the prostrate, multi-noded stems; plants can spread by both rhizomes and stolons as well as through seed production. *B. decumbens* has been widely adopted as a pasture for grazing ruminants due to its high nutritive value and aggressive growth habit that provides a dense ground cover able to suppress weeds. However, the aggressive growth habit can result in pastures becoming monocultures [2,18] by suppressing companion plants such as legumes, thus reducing the potential nutritive value of the pastures for livestock. Legumes with stoloniferous growth habits, such as *Arachis pintoi* (pinto peanut) and *Desmodium heterophyllum* (creeping hetero) and plants with shrubby/tree growth habits (such as *Leucaena leucocephala* and *Stylosanthes guianensis*) have been used with some success as legume companion species *B. decumbens* pastures [18,19,20,21]. The ability of *B. decumbens* to withstand shading has made it an important companion to Leucaena [22], and it has been successfully grown as a pasture in coconut plantations [23,24].

B. decumbens is resilient and is grown over a wide range of soil types including infertile acid soils with low pH (<3.5) and high aluminium levels [4,25] and climates ranging from tropical to sub-tropical. In Western Australia, *B. decumbens* was included in a mixed species pasture mix for the sub-tropics (sandy soils and a maximum annual rainfall of around 600 mm) [26]. This recommendation has now been withdrawn due to outbreaks of photosensitisation in grazing livestock on the mixed pastures. *B. decumbens* is able to withstand short-term drought conditions [2,27,28,29] by re-allocation of biomass and adjustment of growth rates resulting in no significant differences in biomass yield [29] but is highly susceptible to waterlogging [2] and can only tolerate flooding for a few days.

B. decumbens produces more dry matter than most tropical grasses during the dry season [30] and is capable of producing 15–27 mt (metric tonnes) dry matter (DM)/hectare/year [31,32,33]. In a study undertaken in South Sulawesi, Indonesia, *B. decumbens* produced more dry matter than other improved pasture species during the dry season (5700 kg/ha and 1910–3500 kg/ha respectively). It has been suggested that the ability to respond to small amounts of rainfall that occurred in the dry season was due to the extensive root system of *B. decumbens* [29,30]; plants produce new growth rapidly with out-of-season rain events during the dry season and with the break of season. Young vegetative material has been associated with reported outbreaks of photosensitisation among ruminants grazing *B. decumbens* pastures [34,35,36,37] (see Section 3).

B. decumbens responds well to defoliation either through grazing, harvesting or complete defoliation such as burning [23,38]. In Papua New Guinea, *B. decumbens* regrows readily after fire, a practice used by local villagers to hunt animals or as part of the "slash and burn" management of standing grass to prepare new food gardens [39,40,41,42]. *B. decumbens* has been shown to persist better than *Panicum maximum* (guinea grass) under high stocking rates in far north Queensland (Australia) [43].

Brachiaria species, including *B. decumbens*, utilise nitrogen efficiently [2] due to an ability to favour root growth in extreme conditions and to harvest biological nitrogen from bacterial species in the soil [44,45] equivalent to 30–50 kg N/hectare. *B. decumbens* cv. Basilisk has the ability to uptake higher levels of calcium, phosphorus and nitrogen than other *Brachiaria* species in both high and low soil fertility conditions [45]. Data from a number of experiments, extrapolated by Rao *et al.* [45] have indicated that 40% of plant nitrogen content in *B. decumbens* was derived from biologically fixed N_2. *B. decumbens* has shown, in plot trials, an increasing response to the addition of nitrogen fertiliser up to 400 kg N/hectare depending on nitrogen source [46,47], producing more than 40 mt DM/hectare/year at this level. In Brazil, the late summer application of nitrogen (100 kg N/ha) together with 95 days of deferred grazing, were found to produce the greatest biomass on an annual basis [48]. Split applications of nitrogen fertiliser have been recommended, as the growth response is greater in the period immediately following the application of nitrogen [49].

3. ANIMAL PRODUCTION ON *B. DECUMBENS* PASTURES

Brachiaria species have been widely implemented as quality pastures for animal production. Nitrogen content in early regrowth periods (2–5 weeks) ranged from 10–28 g N/kg dry matter decreasing to 7–15 g N/kg DM for 10–12 weeks regrowth [2,21,40,50,51,52]. *In vitro* dry matter digestibility showed similar variation with early growth ranging from 56%–78% up to five weeks of age decreasing to 41.6% to 63.7% as the pasture matures [2,21,30,47,51,53,54]. The estimated nutritive value, and therefore the potential level of liveweight gain, of *B. decumbens* is similar to other tropical pasture species such as *Panicum maximum* (guinea grass), *Digitaria decumbens* (pangola grass) and *Setaria anceps* (setaria) [20,30,51,54,55]. Analytical data on chemical composition and potential digestibility (*in vitro* DM digestibility) of *B. decumbens* have shown high variability between data sets sampled through the growing period. This variability may be due to one or more factors such as season (wet or dry), fertiliser applications (timing and level), species combinations (e.g., presence of legumes), growing conditions (trial plot *vs.* grazing pasture), sampling procedures (plot harvest or picked selection) and analysis methods (*in vivo vs. in vitro* digestibility determination).

Based on estimated nutritive values, animal production from *B. decumbens* pastures would be expected to be comparable to production from other commonly used tropical grass species and from medium quality temperate pastures. This expectation is supported by grazing trials that compare pasture species/composition, stocking rate and growth rates of sheep, goats and cattle; these comparisons have shown that daily and annual live weight gains from grazing *B. decumbens* is comparable toormayexceed growth rates on *P. maximum* pastures—0.46 to 0.78 kg/head/day and 0.49 to 0.61 kg/head/day, respectively [10,18,21,55,56]. Cattle in North Queensland, aged 15 months to three years of age, grazed tropical pastures including *B. decumbens* under a rotational grazing system at stocking rates that ranged from 0.7 to 5 head/ha [47]; annual weight gains were 211 to 950 kg/ha/year [43,47,55,57] with the highest live weight gains achieved on *B. decumbens* pastures. *B. decumbens* persist under high stocking rates and continuous grazing [55,58].

However, conflicting data from trials and from on-farm records indicate that nutritive value, as determined by chemical analysis, may not be a true indicator of potential animal performance. Analysis of growth rates for Brahman cattle in the Markham and Ramu Valleys of Papua New Guinea and in Queensland, Australia showed lower growth rates on *B. decumbens* pastures (0.29–0.37 kg/head/day) in the late wet season/early dry season pastures following rapid pasture growth compared to growth rates during the late dry season (0.57–0.65 kg/head/day) [47,59,60]. Since pastures in the early dry season would be considered to have a high nutritive value following the seasonal rains, these data conflict with performance expectations. Growth rates calculated from farm records over five years, from two different production systems in the Ramu Valley in Papua New Guinea, showed that Brahman weaner cattle on *B. decumbens* pastures (with supplements) averaged 0.05 kg/head/day and 0.2 kg/head/day for 1–100 days and 100–200 days post weaning, respectively [59] irrespective of the production system. The two production systems were (i) born, reared and weaned onto *B. decumbens* pastures or (ii) born onto native pastures and weaned onto *B. decumbens* pastures. Studies in Brazil, however, have shown that sheep adapted to *B. decumbens* pastures gained weight at a similar rate to animals on *P. maximum* but had evidence of liver dysfunction indicated by elevated serum enzyme levels [16]. Naïve sheep (*i.e.*, sheep which had no previous exposure to *B. decumbens* pastures) lost weight rapidly, had elevated serum liver enzyme levels and a 50% mortality rate.

Grazing trials report rates of weight gain and persistence of the pasture under grazing for *B. decumbens* pastures. No reports of production problems, photosensitization or deaths have been found in grazing trial data. However production/disease issues on these pastures have appeared as disease outbreak reports, generally from commercial farms rather than from trials. Recent studies in Brazil have attempted to link live weight gain and grazing history of the livestock species to clinical andsub-clinical disease occurrences [16,41].

4. TOXICITY OF *B. DECUMBENS* PASTURES

Although analyses and animal production studies indicate that *B. decumbens* pastures can produce high live weight gains in grazing livestock, there are many reports of low weight gains, general ill-thrift and sporadic outbreaks of photosensitivity in goats [36,61], sheep [34,62,63,64,65], llama [66], buffaloes [67], deer [68] and cattle [69] across Colombia, Brazil, Australia, Papua New Guinea [59,70], Malaysia and Sri Lanka.

4.1. Photosensitisation and Clinical Biochemistry

Photosensitisation outbreaks are often reported when the pasture is young and actively growing. This includes growth after cultivation or slashing [62], in the early wet season or during the dry season following an out of season rain event [34,36,61]. Photosensitisation outbreaks have often been observed in sheep and goats within 10 days of grazing established *B. decumbens* pastures [61,62,63] and after three weeks in new growth pastures following out of season rainfall events [34,37,40,71,72]. Clinical signs of photosensitisation include progressive weight loss and anorexia, oedema and necrotic tissue, crusting and sloughing of skin in non-pigmented and exposed areas such as ears, face, rump, flank and vulva regions in females as well as visible jaundice [40,62]. Affected animals seek shade, grazing in the evening or early mornings or remaining under trees and shrubs rather than actively grazing [36,40,63]. Data from reported outbreaks indicate that young animals (sheep <12 months and cattle <2 years) are most likely to show clinical signs [41,42,65,73,74] with older animals not exhibiting clinical signs. In these outbreaks, morbidity in sheep varied from 0.42% to 100% and 0.08% to 7.9% for cattle [40,41,42,63,74,75,76].

B. decumbens or signal grass toxicity outbreaks are reported as clinical signs (predominantly secondary hepatogenous photosensitisation) and descriptions of behavioural changes. Secondary photosensitisation is a result of the build-up of phylloerythrin in the skin. Phylloerythrin is a photodynamic molecule produced from the metabolism of plant chlorophyll [77] by microorganisms and normally removed from blood for excretion in bile. Circulating phylloerythrin levels are increased due to hepatic dysfunction and/or bile duct lesions [78]. Skin cells become highly reactive in the presence of visible and UV radiation through increased photoexcitation of these photodynamic molecules [79]. Animals with pigmented skin are less likely to show clinical signs [34,40,63,79,80]; hair and skin pigments protect animals from the impacts of UV exposure. For this reason, areas most affected on animals displaying signs of photosensitivity tend to be those that have minimal hair cover and are unpigmented [79] such as ears, muzzle, vulva and areas around the tail. Clinical signs (*i.e.*, photosensitisation), as the indicator of animals affected by *B. decumbens* toxicity, is likely to be a gross underestimate of the number of animals affected, especially in a pigmented flock or herd.

Biochemical data from clinical and experimentally induced cases suggest that liver dysfunction is a major component of observed subclincal and clinical disease signs. Clinical outbreaks are accompanied by changes in liver enzyme concentrations, particularly serum AST (aspartate aminotransferase) and serum GGT (gamma glutamyltransferase); increased concentrations of these enzymes indicate impaired liver function and have been recorded with other grass species that have also been associated with hepatogenous photosensitisation [16,40,64,66,67,68,72,76,80]. Clinical biochemistry collected from reported outbreaks and from feeding/grazing trails indicates that serum GGT levels are likely to be more highly elevated (compared to normal range information) than serum AST levels. Recorded serum GGT levels of affected animals ranged from 100 U/L to 978 U/L (normal range 20–52 U/L) and serum AST levels ranged from 88.8 U/L to 1810 U/L (normal range 60–300 U/L) [40,64,68,76,80,81]. Similar serum concentrations have been recorded for animals in the same grazing groups but not displaying clinical signs. The correlation between serum GGT levels and possible *B. decumbens* toxicity is higher than for serum AST levels possibly due to the narrow normal range associated with serum GGT levels. GGT is an enzyme located in the cell membranes in the bile ducts and hepatocytes [82]; serum levels increase with membrane damage due to accumulation of bile salts resulting in oxidative stress and decreased antioxidant potential [82,83]. Changes in serum AST could be confirmed only when initial values for each animal or herd/flock had been determined. Animals with dark wool and/or pigmented skin are less likely to display clinical signs but are as likely to have elevated liver enzymes [40,63,80] suggesting that clinical signs are not suitable indicators of the number of animals affected by *B. decumbens* toxicity.

Comparison of serum biochemistry of clinically affected and non-affected animals indicated that, in many cases, serum AST and GGT levels were elevated to the same extent in both groups [64,68,84]. The proportions of affected naïve sheep and goats in three grazing trials on *B. decumbens* pastures in Papua New Guinea determined as clinical signs or through serum biochemistry was 38% and 75%, 15% and 42% and 27% and 82%, respectively [40]. These data suggest that the use of clinical signs to define outbreaks may be flawed, severely underestimating the number of affected animals. Weight loss (with or without clinical signs) of grazing livestock has often been associated with *B. decumbens* pastures where outbreaks have occurred and with animals in experimental feeding trials [16,40,42,80,84]; monitoring of condition score and live weight may be better tools for identifying affected animals.

The proportion of animals clinically affected does, however, appear to be greater for naïve livestock and for young and/or newly weaned livestock [64,74]. Naïve sheep introduced to *B. decumbens* pastures in Brazil had increased morbidity and mortality compared to experienced sheep (33.3% and 11.1% compared to 8.69% and 4.34% respectively) [42,76,80]. Data from reports for other pasture species associated with secondary hepatogenous photosensitisation found that 30%–60% of lambs less than 12 months of age were most affected resulting in case fatalities of up to 90% [64,85,86]. In Papua New Guinea, cattle born and weaned onto *B. decumbens* pastures had low numbers of clinically affected cattle compared to cattle born onto native pastures and weaned onto *B. decumbens* pastures [40] where the annual mortality rate was between 5% and 10%. However, based on farm records, the growth rates recorded in the two management systems were the same suggesting that the observed poor growth rates for livestock grazing *B. decumbens* pastures may be due either to *B. decumbens* toxicity *per se* or to extremely low intakes of pasture, possibly based on previous negative experience for those animals reared on *B. decumbens* pastures. Animals that were born and weaned onto *B. decumbens* pastures camped near the molasses/copra meal supplement and pastures have very limited evidence of grazing suggesting that cattle have developed avoidance behaviour [40].

4.2. Histopathology

Post-mortem examination of animals with *B. decumbens* toxicity indicates that the liver and kidneys are the major sites of cellular damage. Blockage of bile ducts is associated with thickening of bile canaliculi and the presence of birefringent crystals [40,64,65,67] within the bile ducts. Hydropic degeneration of hepatocytes, necrotic cells and the presence of foamy macrophages were found in association with degenerated hepatocytes [42,64,65,67,74,87]. Multifocal cholangitis was common to these cases. Microscopic examination of liver tissue has provided evidence of bile duct proliferation and blockages with birefringent crystals often found in the bile ducts [40,41,64,65,66,67,69,80,87,88]. The severity of damage to hepatocytes is closely linked with the clinical symptom of jaundice [83]. Physical appearance of the liver shows brown/yellow discolouration, a high degree of mottling and an increased lobular pattern [40,42]. The link between *B. decumbens* pastures and the apparent toxicity recorded has been verified through trials where animals have been dosed with rumen contents from affected animals; clinical, biochemical and histopathological patterns were similar to those found in outbreaks and grazing trials [88,89,90]. Although these experiments did not identify the compound(s) responsible for elevated liver enzymes, histopathological changes and photosensitisation, a link between *B. decumbens* consumption and physical/physiological/biochemical observations was confirmed. Similar histopathological signs and the presence of birefringent crystals have been associated with pasture species such as *Panicum schinzii* [85,91,92], *Panicum coloratum* [93,94], *Panicum dichotomiflorum* [95,96], *Panicum miliaceum* [97] and *Tribulus terrestris* [98,99]. As a consequence, identification of possible toxic compounds has been focused across these pasture species as a group. Initial focus was on the possible role of sporidesmin, a toxin produced by the fungus *Pithomyces chartarun* as the physical signs were similar to those observed in animals suffering from facial eczema [100,101].

4.3. The Role of **Pithomyces chartarum**

Clinical signs of animals affected by *B. decumbens* toxicity are similar to those of animals with facial eczema. This disease affects sheep, deer, alpacas and cattle in New Zealand particularly in late summer/autumn [75,102]. Facial eczema is a secondary photosensitisation disease caused by the mycotoxin sporidesmin produced by the saprophytic fungus *Pithomyces chartarum* and was initially thought to be contributing factor to *B. decumbens* toxicity [100,101]. Early studies attempted to reproduce the symptoms observed in animals affected by *Panicum* toxicoses by dosing lambs with commercially available saponins with or without sporidesmin; the saponins used had been identified as components of birefringent bile crystalsthat are commonly associated with secondary photosensitisation. The dosed saponins did not produce liver lesions and sporidesmin produced a weak response only [103]. Further studies on the possible role of sporidesmin in *B. decumbens* toxicity have shown that there are negligible numbers of spores on pastures associated with either outbreaks or grazing studies, thus discounting any role of sporidesmin in this disease [42,64,72,76,104,105,106].

5. IDENTIFICATION AND METABOLISM OF TOXIC COMPOUNDS

Potentially toxic compounds have been isolated from leaf and stem fractions of plants associated with secondary hepatogenous photosensitisation, including *B. decumbens*, and identified as steroidal saponins [8,40,93,96,106,107,108,109,110,111,112,113]. Saponins are surface-acting glycosides that have the ability to produce a stable foam in an aqueous solution [114] and are classed as either triterpenoid or steroid based on structure [114]. Triterpenoid saponins are commonly found in dicotyledons while steroidal saponins are found in monocotyledons [115]. The activity of a saponin is a function of the position and type of sugar component of the molecule [115] with many plant species containing multiple saponins—for example lucerne (alfalfa) roots contain 29 identified saponins. Saponins create pores in cell membranes and are associated with haemolytic activity [114,116]—this has been used as a qualitative test for the presence of saponins. The ability to disrupt blood cells is a consequence of interactions between the saponin and the sterols present in the red blood cell membrane and can be used as an indicator of the presence of saponins [116].

Steroidal saponins are found in the plant as glycosides of sapogenins with the sugar fraction being composed of one or more sugars. Although saponins have been found in a wide range of plants and crops, their role in plant metabolism and function is poorly understood [114,115]. Steroidal saponins have been commercialised as supplements, health foods for human consumption and potential anti-cancer treatments [116,117,118,119,120,121].

Saponins are present in a number of forages that are important for livestock. These include fodder plants such as *Acacia* spp., *Gliricidia sepium* and *Sesbania sesban* as well as legumes including soybean (*Glycine maxima*), lupins (*Lupinus spp*), lucerne (*Medicago sativa*), red clover (*Trifolium pretense*) and ladino clover (*Trifolium repens*) [115]. Saponins commonly found in forages include soyasapogenin, soyasapogenol, diosgenin, dioscin/protodioscin, and yamogenin [115]. Concentrations of saponins in plant tissues are highest in those that are vulnerable to insect, fungal or bacterial attack [115] suggesting that these compounds may act as a plant defense mechanism. It would therefore be expected that saponin concentrations in plant tissues would be highest when the plant is most vulnerable; this could include early growth after prolonged dry spells and flowering and/or seed set. *B. decumbens* spreads by seed distribution and through vegetative growth (stolons). This is supported by studies that found higher protodioscin concentrations at seed set and seed fall [64,112] and other studies that showed highest concentrations in new leaf shoots [35,40,105,108,111,122]. Outbreaks tended to be associated with growth after rain, particularly at the end of the dry season or after prolonged dry periods [34] rather than total rainfall through the growing season [105] suggesting that saponin concentrations were high. However, de Oliveira *et al.* [15] found positive correlations between total saponin concentration and green leaf mass ($p < 0.01$) and total saponin concentration and rainfall ($p < 0.05$). Protodioscin concentration was elevated after the first rains of the season (32.4 to 35.4 mg/g DM) but decreased as the wet season continued (3.9 to 8 mg/g DM) suggesting that initial high concentration could act as a deterrent to grazing animals and attack by insects, bacteria or fungi ensuring survival and establishment.

The main steroidal saponins found in *B. decumbens*, as well as many of the other species associated with secondary hepatogenous photosensitization, have been identified as dichotomin, protodioscin and dioscin [108,110,112]. Diosgenin is produced from acid hydrolysis of either dioscin or protodioscin [123]. Prior to 2000, the saponin content of pasture species including *B. decumbens* was reported as diosgenin and/or yamogenin concentrations (mg/kg DM). As these compounds have been identified as derivatives of the parent saponin "dioscin", through improved analytical methodology, diosgenin equivalents can be calculated from protodioscin concentration using the conversion "(protodioscin mg/kg)/2.5" [8] to allow data comparison. Protodioscin concentrations in pastures associated with outbreaks, expressed as mg/g plant dry matter ranged from 5.2 mg/g DM to 35.4 mg/g DM [64,67,74,76]. Moore *et al.* (2014) [8] used a protodioscin concentration of 2 mg/g or 0.8 mg/g diosgenin equivalent as the critical value for potential toxicity of pastures. From data from planted plot and field pasture trials, protodioscin concentrations ranged from 0 mg/g to 60.25 mg/g plant DM [8,16,40,64,76,80,105,108,111,112,122,124].

Steroidal sapogenins, derived from the parent plant tissue saponins, have been isolated from rumen contents [125,126] and from bile in affected animals and identified primarily as metabolic derivatives of diosgenin [109,127] and yamogenin (the 25 S isomer of diosgenin) [93,108,109,126,128]. Lambs dosed with extracts from *B. decumbens* plant material developed liver lesions, bile duct proliferation with birefringent crystals found in the bile ducts [88,89,126]. Photosensitisation was not observed as the lambs were kept out of direct sunlight. Similarly, geeldikkop, the hepatogenous photosensitization disease associated with steroidal saponins found in *Tribulus terrestris*, was reproduced using crude plant extracts, producing similar liver lesions and birefringent crystals [90,129]. The extracts contained a mixture of saponins including diosgenin and yamogenin [129]. However, lambs dosed with the commercially available saponins diosgenin and sarsasapogenin into the rumen showed neither loss of appetite nor any evidence of liver damage [40,103] suggesting that either the individual saponins may not be hepatotoxic or that the doses were not high enough.

Metabolism of plant saponins found in *B. decumbens* and other species associated with hepatogenous photosensitisation has been elucidated using plant extracts; ingested diosgenin-derived saponins were rapidly hydrolysed by rumen microflora to parent sapogenins [127,129]; diosgenin was converted to epismilagenin, smilagenone, smilagenin and tigogenin and primarily absorbed from the jejunum [127,130,131]. However, these dosing trials did not produce either clinical or biochemical evidence of toxicity, the direct relationship between intake of saponin-containing plant material and clinical or subclinical disease [40,127,131]. The ruminal hydrolysis of plant saponins was confirmed through the analysis of rumen contents of sheep identified as suffering from *B. decumbens* toxicity [125,126]. Cattle dosed with rumen contents from affected sheep showed evidence of liver dysfunction through elevated serum liver enzyme concentrations, providing a further link between saponins in plant material, rumen metabolism and liver dysfunction [89].

Crystals found in the bile ducts of animals suffering from hepatogenous photosensitisation associated with *Panicum miliaceum* and *Tribulus terrestris* have been identified as insoluble calcium salts of the plant saponins [96,99,132,133]. These investigations have provided important information on the possible links between steroidal saponins in plant tissue, rumen metabolism and absorption of the derived compounds and the formation of crystals capable of blocking bile ducts. However, although the birefringent bile crystals have been shown to be calcium salts of saponins derived from diosgenin [90,132,133], dosing experiments using plant extracts or isolated saponins have provided limited evidence of the link between plant steroidal saponins, liver dysfunction, bile crystals and the clinical signs of this disease [40,103,127,129,130]. Similar results have been obtained for dosing experiment (rumen contents or extracted saponins) for the other species associated with secondary hepatogenous photosensitisation and containing steroidal saponins [88,90,99,129]. It is possible that the observed liver dysfunction and associated physical signs may be either the consequence of interactions between saponins or other chemical constituents in the plants or linked to the metabolic activity of rumen microorganisms. Dosing with rumen contents of affected animals has proved more successful in reproducing biochemical and clinical signs

Saponin (particularly protodioscin) concentration varies with plant and leaf age with senescing leaves generally containing lower concentration of saponins than younger leaves [40,111,134]. Data from Brazil and Papua New Guinea have shown that there is a 5–10 fold increase in saponin concentration between new leaves and mature leaves [40,134]. This finding supports the observations that clinical outbreaks tend to be associated with new leaf growth. Ferreira *et al.* [134] found a strong relationship between saponin concentration on young leaves and hours of solar radiation. Ensiling *B. decumbens* has been shown to decrease protodioscin concentrations significantly (2.28 mg/g DM to 0 mg/g DM) over 20 days while conserving *B. decumbens* as hay only reduced protodioscin concentration by 48% [135]. There is no clear trend in saponin concentration with plant age *per se* [76,105,122], with higher concentrations found in plant material 60 days of age or younger but almost stable levels over the following 300 days [105,122]. Brum *et al.* [112] found that saponin concentrations varied with environmental factors, age, stage of growth and cultivation methods. There was no significant relationship in protodioscin concentrations in commercial paddocks sown with *B. decumbens* in Western Australia over a north to south range of 450 km even though some locations had consistently higher concentrations; protodioscin concentrations within sample dates over a two year period ranged from 15 mg/g DM to 60 mg/g DM [8]. Sample collection methods varied from selecting leaf only to whole plant and have been sampled by plucking, hand picking and plot harvest. Samples collected from outbreak sites have been obtained after the outbreak has been detected and may not be representative of plant material consumed by grazing livestock.

Differences in apparent susceptibility (or the possibility of resistance) to *B. decumbens* toxicity [16,64,80] may be due to selective grazing, with animals avoiding plants or plant parts that contain higher concentrations of saponins. Lambs dosed with plant secondary compounds avoided grazing pastures containing the dosed compound [136] and actively sought out species that did not contain the specific compound. Brahman cattle grazing drought stressed *B. decumbens* pastures in the Markham Valley in Papua New Guinea selected older and senescing leaves, leaving new leaves untouched; this suggests an active avoidance of plant components with high saponin concentration [40]. Brahman weaners born onto *B. decumbens* pastures avoided grazing *B. decumbens* pastures after weaning, relying on a molasses/copra meal/mineral supplement [40]; examination of the pasture revealed that very few plants had been consumed and showed evidence of individual leaf selection [40]. The apparent resistance of some animals to pastures containing high levels of saponins may therefore be a consequence of modified grazing behaviour, particularly where supplements or alternative species are available. However, since *B. decumbens* pastures tend towards monocultures by outcompeting companion species, avoidance of plants with high saponin concentrations may be difficult. Avoidance behaviour could provide a possible explanation, on predominantly *B. decumbens* pastures, for poor growth rates and wasting or weight loss often associated with livestock grazing *B. decumbens* pastures since minimising the intake of high risk pastures will reduce nutrient intake and thus reduce growth rates.

Low growth rates from animals consuming *B. decumbens* pastures could also be related to liver dysfunction (as indicated in elevated serum AST/GGT) or to changes in the rumen microbial population. Early studies showed reduced microbial population and cellulolytic activity of sheep grazing *B. decumbens* pastures, resulting in significantly lower levels of volatile fatty acid production [137]. Although saponins have been shown to reduce protozoal numbers in the rumen through rupture of the protozoal cell membrane and reduce methane production and to decrease the numbers of cellulolytic bacteria, the effects depend on the source of the saponins [115,138]. There is some suggestion that rumen bacterial populations adapt to saponins under long term feeding, but this is yet to be confirmed [115]. Grazing trials with naïve and experienced sheep showed that naïve sheep had higher morbidity and mortality rates than experienced sheep [80]. Adaptation of rumen microorganisms may provide grazing livestock with an element of protection from *B. decumbens* toxicity.

6. CONCLUSIONS

B. decumbens is a tropical/sub-tropical pasture species that produces high quality feed for grazing ruminants. The suitability of *B. decumbens* as a quality pasture for ruminant livestock production is compromised by the presence of steroidal saponins including protodioscin, diosgenin and yamogenin. These compounds have been linked to sporadic outbreaks of secondary hepatogenous photosensitisation, particularly in young animals and have been implicated in sub-clinical disease leading to poor growth rates, wasting and anorexia. The occurrence of *B. decumbens* toxicity and its impact on animal productivity may be grossly underestimated since diagnosis is based solely on outbreaks of photosensitisation and recorded as such. Steroidal saponins have been linked to birefringent crystals that potentially block the bile ducts resulting in elevated liver enzymes. Assessment of liver dysfunction is not possible in a production system. Animal behavior changes (such as shelter seeking and grazing avoidance behaviours may be indicative of toxic pastures. The concentration of these compounds is influenced by management, season, stage of growth and environmental conditions. Further research is needed to elucidate the factors influencing saponin concentrations in *B. decumbens*, to study the impact of the saponins from *B. decumbens* on rumen function, and to investigate apparent differences in susceptibility between individual animals.

CONFLICTS OF INTEREST

The authors declare no conflict of interest.

REFERENCES

1. Ndikumana, J.; de Leeuw, P.N. Regional experience with Brachiaria: Sub-Saharan Africa. In Brachiaria: Biology, Agronomy and Improvement; Miles, J.W., Maass, B.L., do Valle, C.B., Eds.; CIAT: Cali, Colombia, 1996; pp. 247–257.
2. Loch, D.S. Brachiaria decumbens (signal grass)—A review with particular reference to Australia. Trop. Grassl. 1977, 11, 141–157.
3. Miles, J.W.; Maass, B.L.; do Valle, C.B. Identifying and naming brachiaria species. In Brachiaria: Biology, Agronomy and Improvement; Vrinda, K., Ed.; CIAT: Cali, Colombia, 1996.
4. Keller-Grein, G.; Maass, B.L.; Hanson, J. Natural variation in Brachiaria and exisiting germplasm collections. In Brachiaria: Biology, Agronomy and Improvement; Miles, J.W., Maass, B.L., do Valle, C.B., Eds.; CIAT: Cali, Colombia, 1996; pp. 16–42.
5. Mackay, J.H.E. Register of Australian herbage plant cultivars. J. Aust. Inst. Agric. Sci. 1974, 40, 91–93.
6. Winter, W. Preliminary evaluation of twelve tropical grasses with legumes in northern Cape York Peninsula. Trop. Grassl. 1976, 10, 15–20.
7. Moore, G.; Sanford, P.; Wiley, T. Signal Grass in Perennial Pastures for Western Australia; Department of Agriculture and Food Western Australia: Perth, Australia, 2006; p. 2.
8. Moore, G. Saponins in Sub-Tropical Grasses in WA; Government of Western Australia: Perth, Australia, 2014; p. 4.
9. Macfarlane, D.; Shelton, H.M. Pastures in Vanuatu; Australian Centre for International Agricultural Research: Bruce, Australia, 1986.
10. Crowder, L.V.; Chaverra, H.; Lotero, J. Productive improved grasses in Colombia. In Proceedings of the 11th International Grassland Congress, Queensland, Australia, 13–23 April 1970; Wageningen Academic Publishers: Wageningen, The Netherlands, 1970; pp. 147–149.

11. Cárdenas, E.; Díaz, F. Tropical Pastures Program Annual Report. Available online: http://pdf.usaid.gov/pdf_docs/PNAAM065.pdf (accessed on 27 June 2015).
12. Robert, O.T. A review of pasture species in Fiji. I. Grasses. Trop. Grassl. 1970, 4, 129.
13. Howes, J.; Campbell, J. Grass and its utilization in Trinidad. Trop. Agric. 1953, 30, 3.
14. Leng, R.A. Research Priorities for Cattle and Sheep Production in Venezuela; University of New England: Armidale, Australia, 1990.
15. De Oliveira, O.C.; de Oliveira, I.P.; Alves, B.J.R.; Urquiaga, S.; Boddey, R.M. Chemical and biological indicators of decline/degradation of Brachiaria pastures in the Brazilian Cerrado. Agric. Ecosyst. Environ. 2004, 103, 289–300.
16. Gracindo, C.V.; Louvandini, H.; Riet-Correa, F.; Barbosa-Ferreira, M.; Castro, M.B. Performance of sheep grazing in pastures of Brachiaria decumbens, Brachiaria brizantha, Panicum maximum and Andropogon gayanus with different protodioscin concentrations. Trop. Anim. Health Prod. 2014, 46, 733–737.
17. Chin, F.Y. Justification, government policy, measures and strategy for the development of grazing reserves in Peninsular Malaysia. In Proceedings of the 1st Meeting of Regional FAO Working Group on Grazing and Feed Resources in S.E. Asia, Serdang, Malaysia, 27 February–3 March 1989; Available online: http://www.fao.org/ag/agp/agpc/doc/publicat/GRASSLAN/79.pdf (accessed on 19 June 2014).
18. Paterson, R.T.; Quiroga, L.; Sauma, G.; Samur, C. Dry season growth of Zebu-Criollo steers with limited access to leucaena. Trop. Anim. Prod. 1983, 8, 138–142.
19. Watson, S.E.; Whiteman, P.C. Grazing studies on the Guadalcanal Plains, Solomon Islands. 2. Effects of pasture mixtures and stocking rate on animal production and pasture components. J. Agric. Sci. 1981, 97, 353–364.
20. Lascano, C.E.; Euclides, V.P.B. Nutritional quality and animal production of Brachiaria pastures. In Brachiaria: Biology, Agronomy and Improvement; Miles, J.W., Maass, B.L., do Valle, C.B., Eds.; CIAT: Cali, Colombia, 1996; pp. 106–123.
21. Galgal, K.K.; Shelton, H.M.; Mullen, B.; Komolong, M. Growth of cattle grazing Leucaena genotypes in the humid lowland tropics of Papua New Guinea. Asian Aust. J. Anim. Sci. 2000, 13, 242–245.
22. Schultze-Kraft, R.; Tietzel, J.K. Brachiara decumbens stapf. In Plant Resources of South East Asia No. 4. Forages; Mannetje, L.T., Jones, R.M., Eds.; PROSEA: Bogor, Indonesia, 1992; pp. 58–59.
23. Rika, I.K. New forage species for coconut plantations in Bali. In Forages for Plantation Crops; ACIAR: Sanur Beach, Bali, 1990.
24. Fisher, M.J.; Cruz, P. Some ecophysiological aspects of Arachis pintoi. In Biology and agronomy of Forage Arachis; CIAT: Cali, Colombia, 1994.
25. Tropical Forages. Available online: http://www.tropicalforages.info/key/Forages/Media/Html/Brachiaria_brizantha.htm (accessed on 18 October 2012).
26. Moore, G.; Yates, R.; Barrett-Lennard, P.; Nichols, P.; Wintle, B. Establishment Guide for Sub-Tropical Grasses—Key Steps to Success; DAFWA: Kensington, Australia, 2013.
27. Bogdan, A. Tropical Pasture and Fodder Plants; Longman: Essex, UK, 1977.
28. Mannetje, L.; Jones, R. Forages. Bogor: Plant Resources of South-East Asia; PROSEA: Bogor, Indonesia, 1992.
29. Guenni, O.; Marín, D.; Baruch, Z. Responses to drought of five Brachiaria species. I. Biomass production, leaf growth, root distribution, water use and forage quality. Plant Soil 2002, 243, 229–241.
30. Bulo, D.; Blair, G.J.; Till, A.R.; Stür, W. Yield and digestibility of forages in East Indonesia II. Grasses. Asian Aust. J. Anim. Sci. 1994, 7, 335–342.
31. Balachandran, N. Pasture research and development in Papua New Guinea. In Proceedings of the Regional Seminar on Pasture Research and Development in the Solomon Islands and Pacific Region, Honiara, Solomon Islands, 26 August–6 September 1977; Available online: http://www.fao.org/ag/agp/agpc/doc/counprof/PDF%20files/PapuaNewGuinea.pdf (accessed on 19 June 2015).

32. Chee, W.C. Productivity and chemical composition of twenty improved tropical grasses in the humid tropics. MARDI Res. Bull. 1980, 8, 163–173.
33. Abdullah, A.S. Natural Toxicants Affecting Animal Health and Production. Available online: http://docrchive.com/document/natural-toxicants-affecting-animal-health-and-production-6645391705595925/ (accessed on 12 June 2015).
34. Graydon, R.J.; Hamid, H.; Zahari, P.; Gardiner, C. Photosensitisation and crystal-associated cholangiohepatopathy in sheep grazing Brachiaria decumbens. Aust. Vet. J. 1991, 68, 234–236.
35. Riet-Correa, F. Intoxicações por plantas e micotoxinas. Doenças de Ruminantes eEqüídeos 2007, 2, 99–219.
36. Wasinghe, D.D.; Balasuriya, B.R.U.; Horadagoda, N.U. A preliminary report on investigation into jaundice and photosensitization among goats and sheep. Sri Lanka Vet. J. 1987, 33, 44–45.
37. Alexander, P.J. A Study of Photosensitization in Sheep, Goats and Cattle Grazing Brachiaria decumbens (Signal Grass); Department of Agriculture, University of Queensland: Queensland, Australia, 1991; p. 54.
38. Gobius, N.R.; Phaikaew, C.; Pholsen, P.; Rodchompoo, O.; Susena, W. Seed yield and its components of Brachiaria decumbens cv. Basilisk, Digitaria milanjiana cv. Jarra and Andropogon gayanus cv. Kent in north-east Thailand under different rates of nitrogen application. Trop. Grassl. 2001, 35, 26–33.
39. Harrington, G.N. Fire effects on a Ugandan savanna grassland. Trop. Grassl. 1974, 8, 87.
40. Low, S.G. Problems Associated with Signal Grass (Brachiaria decumbens) Pastures in the Tropical Lowlands of Papua New Guinea; Department of Animal Science, University of New England: Armidale, Australia, 2000; p. 268.
41. De Lemos, R.A.A. Brachiaria spp. poisoning in ruminants in Mato Grosso do Sul, Brazil. In Poisoning by Plants, Mycotoxins and Related Toxins; CABI: Wallingford, UK, 2011; pp. 129–132.
42. Riet-Correa, B.; Castro, M.B.; de Lemos, R.A.A.; Riet-Correa, G.; Mustafa, V.; Riet-Correa, F. Brachiaria spp. poisoning of ruminants in Brazil. Pesquisa Veterinária Brasileira 2011, 31, 183–192.
43. Winter, W.H.; Edye, L.A.; Megarrity, R.G.; Williams, W.T. Effects of fertilizer and stocking rate on pasture and beef production from sown pastures in Northern Cape York Peninsula 1. Botanical and chemical composition of the pastures. Aust. J. Exp. Agric. Anim. Husb. 1977, 17, 66–74.
44. Boddey, R.M.; Victoria, R.L. Estimation of biological nitrogen fixation associated with Brachiaria and Paspalum grasses using 15N labelled organic matter and fertilizer. Plant Soil 1986, 90, 265–292.
45. Rao, I.; Kerridge, P.; Macedo, M. Nutritional requirements of Brachiaria and adaptation to acid soils. In Biology, Agronomy and Improvement; CIAT and EMBRAPA-CNPGC: Cali, Colombia, 1996.
46. Watson, S.E. Pasture research on the Guadalcanal Plains. In Proceedings of the Regional Seminar on Pasture Research and Development in the Solomon Islands and Pacific Region, Honiara, Solomon Islands, 29 August–6 September 1977; Ministry of Agriculture and Lands: Honiara, Solomon Islands, 1978.
47. Harding, W.A.T.; Grof, B. Effect of fertilizer nitrogen on yield, nitrogen content and animal productivity of Brachiaria decumbens cv. Basilisk on the wet tropical coast of north Queensland. Qld. J. Agric. Anim. Sci. 1978, 34, 11–21.
48. Teixeira, F.A.; Bonomo, P.; Pires, A.J.V.; da Silva, F.F.; Fries, D.D.; da Hora, D.S. Produção anual e qualidade de pastagem de Brachiaria decumbens diferida e estratégias de adubação nitrogenada. Anim. Sci. 2011, 33, 241–248.
49. Watson, S.E. Animal production research on the Guadalcanal Plains. In Proceedings of the Regional Seminar on Pasture Research and Development in the Solomon Islands and Pacific Region, Honiara, Solomon Islands, 29 August–6 September 1977; Ministry of Agriculture and Lands: Honiara, Solomon Islands, 1978.

50. Grieve, C.M.; Osbourn, D.F. The nutritional value of some tropical grasses. J. Agric. Sci. Camb. 1965, 65, 411–417.
51. Tuen, A.; Djajanegara, A. Chemical composition and rumen degradability of forages and browse for goats and sheep in Sarawak. Sustainable animal production and the environment. In Proceedings of the 7th AAAP Animal Science Congress, Bali, Indonesia, 11–16 July 1994; Indonesian Society of Animal Science: Jakarta Selatan, Indonesia, 1994; pp. 503–504.
52. Aregheore, E.M. Nutritive value and utilization of three grass species by crossbred Anglo-Nubian goats in Samoa. Asian Aust. J. Anim. Sci. 2001, 14, 1389–1393.
53. Trujillo, G.M.; Posada, J.G.; Sierra, O. Efecto de la edad de rebrote en la calidad nutriva de Brachiaria decumbens. Pasturas Tropicales Bolet'in 1986, 8, 7–9.
54. Aregheore, E.M. Nutrient characterization and in vitro digestibility of grass and legume/browse species—Based diets for beef cattle in Vanuatu. S. Pac. J. Nat. Appl. Sci. 2006, 24, 20–27.
55. Whiteman, P.C.; Halim, N.R.; Norton, B.W.; Hales, J.W. Beef production from three tropical grasses in south-eastern Queensland. Aust. J. Exp. Agric. 1985, 25, 481–488.
56. Vanuatu Pasture Improvement Project. Available online: http://www.fao.org/ag/agp/agpc/doc/publicat/faobul1/1text.htm (accessed on 21 June 2015).
57. Grof, B.; Harding, W.A. Dry matter yields and animal production of guinea grass (Panicum maximum) on the humid tropical coast of north Queensland. Trop. Grassl. 1970, 4, 85.
58. Tergas, L.E.; Paladines, M.; Kleinheisterkamp, I. Animal productivity and pasture management of Brachiaria decumbens Stapf. in the Colombian Llanos. Trop. Anim. Prod. 1972, 7, 246–256.
59. Low, S.G. Photosensitisation of cattle grazing Signal grass (Brachiaria decumbens) in Papua New Guinea. N. Z. Vet. J.1993, 41, 220–221.
60. Low, S.G.; Shiel, S. Performance of Brahman weaners grazing signal grass (Brachiaria decumbens) in Papua New Guinea. Trop. Grassl. 1996, 30, 155.
61. Opasina, B.A. Photosensitzation jaundice syndrome in West African Dwarf sheep and goats grazed on Brachiaria decumbens. Trop. Grassl. 1985, 19, 120–123.
62. Briton, N.W.; Paltridge, T.B. Preliminary note on photosensitisation of sheep grazed on Brachiaria brizantha. Proc. R. Soc. Qld. 1942, 11, 121–122.
63. Othman, A.M.; Yusuff, M.K.; Rahman, O.A. Jaundice and photosensitization in indigenous sheep of Malaysia grazing on Brachiaria decumbens. Malays. Vet. J. 1983, 7, 254–263.
64. Brum, K.B.; Haraguchi, M.; Lemos, R.; Rite-Correa, F.; Fioravanti, M.C.S. Crystal-associated cholangiopathy in sheep grazing Brachiaria decumbens containing the saponin protodioscin. Pesqui. Vet. Bras. 2007, 27, 39–42.
65. Castro, M.B. Brachiaria spp. poisoning in sheep in Brazil: Experimental and epidemiological findings. In International Symposium on Poisonous Plants—Poisonings by Plants, Mycotoxins and Related Toxins; CABI: Paraiba, Brazil, 2009.
66. Birgel, E.H., Jr.; dos Santos, M.C.; de Ramos, J.A.; Pogliani, F.C.; Birgel, D.B.; Libera, A.M.; Gregory, L.; de Araujo, W.P.; Benesi, F.J. Secondary hepatogenous photosensitization in a llama (Lama glama) bred in the state of Sao Paulo, Brazil. Can. Vet. J. 2007, 48, 323–324.
67. De Oliveira, C. Hepatic photosensitization in buffaloes intoxicated by Brachiaria decumbens in Minas Gerais state, Brazil. Toxicon. 2013, 73, 121–129.
68. Nasir, M.A. Effects of Brachiaria decumbens (signal grass) on Timor deer (Cervus timorensis). In Proceedings of the 7th AAAP Animal Science Congress, Bali, Indonesia, 11–16 July 1994; Indonesian Society of Animal Science: Jakarta Selatan, Indonesia, 1994.
69. Caicedo, J. Hepatic lesions in cattle grazing on Brachiaria decumbens in Mesetas, Meta (Colombia). Revista de la Facultad de Medicina Veterinaria y de Zootecnia 2012, 59, 102–108.
70. Low, S.G. Signal grass—Friend or foe? In Recent Advances in Animal Nutrition in Australia; University of New England: Armidale, Australia, 1997; p. 251.

71. Tergas, L.E. Potential animal production and management of pastures on Ultisol soil in Colombia. Trop. Anim. Prod.1982, 7, 1–8.
72. Othman, A.M.; Haron, S. Hepatotoxicity in indigenous sheep of Malaysia stall-fed with different forms of Brachiaria Decumbens. Pertanika 1988, 11, 57–61.
73. Abas, M.O.; Sharif, H. Deleterious Effects of Brachiaria decumbens (Signal Grass) on Ruminants. MARDI Report; Institut Penyelidikan dan Kemajuan Pertania: Selangor, Malaysia, 1986.
74. Mustafa, V.S. Intoxicacao natural por Brachiaria spp. em ovinos no Brasil Central. Pesqui. Agropec. Bras. 2012, 32, 1272–1280.
75. Baber, R. Photosensitisation: A note of caution in the use of Brachiaria pastures—A review. Trop. Anim. Health Prod.1989, 21, 277–280.
76. Castro, M. Brachiaria spp. poisoning in sheep in Brazil: Experimental and epidemiological findings. In Poisoning by Plants, Mycotoxins and related Toxins; CAB International: London, UK, 2011; pp. 110–117.
77. Cheeke, P.R. Endogenous toxins and mycotoxins in forage grasses and their effects on livestock. J. Anim. Sci. 1995, 73, 909–918.
78. Barrington, G.M. Overview of Photosensitization. Available online: http://www.merckvetmanual.com/mvm/integumentary_system/photosensitization/overview_of_photosensitization.html (accessed on 9 July 2015).
79. Quinn, J.C.; Kessell, A.; Weston, A.L. Secondary plant products causing photosensitization in grazing herbivores: Their structure, activity and regulation. Int. J. Mol. Sci. 2014, 15, 1441–1465.
80. Faccin, T.C.; Riet-Correa, F.; Rodrigues, F.S.; Santos, A.C.; Melo, G.K.; Silva, J.A.; Ferreira, R.; Itavo, C.C.; Lemos, R.A. Poisoning by Brachiaria brizantha in flocks of naïve and experienced sheep. Toxicon 2014, 82, 1–8.
81. Jackson, P.G.G.; Cockcroft, P.D. Appendix 3: Laboratory reference values: Biochemistry. In Clinical Examination of Farm Animals; Jackson, P.G.G., Cockcroft, P.D., Eds.; John Wiley & Sons Inc.: Hoboken, NJ, USA, 2002; pp. 303–305.
82. Kaneko, J.J.; Harvey, J.W.; Bruss, M.L. Clinical Biochemistry of Domestic Animals, 6th ed.; Academic Press: Burlington, VT, USA, 2008.
83. Assumaidaee, A.A.; Zamri-Saad, M.; Jasni, S.; Noordin, M.M. The role of oxidative stress in Brachiaria decumbens toxicity in sheep. Pertanika J. Trop. Agric. Sci. 2010, 33, 151–157.
84. Moreira, C.N.; Moreira, M.; Garcia, E.C.; Neto, S.C.; de Araujo, E.G.; Fioravanti, M.C.S. Bovinos alimentados com capim Brachiaria e Andropogon: Alterações histológicas do fígado e linfonodos. Ciência Anim. Bras. 2009, 10, 206–218.
85. Button, C.; Paynter, D.I.; Shiel, M.J.; Colson, A.R.; Paterson, P.J.; Lyford, R.L. Crystal-associated cholangiohepatopathy and photosensitisation in lambs. Aust. Vet. J. 1987, 64, 176–180.
86. Dollahite, J.; Younger, R.; Jones, L. Photosensitization in lambs grazing Kleingrass. J. Am. Vet. Med. Assoc. 1977, 171, 1264–1265.
87. Driemeier, D.; Colodel, E.M.; Seitz, A.L.; Barros, S.S.; Cruz, C.E. Study of experimentally induced lesions in sheep by grazing Brachiaria decumbens. Toxicon 2002, 40, 1027–1031.
88. Cruz, C.; Driemeier, D.; Pires, V.S.; Schenkel, E.P. Experimentally induced cholangiohepatopathy by dosing sheep with fractionated extracts from Brachiaria decumbens. J. Vet. Diagn. Investig. 2001, 13, 170–172.
89. Noordin, M.; Abdullah, A.S.; Rajion, M. Experimental Brachiaria decumbens toxicity in cattle. Vet. Res. Commun. 1989, 13, 491–494.
90. Kellerman, T.S.; Erasmus, G.L.; Coetzer, J.A.; Brown, J.M.; Maartens, B.P. Photosensitivity in South Africa. VI. The experimental induction of geeldikkop in sheep with crude steroidal saponins from Tribulus terrestris. Onderstepoort J. Vet. Res. 1991, 58, 47–53.
91. Lancaster, M.; Vit, I.; Lyford, R. Analysis of bile crystals from sheep grazing Panicum schinzii (sweet grass). Aust. Vet. J. 1991, 68, 281.

92. Miles, C.O.; Munday, S.C.; Holland, P.T.; Lancaster, M.J.; Wilkins, A.L. Further analysis of bile crystals from sheep grazing Panicum schinzii (sweet grass). Aust. Vet. J. 1992, 69, 34.
93. Patamalai, B.; Hejtmancik, E.; Bridges, C.H.; Hill, D.W.; Camp, B.J. The isolation and identification of steroidal sapogenins in Kleingrass. Vet. Hum. Toxicol. 1990, 32, 314–318.
94. Bridges, C.; Camp, B.J.; Livingston, C.W.; Bailey, E.M. Kleingrass (Panicum coloratum L.) poisoning in sheep. Vet. Pathol. 1987, 24, 525–531.
95. Miles, C.O.; Munday, S.C.; Holland, P.T.; Smith, B.L.; Embling, P.P.; Wilkins, A.L. Identification of a sapogenin glucuronide in the bile of sheep affected by Panicum dichotomiflorum toxicosis. N. Z. Vet. J. 1991, 39, 150–152.
96. Miles, C.O.; Wilkins, A.L.; Munday, S.C.; Holland, P.T.; Smith, B.L.; Lancaster, M.J.; Embling, P.P. Identification of the calcium salt of epismilagenin.β-D-glucuronide in the bile crystals of sheep affected by Panicum dichotomiflorum and Panicum schinzii toxicoses. J. Agric. Food Chem. 1992, 40, 1606–1609.
97. Badiei, K.; Mostaghni, K.; Nazifi, S.; Tafti, K.A.; Ghane, M.; Momeni, S.A. Experimental Panicum miliaceum poisoning in sheep. Small Rumin. Res. 2009, 82, 99–104.
98. McDonough, S.P.; Woodbury, A.H.; Galey, F.D.; Wilson, D.W.; East, N.; Bracken, E. Hepatogenous photosensitization of sheep in California associated with ingestion of Tribulus terrestris (puncture vine). J. Vet. Diagn. Investig. 1994, 6, 392–395.
99. Miles, C.O.; Wilkins, A.L.; Erasmus, G.L.; Kellerman, T.S.; Coetzer, J.A. Photosensitivity in South Africa. VII. Chemical composition of biliary crystals from a sheep with experimentally induced geeldikkop. Onderstepoort J. Vet. Res. 1994, 61, 215–222.
100. Smith, B.; Miles, C. A role for Brachiaria decumbens in hepatogenous photosensitization of ruminants? Vet. Hum. Toxicol. 1993, 35, 256–257.
101. Mullenax, C. Brachiaria decumbens toxicity—Grass or fungus? Vet. Hum. Toxicol. 1991, 33, 464.
102. Smith, B.; Towers, N. Pithomycotoxicosis (facial eczema) in New Zealand and the use of zinc salts for its prevention. Plant Toxicol. 1985, 50, 70–79.
103. Flaøyen, A.; Smith, B.L.; Miles, C.O. An attempt to reproduce crystal-associated cholangitis in lambs by the experimental dosing of sarsasapogenin or diosgenin alone and in combination with sporidesmin. N. Z. Vet. J. 1993, 41, 171–174.
104. Nobre, D.; Andrade, S. Relação entre fotossensibilização em bovinos jovens e a gramínea Brachiaria decumbens Stapf. Biológico. 1976, 42, 249–258.
105. Lima, F.; Haraguchi, M.; Pfister, J.A.; Guimaraes, V.Y.; Andrade, D.D.F.; Ribeiro, C.S.; Costa, G.L.; Araujo, A.L.; Fioravanti, M.C.S. Weather and plant age affect the levels of steroidal saponin and Pithomyces chartarum spores in Brachiaria grass. Int. J. 2013, 2, 45–53.
106. Cruz, C.; Driemeier, D.; Pires, V.S.; Colodel, E.M.; Taketa, A.T.; Schenkel, E.P. Isolation of steroidal sapogenins implicated in experimentally induced cholangiopathy of sheep grazing Brachiaria decumbens in Brazil. Vet. Hum. Toxicol. 2000, 42, 142–145.
107. Lee, S.T.; Stegelmeier, B.L.; Gardner, D.R.; Vogel, K.P. The isolation and identification of steroidal sapogenins in switchgrass. J. Nat. Toxins 2001, 10, 273–281.
108. Meagher, L.P.; Wilkins, A.L.; Miles, C.O.; Collin, R.G.; Fagliari, J.J. Hepatogenous photosensitization of ruminants by Brachiaria decumbens and Panicum dichotomiflorum in the absence of sporidesmin: Lithogenic saponins may be responsible. Vet. Hum. Toxicol. 1996, 38, 271–274.
109. Munday, S.C.; Wilkins, A.L.; Miles, C.O.; Holland, T.P. Isolation and structure elucidation of dichotomin, a furostanol saponin implicated in hepatogenous photosensitization of sheep grazing Panicum dichotomiflorum. J. Agric. Food Chem. 1993, 41, 267–271.
110. Pires, V.S.; Taketa, A.T.C.; Gosmann, G.; Schenkel, E.P. Saponins and sapogenins from Brachiaria decumbens Stapf. J. Braz. Chem. Soc. 2002, 13, 135–139.
111. Barbosa-Ferreria, M. Steroidal saponin protodioscin concentration in different species and cultivars of Brachiaria spp. In Veterinária e Zootecnia, Faculdade de

Medicina Veterinária e Zootecnia; Universidade Estadual Paulista: São Paulo, Brazil, 2011.
112. Brum, K.B.; Haraguchi, M.; Garutti, M.B.; Nobrega, F.N.; Rosa, B.; Fioravanti, M.C.S. Steroidal saponin concentrations in Brachiaria decumbens and B. brizantha at different developmental stages. Ciênc. Rural 2009, 39, 279–281.
113. Lee, S.T.; Mitchell, R.B.; Wang, Z.; Heiss, C.; Gardner, D.R.; Isolation, P.A. Characterization, and quantification of steroidal saponins in switchgrass (Panicum virgatum L.). J. Agric. Food Chem. 2009, 57, 2599–2604.
114. Francis, G. The biological action of saponins in animal systems: A review. Br. J. Nutr. 2002, 88, 587–605.
115. Wina, E.; Muetzel, S.; Becker, K. The impact of saponins or saponin-containing plant materials on ruminant production: A review. J. Agric. Food Chem. 2005, 53, 8093–8105.
116. Sparg, S.G.; Light, M.E.; van Staden, J. Biological activities and distribution of plant saponins. J. Ethnopharmacol. 2004, 94, 219–243.
117. Kostova, I.; Dinchev, D. Saponins in Tribulus terrestris-chemistry and bioactivity. Phytochem. Rev. 2005, 4, 111–137.
118. Raju, J.; Mehta, R. Cancer chemopreventive and therapeutic effects of diosgenin, a food saponin. Nutr. Cancer 2008, 61, 27–35.
119. Shi, J.; Arunasalam, K.; Yeung, D.; Kakuda, Y.; Mittal, G.; Jiang, Y. Saponins from edible legumes: Chemistry, processing, and health benefits. J. Med. Food 2004, 7, 67–78.
120. Wallace, R.J. Antimicrobial properties of plant secondary metabolites. Proc. Nutr. Soc. 2004, 63, 621–629.
121. Man, S.; Gao, W.; Zhang, Y.; Huang, L.; Liu, C. Chemical study and medical application of saponins as anti-cancer agents. Fitoterapia 2010, 81, 703–714.
122. De Lima, F. Analysis of saponin protodioscin by HPLC-ESI-MS in Brachiaria brizantha and Brachiaria decumbens during 13 months. Veterinária e Zootecnia. 2011, 18, 576–579.
123. Dewick, P.M. Medicinal Natural Products—A Biosynthetic Approach, 3rd ed.; John Wiley & Sons: Oxford, UK, 2009.
124. Lee, S.T.; Mitchell, R.B.; Gardner, D.R.; Tokarnia, C.H.; Riet-Correa, F. Measurement of steroidal saponins in Panicum and Brachiaria grasses in the USA and Brazil. In Poisoning by Plants, Mycotoxins, and Related Toxins; CABI: Paraiba, Brazil, 2011.
125. Abdullah, A.S.; Lajis, N.H.; Bremner, J.B.; Davies, N.W.; Mustapha, W.; Rajion, M.A. Hepatotoxic constituents in the rumen of Brachiaria decumbens intoxicated sheep. Vet. Hum. Toxicol. 1992, 34, 154–155.
126. Lajis, N.H.; Abdullah, A.S.; Salim, S.J.; Bremner, J.B.; Khan, M.N. Epi-sarsasapogenin and epi-smilagenin: Two sapogenins isolated from the rumen content of sheep intoxicated by Brachiaria decumbens. Steroids 1993, 58, 387–389.
127. Meagher, L.P.; Smith, B.L.; Wilkins, A.L. Metabolism of diosgenin-derived saponins: implications for hepatogenous photosensitization diseases in ruminants. Anim. Feed Sci. Technol. 2001, 91, 157–170.
128. Wilkins, A.L. GC-MS method for the analysis of steroidal sapogenins in plant and animal samples associated with ovine photosensitisation. In Plant-Associated Toxins—Agricultural, Phytochemical and Ecological Aspects; CAB International: Perth, Australia, 1994.
129. Miles, C.O.; Wilkins, A.L.; Erasmus, G.L.; Kellerman, T.S. Photosensitivity in South Africa. VIII. Ovine metabolism of Tribulus terrestris saponins during experimentally induced geeldikkop. Onderstepoort J. Vet. Res. 1994, 61, 351–359.
130. Flaoyen, A.; Wilkins, A.L. Metabolism of saponins from Narthecium ossifragum—A plant implicated in the aetiology of alveld, a hepatogenous photosensitization of sheep. Vet. Res. Commun. 1997, 21, 335–345.
131. Flaoyen, A.; Wilkins, A.L.; Deng, D.; Brekke, T. Ovine metabolism of saponins: Evaluation of a method for estimating the ovine uptake of steroidal saponins from Narthecium ossifragum. Vet. Res. Commun. 2001, 25, 225–238.

132. Miles, C.O.; Wilkins, A.L.; Munday, S.C.; Flaoyen, A.; Holland, P.T.; Smith, B.L. Identification of insoluble salts of the β-D-glucuronides of episarsasapogenin and epismilagenin in the bile of lambs with alveld and examination of Narthecium ossifragum, Tribulus terrestris, and Panicum miliaceum for sapogenins. J. Agric. Food Chem. 1993, 41, 914–917.
133. Holland, P.T.; Miles, C.O.; Mortimer, P.H.; Wilkins, A.L.; Hawkes, A.D.; Smith, B.L. Isolation of the steroidal sapogenin epismilagenin from the bile of sheep affected by Panicum dichotomiflorum toxicosis. J. Agric. Food Chem.1991, 39, 1963–1965.
134. Ferreira, M. Variation in saponin concentration in Brachiaria brizantha leaves as a function of maturation: Preliminary data. Available online: http://www.alibris.com/search/books/author/International-Symposium-on-Poisonous-Plants (accessed on 27 June 2015).
135. De Lima, F.G.; Lee, S.T.; Pfister, J.A.; Miyagi, E.S.; Costa, G.L.; da Silva, R.D.; Fioravanti, M.C.S. The effect of ensiling and haymaking on the concentrations of steroidal saponin in two Brachiaria grass species. Ciênc. Rural 2015, 45, 858–866.
136. Villalba, J.; Provenza, F.D.; Clemensen, A.K.; Larsen, R.; Juhnke, J. Preference for diverse pastures by sheep in response to intraruminal administrations of tannins, saponins and alkaloids. Grass Forage Sci. 2011, 66, 224–236.
137. Abdullah, A.S.; Nordin, M.M.; Rajion, M.A. Signal grass (Brachiaria decumbens) toxicity in sheep: Changes in motility and pH of reticulo-rumen. Vet. Hum. Toxicol. 1988, 30, 256–258.
138. Patra, A.K.; Saxena, J. The effect and mode of action of saponins on the microbial populations and fermentation in the rumen and ruminant production. Nutr. Res. Rev. 2009, 22, 204–219.

CHAPTER 9

Aflatoxicosis: Lessons from Toxicity and Responses to Aflatoxin B_1 in Poultry

Melissa S. Monson [1], Roger A. Coulombe [2] and Kent M. Reed [1,*]

[1]*Department of Veterinary and Biomedical Sciences, College of Veterinary Medicine, University of Minnesota, St Paul, MN 55108, USA*

[2]*Department of Animal, Dairy and Veterinary Sciences, College of Agriculture, Utah State University, Logan, UT 84322, USA*

ABSTRACT

This review is a comprehensive introduction to the effects of poultry exposure to the toxic and carcinogenic mycotoxin aflatoxin B_1 (AFB_1). The relationship between AFB_1 sensitivity and metabolism, major direct and indirect effects of AFB_1, recent studies of gene expression and transcriptome responses to exposure, and mitigation strategies to reduce toxicity are discussed. Exposure to AFB_1 primarily occurs by consumption of contaminated corn, grain or other feed components. Low levels of residual AFB_1 in poultry feeds can cause reduction in growth, feed conversion, egg production, and compromised immune functions, resulting in significant economic costs to producers. Thus, AFB_1 acts as a "force multiplier" synergizing the adverse effects of microbial pathogens and other agents, and factors detrimental to poultry health. Domestic turkeys (*Meleagris gallopavo*) are one of the most sensitive animals known to AFB_1 due, in large part, to a combination of efficient hepatic bioactivation by cytochromes P450 1A5 and 3A37, and deficient hepatic glutathione-*S*-transferase (GST)-mediated detoxification. Because of their sensitivity, turkeys are a good model to investigate chemopreventive treatments and feed additives for their ability to reduce AFB_1 toxicity. Transcriptome analysis (RNA-seq) of turkey poults (liver and spleen) has identified AFB_1-induced gene expression changes in pathways of apoptosis, carcinogenesis, lipid regulation, antimicrobial activity, cytotoxicity and antigen presentation. Current research focuses on further identifying the molecular mechanisms underlying AFB_1 toxicity with the goal of reducing aflatoxicosis and improving poultry health.

Keywords: turkey; aflatoxin B_1; hepatotoxicity; immunosuppression; feed additives; transcriptome; RNA-seq

1. INTRODUCTION

Dietary exposure to aflatoxins can have severe toxic and carcinogenic effects in humans and animals. The production and metabolism of aflatoxin, symptoms and biomarkers of exposure, and methods to reduce aflatoxicosis have been extensively investigated over the past 50 years [1,2,3,4,5,6,7,8,9]. Most studies have focused on humans, laboratory model species, or agricultural animals, and have identified conserved and species-specific aspects of aflatoxicosis [2,10,11,12,13,14,15]. This review specifically examines the responses of poultry, a particularly sensitive group, to aflatoxin B_1. Aflatoxin metabolism, toxicity, and expression responses in poultry are discussed, along with potential mitigation strategies.

2. ETIOLOGY OF AFLATOXICOSIS

The acute toxicity of dietary aflatoxin was discovered in 1960 when a then unknown disease, termed Turkey "X" Disease, caused the deaths of over 100,000 turkeys (*Meleagris gallopavo*) and other poultry in England [16,17]. Upon examination, the causative agent was identified as imported Brazilian peanut-meal contaminated with aflatoxins [16,18]. Aflatoxins belong to a heterologous group of fungal secondary metabolites called mycotoxins that adversely affect human and animal health. Structurally derivatives of difurocoumarin, aflatoxins are most commonly produced by strains of *Aspergillus flavus*, *A. parasiticus*, and *A. nominus*, although many other *Aspergilli* (including *Emericella* teleomorphs) have aflatoxigenic capabilities [7,19,20,21]. Named according to their blue or green fluorescence under UV light, there are four primary aflatoxins: aflatoxin B_1 (AFB_1), B_2 (AFB_2), G_1 (AFG_1), and G_2 (AFG_2) (Figure 1) [7,21]. Of these, AFB_1 is the most hepatotoxic, the most mutagenic, and the most prevalent worldwide [2,3,21,22].

Livestock, including poultry, are exposed to AFB_1 and other aflatoxins by consuming contaminated feed. Many agricultural feed commodities (corn, cottonseed, peanuts and sorghum) and other foods (figs, tree nuts and spices) are at especially high risk [21,23]. Stress from drought or insect damage can reduce crop resistance to *Aspergilli* and lead to aflatoxin contamination prior to harvest [21,24]. Warm and humid conditions during maturation, harvest, transport or storage, promote *Aspergillus* colonization and subsequent aflatoxin production [21,24]. Temperatures near 30 °C and water activity of 0.99 provide ideal conditions for AFB_1 biosynthesis, although substrate, time, CO_2 levels, and other environmental factors are also important [25,26,27,28,29,30]. Along with primary contamination of crops, aflatoxins can transfer to milk, meat and eggs of livestock and poultry fed the toxins [23,31,32,33,34,35,36,37,38]. Therefore, AFB_1 is a human food safety risk in both plant and animal products.

Animal susceptibility to the acute effects of aflatoxicosis varies widely (Table 1). Domestic turkeys and ducks (*Anas platyrhynchos*) are highly sensitive to both the acute and chronic toxicity of AFB_1 [21,39,40]. Chickens (*Gallus gallus*) are more resistant to acute aflatoxicosis than other poultry species, except during embryonic development (Table 1). Even when exposure does not cause mortality or morbidity, aflatoxicosis contributes directly and indirectly to losses for the poultry industry. While precise numbers are not available, it has been estimated that aflatoxins inflict a loss of at least $143 million each year due to AFB_1-induced hepatotoxicity, reduced performance and secondary infections [23,41].

3. AFB_1 METABOLISM AND SENSITIVITY

3.1. Metabolism

Bioactivation is required for AFB_1 to be toxic and this processing predominantly occurs in hepatocytes [1,4,7,21]. AFB_1 is initially absorbed in the small intestine, especially the duodenum [42]. While bioactivating enzymes with low affinity for AFB_1 are present in the small intestine [43], the majority of the toxin is metabolized in the liver, where AFB_1 is converted by hepatic cytochromes P450 (*CYP*) enzymes into the reactive and electrophilic *exo*-AFB_1-8,9-epoxide (AFBO) (Figure 2) [1,3,4,21,22,40]. An *endo* stereoisomer of the AFBO epoxide can also be produced, but is far less toxic and not relevant to AFB_1 toxicity [1,4].

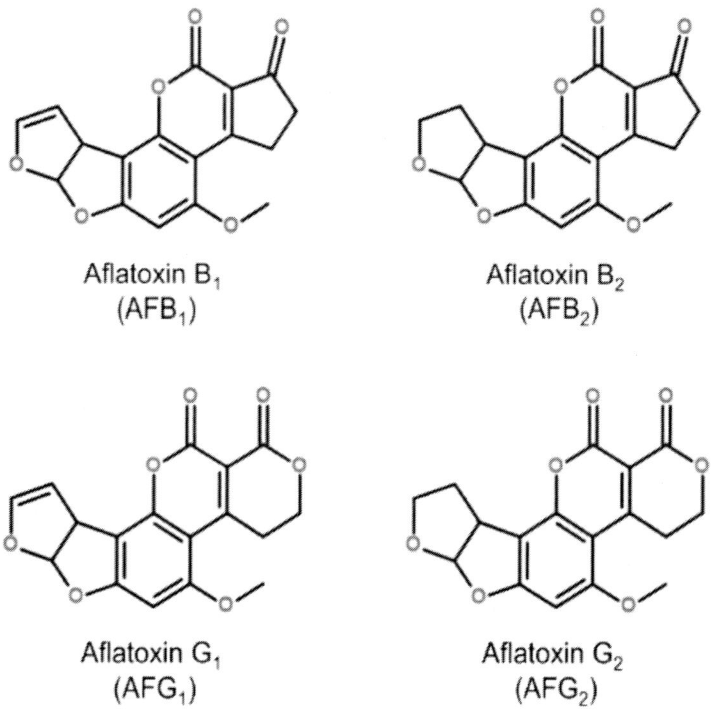

Figure 1. Molecular structures of the four primary aflatoxins.

AFBO is a highly unstable intermediate ($t_{0.5}$ ~ 0.5 s) and quickly reacts to form adducts with DNA, RNA, and proteins [1,4,7], which are then responsible for AFB_1 toxicity. In most mammals, AFBO is primarily detoxified by glutathione *S*-transferase (GST) enzymes that add glutathione (GSH) to form an 8,9-dihydro-8-(*S*-glutathionyl)-9-hydroxy-AFB_1 (AFB_1-SG) adduct (Figure 2) [3,4,7,21,40]. Although AFBO is the most toxic, other metabolites of AFB_1 include aflatoxicol (AFL), aflatoxin M_1 (AFM_1), aflatoxin P_1 (AFP_1), and aflatoxin Q_1 (AFQ_1,) [3,4,7].

Table 1. Comparative acute toxicity of a single oral dose of aflatoxin B_1 (AFB_1).

Species	Age	Oral LD_{50} (mg/kg Body weight) [1]
Baboon	A	2.0–2.2
Cat	A	0.6
Chicken	E	0.3–5.0
Chicken	Y	6.5–18.0
Dog	A	0.5–1.0
Duck	E	0.5–1.0
Duck	N	0.3–0.6
Guinea Pig	Y	1.4–2.0
Hamster	Y	10.2–12.8
Macaque (Cynomolgus)	A	2.2
Macaque (Rhesus)	A	7.8–8.0
Mouse	N	1.5
Mouse	Y	7.3–9.0
Rabbit	Y	0.3–0.5
Rat	N	0.6–1.0
Rat	Y	5.5–7.4
Rat	A	6.3–18.0
Sheep	A	2.0
Swine	Y	0.6
Trout	Y	0.5
Turkey	Y	1.4–3.2

Lethal dose in 50% (LD_{50}), adult (A), embryo (E), neonate (N), young (Y); [1] compiled from [7,10–12,14,22].

3.2. Sensitivity: Mice to Humans

Sensitivity to AFB_1 is determined not only by the rate of P450-mediated bioactivation, but more importantly by the subsequent detoxification of the reactive AFBO intermediate [1,4]. In mice (*Mus musculus*), P450 enzymes such as P450 2A5, 3A11 and 3A13, effectively activate AFB_1 and can produce large amounts of AFBO [4,44,45,46,47,48,49,50]. However, the murine alpha-class GST (GSTA) enzyme mGSTA3 has high affinity for AFBO, making mice extremely resistant to aflatoxicosis [4,45,51,52,53,54,55]. Interestingly, the orthologous hepatic GST enzymes from rats (*Rattus norvegicus*) can have more than 50 times lower activity towards AFBO [45,54,56]. Effective activation by rat P450 3A2 and 2C11 and minimal conjugation by the rat orthologue of mGSTA3 (rGSTA3, Yc1) make the rat far more susceptible to aflatoxicosis

[4,44,45,51,55,57,58,59]. Another rat AFBO-conjugating enzyme rGSTA5 (Yc2) has high affinity for AFBO, but expression is limited to neonatal tissue, some female adults and after antioxidant-induction [59,60,61,62]. Therefore even within rodents, the metabolism and sensitivity to AFB_1 is species-specific.

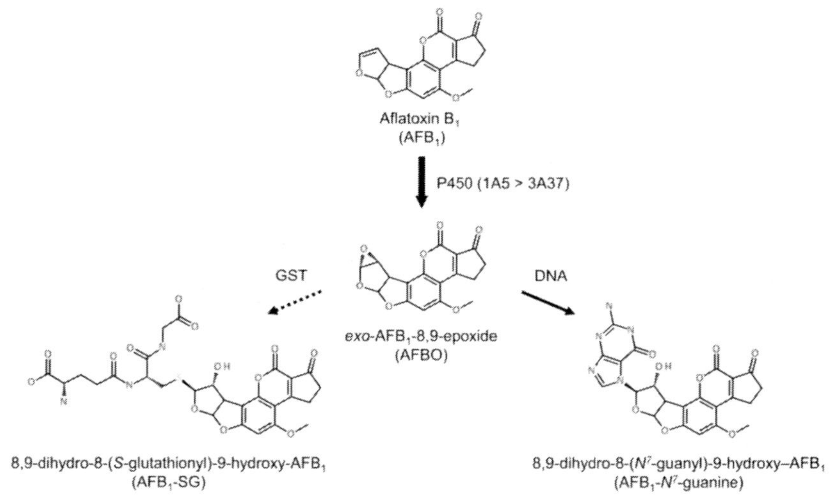

Figure 2. Metabolites and enzymes involved in aflatoxin B_1 (AFB_1) metabolism in the turkey liver. Arrow width shows reaction efficiency (wide > narrow), while a dashed line indicates that a reaction does not occur in all birds. Cytochrome P450 (*CYP*) enzymes effectively interact with AFB_1. Domestic turkey glutathione *S*-transferase (GST) enzymes cannot conjugate *exo*-AFB_1-8,9-epoxide (AFBO), although wild and heritage turkey GST enzymes can have activity.

In humans (*Homo sapiens*), multiple hepatic P450 enzymes can activate AFB_1, but P450 1A2 and 3A4 are the primary producers of AFBO [4,43,63,64,65,66,67]. Human P450 3A4 is active at high AFB_1 concentrations, while P450 1A2 has high affinity at low biologically-relevant concentrations [64,66]. Overall, human P450 enzymes produce far less AFBO than rodents, even at high concentrations of AFB_1 (1/4 that of rats, 1/8 that of mice) [48]. Although lower levels of AFBO reduce toxicity, the epoxide is less effectively detoxified in humans. Hepatic GST enzymes in humans can have over 3000 fold less activity against AFBO than murine GST [56]. Human enzymes hGSTA1 and hGSTA2 have little AFBO-conjugating activity [57,59,68]. Instead, some AFB_1-SG adducts are produced by the mu-class GST (GSTM) enzyme hGSTM1 [59,68,69,70,71].

Although hGSTM1 forms more AFB_1-SG adducts from the *endo* stereoisomer, it is able to detoxify mutagenic *exo*-AFBO [68].

3.3. Sensitivity: Poultry

Poultry are sensitive to even low levels of AFB_1, and among species of agricultural importance, the order of sensitivity is ducks > turkeys > Japanese quail (*Coturnix japonica*) > chickens [7,21,72,73,74,75]. Therefore, lower concentrations of $AFB_1$are lethal to turkeys and ducks and more adversely affect production and health in these species (Table 2). In domestic turkey, efficient production of AFBO contributes to sensitivity. We have conducted considerable research to determine the P450 enzymes responsible for AFB_1 bioactivation and metabolism in turkey livers. This work revealed two turkey P450 enzymes, encoded by *CYP1A5* and *CYP3A37*, predominantly responsible for converting AFB_1 into AFBO *in vitro* and *in vivo* [76,77,78,79]. P450 1A5 has high-affinity (high V_{max}, K_{cat}; low K_m) and catalyzes the production of both *exo*-AFBO and the detoxified metabolite AFM_1 according to traditional Michaelis-Menten kinetics [76,77]. P450 3A37 is the lower affinity catalyst, exhibiting apparent subunit allostery conforming to Hill enzyme kinetics and producing *exo*-AFBO and AFQ_1[76,77].

We used polyclonal anti-peptide antibodies in a series of immunoinhibition experiments to determine the relative importance of these P450 enzymes in AFB_1 bioactivation in turkey livers and liver microsomes. Turkey P450 1A5 is the predominant catalyst (>98%) at low (<50 μM) pharmacologically-relevant AFB_1 concentrations commonly seen in the livers of exposed turkeys, while P450 3A37 predominates at much higher AFB_1 concentrations not likely to be achieved in tissues under *in vivo* conditions [78]. Another P450 enzyme, likely orthologous to mammalian P450 2A6, also converts AFB_1 to AFBO in turkeys, chickens, ducks and quail [80,81,82]. However, the specific gene encoding this protein has not been identified in any poultry species [80,81,82,83]. These same studies implicated a P450 1A1 orthologue in AFBO production, which is likely encoded by *CYP1A5* in turkeys [78,80,81,82].

The activity of P450 enzymes in the turkey, and therefore AFBO production, is inversely related to age, with extreme sensitivity to AFB_1 in young poults [39,77,84,85]. A similar age-dependent sensitivity occurs in chickens [86,87,88]. Hepatic microsomal P450 enzymes from turkeys and quail produce 2 to 4 times more AFBO than ducks or chickens [74]. A later study confirmed the highest AFBO production in turkey, with intermediate levels in duck and quail, and the lowest in chicken [83]. AFBO production correlates with the sensitivity of poultry, except for the duck. In addition, the levels of AFL produced by liver cytosol are highest in turkey, followed by ducks, chickens and quail [74]. The sensitivity of ducks may be more linked to AFL production and toxicity or to oxidative stress [74,83,89].

Table 2. Minimum AFB_1 concentrations with major effects in poultry.

Species	Minimum Dietary Contamination Level (ppb) to Cause		
	100% Lethality	Gross Hepatic Lesions	Impaired Production
Chicken	NR (>4000)	800 [1]	800 [1]
Duck	1000 [2]	500 [2]	500 [2]
Goose	4000 [2]	500 [2]	700 [3]
Pheasant	4000 [2]	500 [2]	1000 [2]
Quail (Bobwhite)	ND	ND	700 [3]
Turkey	800 [1]	400 [1]	400 [1]

Compiled from studies that examined aflatoxicosis in multiple poultry species; parts per billion (ppb); not determined (ND); not reached (NR); [1] [39]; [2] [75]; [3] [72].

Like the cytochrome P450 enzymes, hepatic tGSTA enzymes contribute to the severity of aflatoxicosis in the turkey. Six *tGSTA* genes have been amplified, cloned and heterologously expressed turkey liver (*GSTA1.1, GSTA1.2, GSTA1.3, GSTA2, GSTA3,* and *GSTA4*) [90,91]. Hepatic cytosolic GST enzymes from domestic turkey have essentially no AFBO-conjugating activity [40,91]. However, tGSTA enzymes can conjugate AFBO to GSH *in vitro* using recombinant proteins in an *E. coli* expression system, suggesting that gene silencing mechanisms or post-transcriptional modifications are likely responsible for their lack of function *in vivo* [91]. The high sensitivity of the domestic turkey to AFB_1 appears to be due to an unfortunate combination of efficient P450 enzymes and dysfunctional GST enzymes that allow accumulation of AFB_1adducts in the liver.

The effects of AFB_1 exposure on North American wild turkeys are similar to, but less severe than those seen in domestic poultry [92]. This difference in response may be the result of genetic changes that occurred during domestic selection, but could also be due to the separate genetic background of the domestic turkey. Although members of the same species, domestic turkeys were originally derived from the Mexican subspecies (*Meleagris gallopavo gallopavo*) of wild turkey native to Central America; the Eastern subspecies from North America (*Meleagris gallopavo silvestris*) was likely involved in later crosses [93]. Wild turkeys were originally exported to Europe from Mexico in the 1500's and then reintroduced to North America in the 1600's. Reintroduced turkeys were selectively bred, first forming the heritage breeds, and then developed into the modern commercial breeds. Current commercial production predominately utilizes the Broad Breasted White. Selection for production traits in poultry is known to increase metabolic disorders [94] and decrease immune functions [95,96,97,98]. Likewise, detoxification capabilities could be reduced as an unintended side-effect of selection in domestic birds.

Eastern wild turkeys are more resistant to aflatoxicosis than their domestic relatives [92]. AFBO detoxification is likely a main contributor to the differences seen among turkeys. Consistent with this hypothesis, hepatic cytosolic GST enzymes from both the Eastern and Rio Grande (*Meleagris gallopavo intermedia*) subspecies of wild turkey have activity against AFBO [99], unlike their domestic counterparts. Wild turkey GSTA enzymes heterologously expressed *in vitro* also had AFBO-conjugating activity [99]. Like wild strains, the Royal Palm turkey, a heritage breed, retains hepatic and *in vitro* activity [99]. tGSTM enzymes were also amplified, cloned and heterologously expressed from the livers of wild and domestic turkeys, but had no measurable affinity toward AFBO, and therefore are not thought to be involved in AFB_1 detoxification [100].

4. EFFECTS OF AFB_1 EXPOSURE

4.1. AFB_1 Adducts

Formation of AFB_1 adducts is detrimental to cellular processes. AFBO can react with DNA or RNA to form *trans*-8,9-dihydro-8-(N7-guanyl)-9-hydroxy-AFB_1 (AFB_1-N7-guanine) adducts (Figure 2) [1,3,4,7,21]. Some AFB_1-N7-guanine adducts convert into stable AFB_1-formamidopyrimidine (AFB_1-FAPY) adducts [1,3,7,101]. These AFB_1-DNA and AFB_1-RNA adducts can inhibit transcription and translation, induce DNA mutations during DNA repair and replication, and even initiate apoptosis or carcinogenesis [1,3,4,21,44,102]. AFBO can hydrolyze into AFB_1-8,9-dihydrodiol (AFB_1-diol) and through a series of reactions generate adducts with lysine residues in proteins [4,89,103]. AFB_1-lysine adducts can cause toxicity by impairing protein stability and function [4,89,103]. Although not as well studied, AFB_1 adducts have similar effects on poultry. In chicken primary hepatocytes, the interaction between AFBO and DNA, RNA and proteins has been verified and shown to strongly inhibit synthesis of these macromolecules [104].

4.2. Mutagenicity

Binding of AFBO to DNA introduces G-T transversion mutations in hepatic DNA [4,65,105]. The high incidence of G-T transversions results from AFB_1-FAPY adducts predisposing bypass DNA synthesis machinery to make G-T changes [101]. A G-T transversion in codon 249 of the p53 tumor suppressor has been identified in many human liver cancers and may mechanistically contribute to cancer formation [7,65,71,101,105,106,107]. Chronic AFB_1 exposure, especially in combination with hepatitis-B infections, severely increases the risk of hepatocellular carcinoma in humans [1,3,4,5,6,9,21,105,108]. Dietary AFB_1 is known to have hepatocarcinogenic effects in other mammals, especially the sensitive rat [4,109,110,111]. Based on this evidence from humans and other mammals, AFB_1 is classified as a group I carcinogen by the International Agency for Research on Cancer [4,105]. AFBO is highly mutagenic in poultry, although adenoma and hepatocellular carcinoma have only been reported in ducks [82,112,113,114]. The potential synergistic effect of hepatitis-B virus and AFB_1has not been consistently reproduced in these studies [112,113,114].

4.3. Production Losses

Beyond its mutagenic effects, AFB_1 negatively affects production values, resulting in economic losses for the poultry industry. Dietary exposure to AFB_1 and other aflatoxins leads to lower weight gain and absolute body weights in both chickens and turkeys [34,39,84,88,115, 116,117,118,119,120,121,122,123,124]. Reduced feed intake [34,115,118,122,123,124,125] and decreased efficiency of nutrient usage [34,39,84,88,115,122,125] both contribute to this impaired growth during aflatoxicosis. AFB_1 lowers feed conversion causing poultry to require more feed to produce muscle (broilers and turkeys) [39,84,88,115,122] and eggs (layers) [125]. Although less severe, AFB_1 also reduces feed intake and weight gain in wild turkeys [92]. Similarly, feed consumption was decreased in AFB_1-exposed quail, but body weight and feed conversion were unaffected [126]. In ducks fed AFB_1, both feed intake and weight gain were reduced but without affecting feed efficiency [127].

Exposure to aflatoxins lowers reproductive performance in poultry. In layers fed AFB_1, age to maturity is increased [34] and egg production is reduced [34,121,125,128,129,130,131]. Egg quality parameters such as total weight, shape, albumin or yolk percentage, and shell thickness in chickens and quail can be adversely affected by AFB_1, although the effects were variable among studies [34,125,126,128,131,132,133]. The declines in poultry production traits are often indirect effects of AFB_1 reducing the metabolic potential of the liver. For example, impaired hepatic protein production likely contributes to AFB_1-induced changes within eggs, as the liver is the chief site of synthesis of proteins and lipids incorporated into the egg yolk.

4.4. Hepatotoxicity

Critical to protein synthesis, enzymatic metabolism and detoxification processes, the liver is the primary site of AFB_1activation and therefore toxicity [1,4,7,21]. Aflatoxicosis in poultry is characterized by an enlarged, pale, and friable liver [10,34,38,39,84, 116,117,120,121,134,135] Although relative liver weight can initially decrease [117], longer exposure to dietary AFB_1 raises the relative weight of the liver and causes pale or yellowed pigmentation [10,84,88,115,116,117,120,121,122,123,124,135,136]. At the cellular level, increased vacuolation of AFB_1-exposed hepatocytes allows high levels of lipids to accumulate [10,34,39,84,116,120,121,126]. Steatosis is therefore responsible for the changes in liver color and size during aflatoxicosis.

Both acute and chronic AFB_1 consumption by poultry cause other hepatic lesions. Common histopathological signs of AFB_1-induced liver damage include focal necrotic hepatocytes or hemorrhages [10,34,39,84, 116,121,137]. Acute damage initiates inflammatory responses and leads to leukocyte infiltration and proliferation in the liver [10,34,84, 112,116,138]. Short-term exposure to higher dose can cause morbidity and mortality from extensive liver damage [3,21]. In poultry, chronic AFB_1 consumption is mutagenic and leads to remodeling of liver tissues. Hyperplasia of bile duct epithelial cells or oval cells develops first, followed by periportal fibrosis and nodular tissue regeneration [10,34,39,84,112,116,121,126,134,139].

AFB_1 adducts with biomolecules cause damage to hepatocytes that impairs metabolic functions of the liver during AFB_1exposure. This is exemplified by AFB_1-reduced total serum protein levels, as the liver is responsible for production of most circulating proteins [88,115,117,118,119,123,127,129,140,141]. Aflatoxicosis negatively affects albumin, globulin, cholesterol, and triglyceride levels in serum [88,115,117,119,127,129,136,140,141,142]. Multiple blood coagulation factors are produced in the liver; the activity of these clotting factors and the serum levels of fibrinogen are diminished by AFB_1 in both chickens and turkeys [142,143,144,145,146]. Hepatic protein concentrations also decrease in AFB_1-fed chickens [88]. Protein content likely declines because AFB_1-DNA adducts inhibit transcription or translation and AFB_1-

lysine adducts result in protein degradation or excretion. Reduced synthesis of enzymes in the liver would have systemic effects on poultry metabolism. For example, the decreased hepatic fatty acid synthesis observed in AFB_1-exposed chickens [147] could be responsible for lower production of serum cholesterol and triglycerides [129,140].

4.5. Immunotoxicity

The avian immune system relies on the bursa of Fabricius, thymus and spleen to produce mature or active leukocytes. Even at low dietary concentrations, AFB_1 can damage these immune tissues and suppress innate and adaptive immune responses [3,7,148]. AFB_1 consumption during growth can lead to immune tissue atrophy, reducing relative weights of the bursa, spleen and thymus [120,122,134, 139,149,150,151,152]. Increases in relative spleen weight have also been observed during aflatoxicosis [117,120,153,154,155,156]. In young chickens, tissue changes are concomitant to the development of histopathological lesions. AFB_1 exposure causes visible congestion in the spleen and thymus, while nuclear debris accumulates in the thymus and bursa [139,149,150,156]. In AFB_1-exposed chicks, vacuoles increase in the lymphoid follicles of the bursa and the white pulp of the spleen, especially the T-cell rich periarteriolar lymphoid sheaths [139,150,156].

At the cellular level, innate and adaptive cell-mediated immune functions are impaired by AFB_1 exposure in poultry [148]. *In vivo* exposure to AFB_1 and other aflatoxins has been shown to decrease phagocytic activity in chicken leukocytes, including heterophils [157,158], macrophages [159,160,161], and monocytes [162]. Reduced phagocytosis was demonstrated *in vitro* for peritoneal macrophages isolated from both chickens and turkeys, although microsomally activated AFB_1 was required [163,164]. In contrast, the phagocytic activity of thrombocytes was not affected in chickens fed an aflatoxin-contaminated diet [165]. Dietary AFB_1 inhibits T lymphocyte activation in both chickens and turkeys as evidenced by delayed hypersensitive skin tests and graft-versus-host response tests [39,84,116,122,159,166,167].

AFB_1 can induce circulating lymphocytopenia [34,159,166,168] and cause lymphoid depletion in the bursa, spleen and thymus [34,134,139,149,150,159,166]. This likely results from increased apoptosis of splenocytes, thymocytes and bursal B-cells as seen in young chickens during aflatoxicosis [139,149,156,169,170]. In the chicken, both $CD4^+$ and $CD8^+$ T lymphocytes in the spleen, thymus, peripheral blood, and even the ileum can be affected by AFB_1 exposure [149,150,156,168,171,172]. Oxidative stress and DNA damage from AFB_1 are likely responsible for initiating apoptotic processes in lymphocytes [169,170]. Together these losses reduce the adaptive immune potential of poultry fed AFB_1.

Although some studies only observed effects on cell-mediated immunity [39,84,116], others found that AFB_1 can similarly diminish innate and adaptive humoral immune capabilities. Total serum complement is decreased by feeding AFB_1 to chickens, ducks and turkeys [115,127,173,174,175]. Antibody titers are often reduced, whether measured as total serum levels of IgA, IgG and IgM [139,141,167], as production of specific antibodies in response to sheep red blood cells [122,152,159] or as exposure response to infectious bronchitis virus, infectious bursal disease virus, Newcastle disease virus or *Pasteurella multocida* [128,176,177]. AFB_1 can also impair the effectiveness of vaccination for these poultry diseases [128,158,176,177,178].

As a consequence of AFB_1-driven immunosuppression, exposed poultry have lower resistance to secondary infections [148]. Poultry with aflatoxicosis are more susceptible to the bacterial diseases, fowl cholera [151,176,179,180,181] and salmonellosis [182,183,184]. Exposure to AFB_1 can increase the severity of the protozoan disease, cecal coccidiosis [185,186,187,188] and the fungal infection, crop mycosis [189]. Similarly, AFB_1 consumption decreases resistance to viral pathogens, including infectious bronchitis virus [128,177], infectious bursal disease virus [128,176,177,190,191], Marek's disease virus [185,192], and Newcastle disease virus [128,158,177,178,193]. These secondary infections dramatically increase the economic losses attributed to AFB_1 exposure. Thus, AFB_1 is a potent immunotoxin and acts as a synergistic "force multiplier" that can enhance the incidence and impacts of avian diseases.

4.6. Intestinal Toxicity

Nutrients are primarily absorbed by epithelial cells in the small intestine and these cells facilitate uptake of AFB_1. Most AFB_1 is transferred directly into the blood stream; however, as in mammals [43], some may be metabolized in intestinal tissues. Therefore in poultry, both local and systemic effects of AFB_1 exposure likely occur in the small intestine, but these are not well characterized. Dietary exposure to AFB_1 can lower the unit weight (length/weight) of the duodenum and jejunum [194,195] and affect tissue morphology. In chickens, AFB_1 has been shown to raise crypt depth in the jejunum [196], decrease villus height in the duodenum [197], and lower the ratio between villus height/crypt depth in all three sections of the small intestine [197]. However, these histopathological effects may not be pervasive [198].

The severity of aflatoxicosis in poultry can be affected by nutrition [7]. Deficiencies in some dietary vitamins raised AFB_1sensitivity in chickens, although feeding vitamins in excess was not protective [199,200]. Dietary supplementation with tryptophan also increased hepatotoxicity of AFB_1 [201]. Conversely, diets high in fat or protein may be beneficial to chickens and turkeys fed AFB_1 [202,203,204].

Direct investigations of AFB_1-effects on absorption or retention of individual nutrients have had variable results [34,194,196,205,206,207,208]. Exposure to AFB_1 decreases the apparent metabolizable energy poultry can obtain from their diet [34,195,196,205,206,207]. Therefore, increased dietary nutrients are needed to compensate for impaired uptake. It is currently unclear how much reduced nutrient uptake in the intestine contributes to AFB_1 effects on growth and feed efficiency in poultry.

4.7. Embryotoxicity

Exposure during development recapitulates many of signs of aflatoxicosis seen in hatched chicks and poults. Although most studies are carried out by *in ovo* AFB_1 injections, embryonic exposure to the toxin is a known risk to poultry. AFB_1 and its metabolites can be transferred from the laying hen into the albumin and yolk of the egg [31,34,37,128,131,133,209]. $AFM_1$is a common metabolite detected in eggs and, while not as carcinogenic as AFB_1, is acutely toxic [4,37,133]. Contamination of unfertilized shell eggs is therefore a food safety risk when used for human consumption.

Transfer of aflatoxins into embryonated eggs is also a concern for poultry producers. In experimental settings, *in ovo* exposure of chickens or turkeys to AFB_1 caused DNA damage in the embryonic liver and increased embryo mortality [131,161,210,211,212,213,214,215]. When introduced into the maternal diet to simulate the natural route of embryonic exposure, AFB_1 caused reduced hatchability [130,131,161]. Consistent with studies in hatched poultry, turkeys may be more sensitive to embryotoxic effects than chickens [215]. Since embryonic liver has an active protein from the *CYP1A* family [216], hepatotoxicity in turkey embryos is likely mediated by the same P450 1A5 as in poults.

Embryonic AFB_1 exposure can lead to morphological defects [212], such as abnormal area opaca cells [210,214], skeletal defects in the tibia growth plate [213], and inhibition of bursal follicle development [210,214]. These mutagenic effects can reduce embryo viability and adversely affect hatched progeny. *In ovo* AFB_1-driven immunosuppression has the potential to increase the incidence of infectious disease in young poultry and negatively affect their health and productivity. Whether *in ovo* injection or maternal feeding, chickens exposed to AFB_1 during embryogenesis have compromised cellular and humoral immune functions post hatch [160,161,211,217,218].

5. GENE EXPRESSION AND AFB_1

5.1. P450 and GST Enzymes

Gene expression can improve our understanding of responses to AFB_1 and provide targets to modulate the mechanisms of toxicity. Both cellular responses to toxicity and AFB_1 inhibition of transcription and translation will affect gene expression [102]. For example, AFB_1 down-regulates *p53* expression in human cells [67] and in tissue from hepatocellular carcinomas in rats [219], likely due to mutations introduced by AFB_1-DNA adducts. Expression of AFB_1-metabolizing genes can also be affected by AFB_1. Exposure in the rat liver to AFB_1 can increase expression of P450 (*CYP3A, CYP4F1*), and *rGSTM2* genes (Yb2) [220]. Another study in rats identified effects on multiple *P450* and *GST* genes, with greatest up-regulation in *rGSTA5* and pi-class GST (GSTP) *rGSTP1* [221]. In chickens, dietary AFB_1 up-regulated hepatic expression of *CYP1A1* and *CYP2H1*, and down-regulated expression of epoxide hydrolase (*EH*) and *GSTA*, although the specific *GSTA* gene target was not identified [123,124]. Other *P450* family members were significantly down-regulated [124]. Interestingly, a recent study in broilers observed the opposite changes in the liver, down-regulation of *CYP1A1* and up-regulation of *EH* and *GST* [115].

5.2. Cytokines and the MHC

AFB_1 is known to initiate hepatic inflammation, and correspondingly, was shown to affect expression of pro-inflammatory cytokines in both mammals and poultry [124,171,222,223,224,225,226]. Splenic expression of interleukins 1 beta (*IL1*β), 6 (*IL6*), 10 (*IL10*), interferon gamma (*IFN*-γ) and tumor necrosis factor alpha (*TNF*-α) in pigs (*Sus scrofa*) were increased by AFB_1 exposure [225]. Expression of *IFN*-γ and *TNF*-α increased in rats during aflatoxicosis, and protein levels of IL-1, IL-2, and IL-6 were also modulated [224,226]. In chickens, dietary AFB_1 increased hepatic expression of *IL6* [115,124]. However, in another experiment [123], expression of the IL6 receptor (*IL6R*) and IL10 receptor beta (*IL10RB*) were reduced in the liver. Exposure to AFB_1 reduced ileac expression of *IL2*, interleukin 4 (*IL4*), *IL6, IL10,* interleukin 17 (*IL17*), and *IFN*-γ in chickens [171,172]. Lipopolysaccharide-induced TNF factor (*LITAF*) expression also decreased in the intestine [171,172]. This gene is used as a marker for TNF activity since *TNF*-α has not been identified and may not be present in birds. Lastly, serum protein levels of IL-2 and IFN-γ were lower in chickens exposed to AFB_1 [168]. Down-regulation of these cytokines and their receptors in poultry may result in decreased T lymphocyte activation and proliferation.

Both up- and down-regulation of major histocompatibility complex (MHC) class I genes in response to AFB_1 exposure was observed in pigs [227], rats [221], and rainbow trout (*Oncorhynchus mykiss*) [228]. The MHC is a highly polymorphic genomic region that contains genes encoding proteins essential to innate and adaptive immune functions. For example, MHC class I and class II molecules are necessary for antigen presentation to T lymphocytes and are common to the galliform MHC [229,230,231,232,233,234,235,236,237]. In turkeys and chickens, the MHC is composed of 2 genetically unlinked regions, the *B*-locus (*MHC-B*) and *Y*-locus (*MHC-Y*), co-located on a single microchromosome (GGA16 or MGA18, respectively) [229,230,231,232,233,234,235,236,237]. Although some genes are well characterized, the functions and expression patterns of many poultry MHC genes are still unknown. In the turkey, expression of multiple MHC genes significantly increased in response to AFB_1 exposure [238].

5.3. Moving towards Transcriptomics

Few studies have examined gene expression changes across the entire transcriptome. Although the specific genes affected by AFB_1 vary between species, exposure to the toxin has an up-regulatory effect on expression of damage responses, or when those fail, pathways of carcinogenesis. In rats, AFB_1 enhanced hepatic expression of genes involved in xenobiotic detoxification, cell cycle regulation, oxidative stress, DNA damage repair, tumor development and amino acid metabolism [221]. For example, E2F transcription factor 1 (*E2F1*) and many genes downstream of *E2F1* were up-regulated by AFB_1exposure; E2F1 regulates DNA replication and apoptosis, and could contribute to carcinogenesis [221]. Aflatoxicosis in pigs affects expression of many of the same pathways, with greatest effects on metabolism, cell cycle, DNA damage responses and apoptotic processes in one study [239] and on protein degradation, metabolism, apoptosis, and immune responses in another [227]. Hepatocellular carcinomas from AFB_1-exposed rainbow trout showed changes in cell cycle, metabolism, immune and acute phase response genes when compared to adjacent non-cancerous liver tissue

[228]. Two of these studies utilized high throughput RNA sequencing (RNA-seq), rather than microarrays, to characterize transcriptome changes after AFB_1 exposure [221,239].

5.4. Poultry Transcriptomics and AFB_1

AFB_1 effects on the transcriptome have been investigated in chickens using microarray [123]. Exposure to AFB_1 affected hepatic expression of genes associated with fatty acid metabolism, development, detoxification, coagulation, immunity and cell proliferation [123]. Among these, the greatest proportions of differentially expressed genes were involved in cell proliferation and metabolism. For example, aflatoxicosis had the greatest effect on expression of the cell signaling inhibitor Dickkopf homolog 3 (*DKK3*) (down-regulated) and the metabolic glycogen synthase 1 (*GYS1*) (up-regulated).

More recently, RNA-seq has been applied to the domestic turkey transcriptome to elucidate hepatic and splenic responses to dietary AFB_1 [227,238,240]. Similar to expression changes in other species, transcriptome analysis in the turkey found that more genes were up-regulated than down-regulated by AFB_1 [227,238,240]. In one study of the liver, more than 80 genes belonging to cancer or focal adhesion pathways were up-regulated during aflatoxicosis, whereas smaller numbers of genes involved in cell signaling, cytoskeleton, and cell cycle were also up-regulated [227]. Conversely, the greatest numbers of down-regulated genes were involved in complement and coagulation.

Our investigation of AFB_1-hepatotoxicity in the turkey by RNA-seq found 313 transcripts significantly affected by AFB_1exposure, with up-regulation of genes involved in apoptosis, cancer, and cell cycle regulation, and down-regulation of lipid metabolism [240]. Greatest up-regulation was observed for keratin 20 (*KRT20*), cell-death activator CIDE-3 (*CIDEC*), and E3 ubiquitin-protein ligase Mdm2 (*MDM2*). Alpha-2-macroglobulin (*A2M*) was the most down-regulated gene. Exposure to AFB_1 also turned on expression of HEPACAM family member 2 (*HEPACAM2*), a protein important in modulating cell adhesion and migration, and *S*-adenosylmethionine synthase isoform type-2 (*MAT2A*), important in methylation pathways.

Utilizing spleen samples from the same AFB_1-challenge trial, immunotoxicity in domestic turkey was also examined through RNA-seq [238]. Exposure to AFB_1 induced significant expression changes in 391 *de novo* assembled transcripts; the greatest up-regulation was seen in E3 ubiquitin-protein ligase CBL-B (*CBLB*) and ubiquitin specific peptidase 40 (*USP40*). Of the significantly altered transcripts, 27.6% encoded proteins with known immune functions representing both innate and adaptive responses. Antimicrobial genes, including beta-defensin 1 (*THP1*) and 2 (*THP2*), were down-regulated, while cytotoxic and antigen presentation genes, such as granzyme A (*GZMA*), perforin 1 (*PRF1*), MHC class IA and class IIB, were up-regulated in AFB_1-exposed tissue.

These studies were also designed to evaluate the ability of a *Lactobacillus*-based oral probiotic to reduce AFB_1-effects on the liver and spleen [238,240]. In the same challenge trial, domestic turkeys were exposed to probiotics alone or in combination with AFB_1. Addition of probiotics during AFB_1 exposure mitigated AFB_1-induced expression changes in genes such as serine/arginine repetitive matrix protein 1 (*SRRM1*), 28S ribosomal RNA (*RNA28S*), and ISG12-2 protein-like (*ISG12-2*) (Figure 3A) [240]. In the spleen, probiotics had greater ameliorating properties, significantly reducing AFB_1-effects on multiple immune genes, including *THP2, GZMA,* and *PRF1* (Figure 3B) [238]. However, probiotics were unable to reverse most AFB_1-induced expression changes and even had synergistic effects with AFB_1. For example, combined treatment increased differential expression in apolipoprotein A-IV (*APOA4*) in the liver and interestingly both *RNA28S* and *ISG12-2* in the spleen (Figure 3). Therefore, oral probiotics modulated expression in both tissues, but did not restore normal transcriptome profiles.

A.

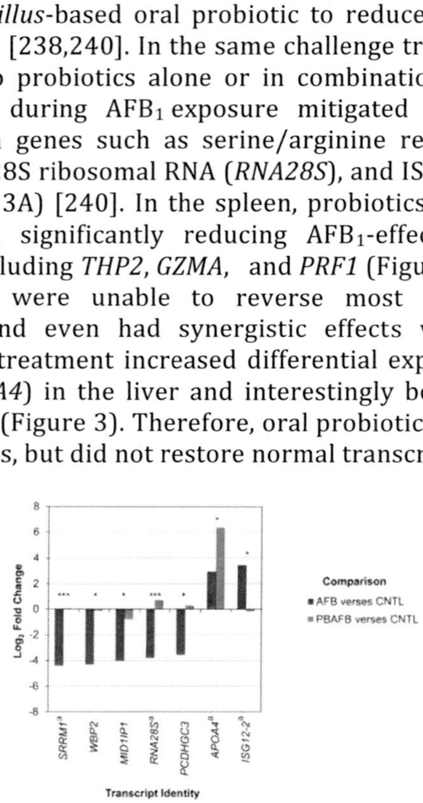

B.

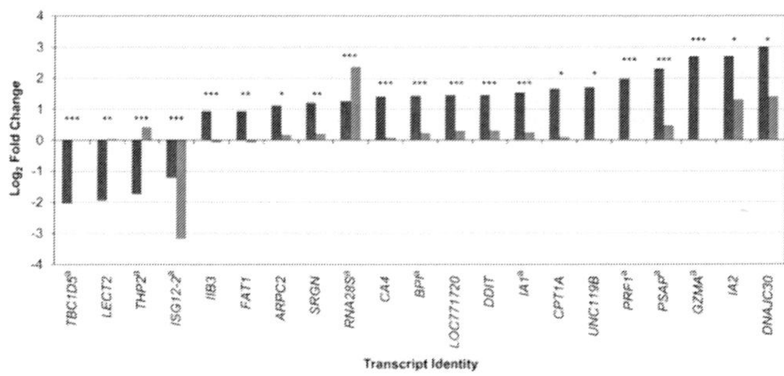

Figure 3. Oral probiotics mitigate gene expression changes induced by aflatoxin B_1(AFB_1). **(A)** Liver. **(B)** Spleen. Graphs show annotated transcripts from each tissue with significant differential expression in the AFB_1-treated group (AFB) *versus* the control group (CNTL) and in the probiotic + AFB_1 group (PBAFB) verses AFB. Bars illustrate log_2 fold change in AFB verses CNTL (**blue**) or PBAFB verses CNTL (**green**). Significance of the probiotics (PBAFB verses AFB) is represented by the number of asterisks (*: 0.05 > *p*-value > 0.01, **: 0.01 > *p*-value > 0.001, ***: *p*-value < 0.001). Genes with multiple significant transcripts are indicated by [a]; only the most significant transcript in PBAFB verses AFB is shown. Apolipoprotein A-IV (*APOA4*), actin-related protein 2/3 complex subunit 2-like (*ARPC2*), bactericidal permeability-increasing protein (*BPI*), carbonic anhydrase IV (*CA4*), carnitine O-palmitoyltransferase 1, liver isoform (*CPT1A*), DNA-damage-inducible transcript 4 (*DDIT*), DnaJ homolog subfamily C member 30 (*DNAJC30*), FAT tumor suppressor homolog 1 (*FAT1*), granzyme A (*GZMA*), MHC class I antigen alpha chain 1 (*IA1*), MHC class I antigen alpha chain 2 (*IA2*), MHC class II antigen beta chain 3 (*IIB3*), ISG12-2 protein-like (*ISG12-2*), leukocyte cell-derived chemotaxin-2 (*LECT2*), uncharacterized LOC771720 (*LOC771720*), MID1 interacting protein 1 (*MID1IP1*), protocadherin gamma subfamily C3 (*PCDHGC3*), perforin 1 (*PRF1*), presaposin (*PSAP*), 28S ribosomal RNA gene (*RNA28S*), serglycin (*SRGN*), serine/arginine repetitive matrix protein 1 (*SRRM1*), TBC1 domain family member 5 (*TBC1D5*), beta-defensin 2 (*THP2*), unc-119 homolog B (*UNC119B*), and WW domain-binding protein 2 (*WBP2*).

6. STRATEGIES TO REDUCE AFB_1 TOXICITY

6.1. Chemical Detoxification

Since prevention of AFB_1 contamination is often impractical, methods to chemically detoxify this mycotoxin have been intensely investigated [7]. Several candidate chemicals have been examined for their ability to detoxify AFB_1 in crops such as grain, rice, corn, and cottonseed. These include ammonium hydroxide [241,242,243,244,245,246], calcium hydroxide [243,247], hydrogen peroxide [243,248], sodium hydroxide [243,249], and sodium hypochlorite [243,250] all of which reduce AFB_1 concentrations through hydrolysis and produce a degraded form with reduced or no toxicity. However, most of these chemicals are of themselves, hazardous and safe use is often expensive and may decrease the nutrient value of feed components. In addition, it is not possible to fully preclude low levels of AFB_1 in feed production, especially in crops that are heavily contaminated, severely reducing returns for producers.

6.2. Feed Additives

Given that eliminating all potential for exposure to AFB_1 is not feasible, feed additives have been examined for their ability to protect poultry from aflatoxicosis [7,21]. Some additives, such as selenium supplementation, attempt to boost detoxification, metabolic, or immune functions to counteract the effects of AFB_1 [139,149,150, 168,170,171,183,251]. However, most additives have been investigated for their potential to reduce AFB_1 uptake by the intestine [21]. Many natural absorbents have been shown to decrease the effects of AFB_1 in poultry, including super-activated charcoal [153], zeolites like hydrated sodium calcium aluminosilicate [115,143,198,252,253,254,255,256,257], clinoptilolite [134,138], and sodium bentonite [158,258]. Antioxidants like butylated hydroxytolouene (BHT) [137,259,260,261,262,263] and turmeric [124,264] can also mitigate the severity of aflatoxicosis.

6.3. Probiotics

Many Gram-positive bacteria, including *Streptococcus*, *Enterococcus*, *Lactococcus*, and *Berevibacillus*, can bind AFB_1 *in vitro* [265,266, 267,268,269]. However, most research has focused on probiotic strains of *Lactobacillus*, *Bifidobacterium*, and *Propionibacterium* [42,266, 267,270,271,272,273,274,275,276,277,278,279,280,281,282,283,284,28 5,286,287,288]. Interactions between AFB_1 and *Lactobacillus rhamnosus* GG (LGG), *L. rhamnosus* LC-705 (LC-705), *Propionibacterium freudenrieichii* strain *shermanii* JS (PJS) or mixtures of these strains have been shown to be especially effective [42,270,271,272, 273,274,276,277,278,279,280,284,285,286,288]. A mixture of these strains was also utilized in our transcriptomic analyzes [238,240]. As gastrointestinal commensals or cultures used in cheese-making, yogurt and other dairy products, the safety of these lactic acid bacterial strains is well-established and easily applicable as potential chemopreventatives.

Strains LGG and LC-705 can sequester up to 80% of AFB_1 introduced into growth media [272,280]. AFB_1 interacts with the thick peptidoglycan layer characteristic of the Gram-positive bacterial cell wall [282,285]. When bound to AFB_1 *in vitro*, LGG and a mixture of LC-705 and PJS interacted less with intestinal mucus [276]. Incubation of LGG with AFB_1 reduced LGG adhesion to a Caco-2 intestinal cell monolayer [284] and decreased transport of AFB_1 across the monolayer [278]. These *in vitro* models suggest the probiotic and the toxin would be excreted together *in vivo* and thereby decrease the effective dose of AFB_1.

Reduced AFB_1 toxicity has been demonstrated in both mice [270] and in rats [277] after addition of dietary LGG. In humans, probiotics enhanced excretion of AFB_1 [289,290]. Furthermore, addition of LGG, LC-705 or PJS *ex vivo* through injection of AFB_1 into the lumen of the chicken duodenum significantly reduced AFB_1 uptake (by 74%, 63% and 37%, respectively) [274]. A mixture of LC-705 and PJS caused a 40% reduction in AFB_1 absorption into chicken duodenal tissue in a repeat experiment [42]. Therefore, a probiotic mixture including LGG or LC-705 could be an effective preventative for aflatoxicosis if added to poultry feeds.

6.4. Selection for Resistance

Another option to minimize the adverse effects of AFB_1 is to increase the resistance of domestic poultry [7]. Selection for AFB_1 resistant lines of chicken [291,292,293,294,295] and quail [296,297,298,299] has been investigated; however, selection studies have not been performed in the far more sensitive domestic turkey. Improvement in AFB_1 resistance is highly dependent on starting population [299] and selection is most effective during AFB_1 challenge since correlations of phenotypic measures like weight gain or blood parameters are most informative during exposure [291,292,298]. The requirement for concurrent aflatoxicosis makes phenotypic selection difficult to implement in a commercial setting. However, understanding the molecular mechanism of aflatoxicosis in domestic turkey and identifying the genetic differences underlying decreased sensitivity in wild turkeys could allow targeted genetic selection for resistant alleles without constant AFB_1-exposure. Due to the similarity of symptoms of aflatoxicosis across species, potential genetic targets may be translatable to other poultry species.

7. CONCLUSIONS: SUGGESTED AREAS FOR FURTHER RESEARCH

Future gene expression analyzes can provide insight into the mechanisms of aflatoxicosis and methods to reduce its effects. Our characterization of AFB_1-induced changes in the turkey liver and spleen transcriptomes identified genes responding to aflatoxicosis and host responses to toxicity. Since both proliferation and apoptosis occur during aflatoxicosis, the expression of cell cycle genes in the liver needs to be quantified alongside measures of apoptotic and mitotic cells. Similarly, gene expression in immune tissues such as the spleen should be measured concurrent with lymphocyte numbers, activation or apoptotic state to better clarify gene functions. Investigation of individual cell types could also detect cell-specific gene modulation. For example, the effects of AFB_1 on expression of immune genes could be measured in heterophils or T lymphocyte subsets (e.g., $CD4^+$ verses $CD8^+$ T cells). Gene expression in cells of the bile duct, known targets of AFB_1 mutagenesis, could be used to characterize hyperplasia. Expression patterns, as determined from mRNA, do not always directly correlate with protein levels or stability; therefore, proteomics could confirm effects on cell cycle regulators or immune mediators and provide a measure for AFB_1 inhibition of protein synthesis.

Analyses of systemic responses to AFB_1 in other tissues, such as the bursa, thymus, kidney or small intestine are needed to fully elucidate AFB_1 effects. Intestinal epithelial cells are directly exposed to AFB_1 during absorption and the potential prevention of AFB_1 uptake by feed additives also occurs within the small intestine. Therefore, investigation of intestinal transcriptome responses to AFB_1 is a priority. The utility of a *Lactobacillus*-based probiotics as preventative for aflatoxicosis requires further examination *in vivo*. Higher concentrations or different compositions of dietary probiotics should be examined for their ability to protect poultry from the development of hepatic lesions and in restoring gene expression profiles. Furthermore, no definitive conclusions can be made regarding the effects of probiotics without characterizing the intestinal microbiota. Shifts in bacterial population structure of the microbiome could be investigated by 16S NGS sequencing for direct comparison to the host effects shown by RNA-seq.

Comparative analysis of transcriptome responses of poultry to AFB_1 could help resolve differences in their sensitivity. RNA-seq studies investigating both domestic and wild turkey using an *in ovo* exposure model [300] and dietary challenge of poults [301] are currently underway. Preliminary data from embryonic exposures illustrates conserved effects on cell cycle regulators and variation in metabolic and anti-oxidant enzymes [300]. Genes and pathways identified in these studies will provide targets for selection efforts to improve resistance to aflatoxicosis in domestic poultry.

ACKNOWLEDGMENTS

The authors gratefully acknowledge support from the United States Department of Agriculture, Agriculture and Food Research Initiative Grants 2005-01326, 2007-35205-17880, 2009-35205-05302, and 2013-

01043 from the USDA National Institute of Food and Agriculture Animal Genome Program. Support from the Minnesota and Utah Agricultural Experiments Stations is also acknowledged.

AUTHOR CONTRIBUTIONS

The work presented is a part of a Ph.D. thesis; the information was collected and summarized by Melissa Monson who took the lead on writing the manuscript. Kent Reed and Roger Coulombe supervised the work and edited the manuscript.

CONFLICTS OF INTEREST

The authors declare no conflict of interest.

REFERENCES

1. Bedard, L.L.; Massey, T.E. Aflatoxin B1-induced DNA damage and its repair. Cancer Lett. 2006, 241, 174–183.
2. Bennett, J.W.; Klich, M. Mycotoxins. Clin. Microbiol. Rev. 2003, 16, 497–516.
3. Coulombe, R.A., Jr. Biological action of mycotoxins. J. Dairy Sci. 1993, 76, 880–891.
4. Eaton, D.L.; Gallagher, E.P. Mechanisms of aflatoxin carcinogenesis. Annu. Rev. Pharmacol. Toxicol. 1994, 34, 135–172.
5. Groopman, J.D.; Kensler, T.W. Role of metabolism and viruses in aflatoxin-induced liver cancer. Toxicol. Appl. Pharmacol. 2005, 206, 131–137.
6. Kensler, T.W.; Roebuck, B.D.; Wogan, G.N.; Groopman, J.D. Aflatoxin: A 50-Year odyssey of mechanistic and translational toxicology. Toxicol. Sci. 2011, 120, S28–S48.
7. Leeson, S.; Diaz, G.J.; Summers, J.D. Aflatoxins. In Poultry Metabolic Disorders and Mycotoxins; University Books: Guelph, Canada, 1995.
8. Yabe, K.; Nakajima, H. Enzyme reactions and genes in aflatoxin biosynthesis. Appl. Microbiol. Biotechnol. 2004, 64, 745–755.
9. Wogan, G.N.; Kensler, T.W.; Groopman, J.D. Present and future directions of translational research on aflatoxin and hepatocellular carcinoma. A review. Food Addit. Contam. 2012, 29, 249–257.
10. Newberne, P.M.; Butler, W.H. Acute and chronic effects of aflatoxin on the liver of domestic and laboratory animals: A review. Cancer Res. 1969, 29, 236–250.
11. Patterson, D.S.P. Metabolism as a factor in determining the toxic action of the aflatoxins in different animal species. Food Cosmet. Toxicol. 1973, 11, 287–294.
12. Pier, A.C. Aflatoxicosis and Immunosuppression in Mammalian Animals. In Aflatoxin in Maize: A Proceedings of the Workshop; Zuber, M.S., Lillehoj, E.B., Renfro, B.L., Eds.; CIMMYT: El Batan, Mexico, 1986.
13. Robens, J.F.; Richard, J.L. Aflatoxins in animal and human health. Rev. Environ. Contam. Toxicol. 1992, 127, 69–94.

14. Wogan, G.N. Chemical nature and biological effects of the aflatoxins. Bacteriol. Rev. 1966, 30, 460–470.
15. Wogan, G.N. Aflatoxins as risk factors for hepatocellular carcinoma in humans. Cancer Res. 1992, 52, 2114s–2118s.
16. Blount, W.P. Turkey "X" disease. Turkeys 1961, 9, 77.
17. Wannop, C.C. The histopathology of turkey "X" disease in Great Britain. Avian Dis. 1961, 5, 371–381.
18. Nesbitt, B.F.; O'Kelly, J.; Sargeant, K.; Sheridan, A. Aspergillus flavus and turkey X disease. Toxic metabolites of Aspergillus flavus. Nature 1962, 195, 1062–1063.
19. Cary, J.W.; Klich, M.A.; Beltz, S.B. Characterization of aflatoxin-producing fungi outside of Aspergillus section Flavi. Mycologia 2005, 97, 425–432.
20. Frisvad, J.C.; Skouboe, P.; Samson, R.A. Taxonomic comparison of three different groups of aflatoxin producers and a new efficient producer of aflatoxin B1, sterigmatocystin and 3-O-methylsterigmatocystin, Aspergillus rambellii sp. nov. Syst. Appl. Microbiol. 2005, 28, 442–453.
21. Rawal, S.; Kim, J.E.; Coulombe, R. Aflatoxin B1 in poultry: Toxicology, metabolism and prevention. Res. Vet. Sci. 2010, 89, 325–331.
22. Cullen, J.M.; Newberne, P.M. Acute Hepatotoxicity of Aflatoxins. In The Toxicology of Aflatoxins: Human Health, Veterinary, and Agricultural Significance; Eaton, D.L., Groopman, J.D., Eds.; Academic Press: London, UK, 1993.
23. CAST. Mycotoxins: Risks in Plant, Animal and Human Systems; No. 139; Council for Agricultural Science and Technology: Ames, IA, USA, 2003.
24. Cotty, P.J.; Jaime-Garcia, R. Influences of climate on aflatoxin producing fungi and aflatoxin contamination. Int. J. Food Microbiol. 2007, 119, 109–115.
25. Gqaleni, N.; Smith, J.E.; Lacey, J.; Gettinby, G. Effects of temperature, water activity, and incubation time on production of aflatoxins and cyclopiazonic acid by an isolate of Aspergillus flavus in surface agar culture. Appl. Environ. Microbiol. 1997, 63, 1048–1053.
26. Medina, A.; Rodriguez, A.; Magan, N. Effect of climate change on Aspergillus flavus and aflatoxin B1 production. Front. Microbiol. 2014, 5, 348.
27. Mousa, W.; Ghazali, F.M.; Jinap, S.; Ghazali, H.M.; Radu, S. Modeling growth rate and assessing aflatoxins production by Aspergillus flavus as a function of water activity and temperature on polished and brown rice. J. Food Sci. 2013, 78, M56–M63.
28. Schindler, A.F.; Palmer, J.G.; Eisenberg, W.V. Aflatoxin production by Aspergillus flavus as related to various temperatures. Appl. Microbiol. 1967, 15, 1006–1009.
29. Schroeder, H.W.; Hein, H. Aflatoxins: Production of the toxins in vitro in relation to temperature. Appl. Microbiol.1967, 15, 441–445.
30. Trenk, H.L.; Hartman, P.A. Effects of moisture content and temperature on aflatoxin production in corn. Appl. Microbiol. 1970, 19, 781–784.
31. Aly, S.A.; Anwer, W. Effect of naturally contaminated feed with aflatoxins on performance of laying hens and the carryover of aflatoxin B1 residues in table egg. Pakistan J. Nutr. 2009, 8, 181–186.
32. Chen, C.; Pearson, A.M.; Coleman, T.H.; Gray, J.I.; Pestka, J.J.; Aust, S.D. Tissue deposition and clearance of aflatoxins from broiler chickens fed a contaminated diet. Food Chem. Toxicol. 1984, 22, 447–451.
33. Galvano, F.; Galofaro, V.; Galvano, G. Occurrence and stability of aflatoxin M1 in milk and milk products: A worldwide review. J. Food Prot. 1996, 59, 1079–1090.
34. Pandey, I.; Chauhan, S.S. Studies on production performance and toxin residues in tissues and eggs of layer chickens fed on diets with various concentrations of aflatoxin AFB1. Br. Poult. Sci. 2007, 48, 713–723.
35. Purchase, I.F.H. Aflatoxin residues in food of animal origin. Food Cosmet. Toxicol. 1972, 10, 531–544.
36. Richard, J.L.; Stubblefield, R.D.; Lyon, R.L.; Peden, W.M.; Thurston, J.R.; Rimler, R.B. Distribution and clearance of aflatoxins B1 and M1 in turkeys fed diets containing 50 or 150 ppb aflatoxin from naturally contaminated corn. Avian Dis. 1986, 30, 788–793.

37. Wolzak, A.; Pearson, A.M.; Coleman, T.H.; Pestka, J.J.; Gray, J.I. Aflatoxin deposition and clearance in the eggs of laying hens. Food Chem. Toxicol. 1985, 23, 1057–1061.
38. Wolzak, A.; Pearson, A.M.; Coleman, T.H.; Pestka, J.J.; Gray, J.I.; Chen, C. Aflatoxin carryover and clearance from tissues of laying hens. Food Chem. Toxicol. 1986, 24, 37–41.
39. Giambrone, J.J.; Diener, U.L.; Davis, N.D.; Panangala, V.S.; Hoerr, F.J. Effects of aflatoxin on young turkeys and broiler chickens. Poult. Sci. 1985, 64, 1678–1684.
40. Klein, P.J.; Buckner, R.; Kelly, J.; Coulombe, R.A., Jr. Biochemical basis for the extreme sensitivity of turkeys to aflatoxin B1. Toxicol. Appl. Pharmacol. 2000, 165, 45–52.
41. CAST. Mycotoxins: Economic and Health Risks; No. 116; Council for Agricultural Science and Technology: Ames, IA, USA, 1989.
42. Gratz, S.; Mykkänen, H.; el-Nezami, H. Aflatoxin B1 binding by a mixture of Lactobacillus and Propionibacterium.: In vitro versus ex vivo. J. Food Prot. 2005, 68, 2470–2474.
43. Guengerich, F.P.; Johnson, W.W.; Ueng, Y.F.; Yamazaki, H.; Shimada, T. Involvement of cytochrome P450, glutathione S-transferase, and epoxide hydrolase in the metabolism of aflatoxin B1 and relevance to risk of human liver cancer. Environ. Health Perspect. 1996, 104, 557–562.
44. Godoy, H.M.; Neal, G.E. Some studies of the effects of aflatoxin B1 in vivo and in vitro on nucleic acid synthesis in rat and mouse. Chem. Biol. Interact. 1976, 13, 257–277.
45. Monroe, D.H.; Eaton, D.L. Comparative effects of butylated hydroxyanisole on hepatic in vivo DNA binding and in vitro biotransformation of aflatoxin B1 in the rat and mouse. Toxicol. Appl. Pharmacol. 1987, 90, 401–409.
46. Pelkonen, P.; Lang, M.A.; Wild, C.P.; Negishi, M.; Juvonen, R.O. Activation of aflatoxin B1 by mouse CYP2A enzymes and cytotoxicity in recombinant yeast cells. Eur. J. Pharmacol. 1994, 292, 67–73.
47. Pelkonen, P.; Lang, M.A.; Negishi, M.; Wild, C.P.; Juvonen, R.O. Interaction of aflatoxin B1 with cytochrome P450 2A5 and its mutants: Correlation with metabolic activation and toxicity. Chem. Res. Toxicol. 1997, 10, 85–90.
48. Ramsdell, H.S.; Eaton, D.L. Species susceptibility to aflatoxin B1 carcinogenesis: Comparative kinetics of microsomal biotransformation. Cancer Res. 1990, 50, 615–620.
49. Yanagimoto, T.; Itoh, S.; Sawada, M.; Hashimoto, H.; Kamataki, T. Molecular cloning and functional expression of a mouse cytochrome P-450 (Cyp3a-13): Examination of Cyp3a-13 enzyme to activate aflatoxin B1 (AFB1). Biochim. Biophys. Acta 1994, 1201, 405–410.
50. Yanagimoto, T.; Itoh, S.; Sawada, M.; Kamataki, T. Mouse cytochrome P450 (Cyp3a11): Predominant expression in liver and capacity to activate aflatoxin B1. Arch. Biochem. Biophys. 1997, 340, 215–218.
51. Buetler, T.M.; Slone, D.; Eaton, D.L. Comparison of the aflatoxin B1-8,9-epoxide conjugating activities of two bacterially expressed alpha class glutathione S-transferase isozymes from mouse and rat. Biochem. Biophys. Res. Commun. 1992, 188, 597–603.
52. Hayes, J.D.; Judah, D.J.; Neal, G.E.; Nguyen, T. Molecular cloning and heterologous expression of a cDNA encoding a mouse glutathione S-transferase Yc subunit possessing high catalytic activity for aflatoxin B1-8,9-epoxide. Biochem. J.1992, 285, 173–180.
53. Ilic, Z.; Crawford, D.; Vakharia, D.; Egner, P.A.; Sell, S. Glutathione-S-transferase A3 knockout mice are sensitive to acute cytotoxic and genotoxic effects of aflatoxin B1. Toxicol. Appl. Pharmacol. 2010, 242, 241–246.
54. Monroe, D.H.; Eaton, D.L. Effects of modulation of hepatic glutathione on biotransformation and covalent binding of aflatoxin B1 to DNA in the mouse. Toxicol. Appl. Pharmacol. 1988, 94, 118–127.
55. Quinn, B.A.; Crane, T.L.; Kocal, T.E.; Best, S.J.; Cameron, R.G.; Rushmore, T.H.; Farber, E.; Hayes, M.A. Protective activity of different hepatic cytosolic glutathione S-

transferases against DNA-binding metabolites of aflatoxin B1. Toxicol. Appl. Pharmacol. 1990, 105, 351–363.

56. Slone, D.H.; Gallagher, E.P.; Ramsdell, H.S.; Rettie, A.E.; Stapleton, P.L.; Berlad, L.G.; Eaton, D.L. Human variability in hepatic glutathione S-transferase-mediated conjugation of aflatoxin B1-epoxide and other substrates. Pharmacogenetics 1995, 5, 224–233.
57. Buetler, T.M.; Bammler, T.K.; Hayes, J.D.; Eaton, D.L. Oltipraz-mediated changes in aflatoxin B1 biotransformation in rat liver: Implications for human chemointervention. Cancer Res. 1996, 56, 2306–2313.
58. Imaoka, S.; Ikemoto, S.; Shimada, T.; Funae, Y. Mutagenic activation of aflatoxin B1 by pulmonary, renal, and hepatic cytochrome P450s from rats. Mutat. Res. 1992, 269, 231–236.
59. Johnson, W.W.; Ueng, Y.F.; Widersten, M.; Mannervik, B.; Hayes, J.D.; Sherratt, P.J.; Ketterer, B.; Guengerich, F.P. Conjugation of highly reactive aflatoxin B1 exo-8,9-epoxide catalyzed by rat and human glutathione transferases: Estimation of kinetic parameters. Biochemistry 1997, 36, 3056–3060.
60. Hayes, J.D.; Judah, D.J.; McLellan, L.I.; Kerr, L.A.; Peacock, S.D.; Neal, G.E. Ethoxyquin-induced resistance to aflatoxin B1 in the rat is associated with the expression of a novel alpha-class glutathione S-transferase subunit, Yc2, which possesses high catalytic activity for aflatoxin B1–8,9-epoxide. Biochem. J. 1991, 279, 385–398.
61. Hayes, J.D.; Nguyen, T.; Judah, D.J.; Petersson, D.G.; Neal, G.E. Cloning of cDNAs from fetal rat liver encoding glutathione S-transferase Yc polypeptides. The Yc2 subunit is expressed in adult rat liver resistant to the hepatocarcinogen aflatoxin B1. J. Biol. Chem. 1994, 269, 20707–20717.
62. Hayes, J.D.; Pulford, D.J.; Ellis, E.M.; McLeod, R.; James, R.F.; Seidegård, J.; Mosialou, E.; Jernström, B.; Neal, G.E. Regulation of rat glutathione S-transferase A5 by cancer chemopreventive agents: Mechanisms of inducible resistance to aflatoxin B1. Chem. Biol. Interact. 1998, 111–112, 51–67.
63. Forrester, L.M.; Neal, G.E.; Judah, D.J.; Glancey, M.J.; Wolf, C.R. Evidence for involvement of multiple forms of cytochrome P-450 in aflatoxin B1 metabolism in human liver. Proc. Natl. Acad. Sci. USA 1990, 87, 8306–8310.
64. Gallagher, E.P.; Wienkers, L.C.; Stapleton, P.L.; Kunze, K.L.; Eaton, D.L. Role of human microsomal and human complementary DNA-expressed cytochromes P4501A2 and P4503A4 in the bioactivation of aflatoxin B1. Cancer Res.1994, 54, 101–108.
65. Macé, K.; Aguilar, F.; Wang, J.S.; Vautravers, P.; Gómez-Lechón, M.; Gonzalez, F.J.; Groopman, J.; Harris, C.C.; Pfeifer, A.M. Aflatoxin B1-induced DNA adduct formation and p53 mutations in CYP450-expressing human liver cell lines. Carcinogenesis 1997, 18, 1291–1297.
66. Ramsdell, H.S.; Parkinson, A.; Eddy, A.C.; Eaton, D.L. Bioactivation of aflatoxin B1 by human liver microsomes: Role of cytochrome P450 IIIA enzymes. Toxicol. Appl. Pharmacol. 1991, 108, 436–447.
67. Van Vleet, T.R.; Watterson, T.L.; Klein, P.J.; Coulombe, R.A., Jr. Aflatoxin B1 alters the expression of p53 in cytochrome P450-expressing human lung cells. Toxicol. Sci. 2006, 89, 399–407.
68. Raney, K.D.; Meyer, D.J.; Ketterer, B.; Harris, T.M.; Guengerich, F.P. Glutathione conjugation of aflatoxin B1 exo- and endo-epoxides by rat and human glutathione S-transferases. Chem. Res. Toxicol. 1992, 5, 470–478.
69. Gross-Steinmeyer, K.; Stapleton, P.L.; Tracy, J.H.; Bammler, T.K.; Strom, S.C.; Eaton, D.L. Sulforaphane- and phenethyl isothiocyanate-induced inhibition of aflatoxin B1-mediated genotoxicity in human hepatocytes: Role of GSTM1 genotype and CYP3A4 gene expression. Toxicol. Sci. 2010, 116, 422–432.
70. Liu, Y.H.; Taylor, J.; Linko, P.; Lucier, G.W.; Thompson, C.L. Glutathione S-transferase mu in human lymphocyte and liver: Role in modulating formation of carcinogen-derived DNA adducts. Carcinogenesis 1991, 12, 2269–2275.
71. McGlynn, K.A.; Rosvold, E.A.; Lustbader, E.D.; Hu, Y.; Clapper, M.L.; Zhou, T.; Wild, C.P.; Xia, X.L.; Baffoe-Bonnie, A.; Ofori-Adjei, D.; et al. Susceptibility to hepatocellular

carcinoma is associated with genetic variation in the enzymatic detoxification of aflatoxin B1. Proc. Natl. Acad. Sci. USA 1995, 92, 2384–2387.
72. Arafa, A.S.; Bloomer, R.J.; Wilson, H.R.; Simpson, C.F.; Harms, R.H. Susceptibility of various poultry species to dietary aflatoxin. Br. Poult. Sci. 1981, 22, 431–436.
73. Gumbmann, M.R.; Williams, S.N.; Booth, A.N.; Vohra, P.; Ernst, R.A.; Bethard, M. Aflatoxin susceptibility in various breeds of poultry. Proc. Soc. Exp. Biol. Med. 1970, 134, 683–688.
74. Lozano, M.C.; Diaz, G.J. Microsomal and cytosolic biotransformation of aflatoxin B1 in four poultry species. Br. Poult. Sci. 2006, 47, 734–741.
75. Muller, R.D.; Carlson, C.W.; Semeniuk, G.; Harshfield, G.S. The response of chicks, ducklings, goslings, pheasants and poults to graded levels of aflatoxins. Poult. Sci. 1970, 49, 1346–1350.
76. Rawal, S.; Mendoza, K.M.; Reed, K.M.; Coulombe, R.A., Jr. Structure, genetic mapping, and function of the cytochrome P450 3A37 gene in the turkey (Meleagris gallopavo). Cytogenet. Genome Res. 2009, 125, 67–73.
77. Rawal, S.; Yip, S.S.; Coulombe, R.A., Jr. Cloning, expression and functional characterization of cytochrome P450 3A37 from turkey liver with high aflatoxin B1 epoxidation activity. Chem. Res. Toxicol. 2010, 23, 1322–1329.
78. Rawal, S.; Coulombe, R.A., Jr. Metabolism of aflatoxin B1 in turkey liver microsomes: The relative roles of cytochromes P450 1A5 and 3A37. Toxicol. Appl. Pharmacol. 2011, 254, 349–354.
79. Yip, S.S.; Coulombe, R.A. Molecular cloning and expression of a novel cytochrome P450 from turkey liver with aflatoxin B1 oxidizing activity. Chem. Res. Toxicol. 2006, 19, 30–37.
80. Diaz, G.J.; Murcia, H.W.; Cepeda, S.M. Bioactivation of aflatoxin B1 by turkey liver microsomes: Responsible cytochrome P450 enzymes. Br. Poult. Sci. 2010, 51, 828–837.
81. Diaz, G.J.; Murcia, H.W.; Cepeda, S.M. Cytochrome P450 enzymes involved in the metabolism of aflatoxin B1 in chickens and quail. Poult. Sci. 2010, 89, 2461–2469.
82. Diaz, G.J.; Murcia, H.W.; Cepeda, S.M.; Boermans, H.J. The role of selected cytochrome P450 enzymes on the bioactivation of aflatoxin B1 by duck liver microsomes. Avian Pathol. 2010, 39, 279–285.
83. Murcia, H.W.; Díaz, G.J.; Cepeda, S.M. Enzymatic activity in turkey, duck, quail and chicken liver microsomes against four human cytochrome P450 prototype substrates and aflatoxin B1. J. Xenobiotics 2011, 1, 17–21.
84. Giambrone, J.J.; Diener, U.L.; Davis, N.D.; Panangala, V.S.; Hoerr, F.J. Effects of purified aflatoxin on turkeys. Poult. Sci. 1985, 64, 859–865.
85. Klein, P.J.; van Vleet, T.R.; Hall, J.O.; Coulombe, R.A., Jr. Biochemical factors underlying the age-related sensitivity of turkeys to aflatoxin B1. Comp. Biochem. Physiol. C Toxicol. Pharmacol. 2002, 132, 193–201.
86. Hussain, Z.; Khan, M.Z.; Khan, A.; Javed, I.; Saleemi, M.K.; Mahmood, S.; Asi, M.R. Residues of aflatoxin B1 in broiler meat: Effect of age and dietary aflatoxin B1 levels. Food Chem. Toxicol. 2010, 48, 3304–3307.
87. Lanza, G.M.; Washburn, K.W.; Wyatt, R.D. Variation with age in response of broilers to aflatoxin. Poult. Sci. 1980, 59, 282–288.
88. Quezada, T.; Cuéllar, H.; Jaramillo-Juárez, F.; Valdivia, A.G.; Reyes, J.L. Effects of aflatoxin B1 on the liver and kidney of broiler chickens during development. Comp. Biochem. Physiol. C Toxicol. Pharmacol. 2000, 125, 265–272.
89. Barraud, L.; Douki, T.; Guerret, S.; Chevallier, M.; Jamard, C.; Trepo, C.; Wild, C.P.; Cadet, J.; Cova, L. The role of duck hepatitis B virus and aflatoxin B1 in the induction of oxidative stress in the liver. Cancer Detect. Prev. 2001, 25, 192–201.
90. Kim, J.E.; Bauer, M.M.; Mendoza, K.M.; Reed, K.M.; Coulombe, R.A., Jr. Comparative genomics identifies new alpha class genes within the avian glutathione S-transferase gene cluster. Gene 2010, 452, 45–53.

91. Kim, J.E.; Bunderson, B.R.; Croasdell, A.; Coulombe, R.A., Jr. Functional characterization of alpha-class glutathione S-transferases from the turkey (Meleagris gallopavo). Toxicol. Sci. 2011, 124, 45–53.
92. Quist, C.F.; Bounous, D.I.; Kilburn, J.V.; Nettles, V.F.; Wyatt, R.D. The effect of dietary aflatoxin on wild turkey poults. J. Wildl. Dis. 2000, 36, 436–444.
93. Reed, K.M. Turkey. In Genome Mapping and Genomics in Domestic Animals; Cockett, N.E., Kole, C., Eds.; Springer-Verlag: Berlin, Germany, 2009.
94. Julian, R.J. Production and growth related disorders and other metabolic diseases of poultry—A review. Vet. J. 2005, 169, 350–369.
95. Bayyari, G.R.; Huff, W.E.; Rath, N.C.; Balog, J.M.; Newberry, L.A.; Villines, J.D.; Skeeles, J.K.; Anthony, N.B.; Nestor, K.E. Effect of the genetic selection of turkeys for increased body weight and egg production on immune and physiological responses. Poult. Sci. 1997, 76, 289–296.
96. Li, Z.; Nestor, K.E.; Saif, Y.M.; Luhtala, M. Flow cytometric analysis of T lymphocyte subpopulations in large-bodied turkey lines and a randombred control population. Poult. Sci. 2000, 79, 219–223.
97. Li, Z.; Nestor, K.E.; Saif, Y.M.; Anderson, J.W.; Patterson, R.A. Effect of selection for increased body weight in turkeys on lymphoid organ weights, phagocytosis, and antibody responses to fowl cholera and Newcastle disease-inactivated vaccines. Poult. Sci. 2001, 80, 689–694.
98. Van der Most, P.J.; de Jong, B.; Parmentier, H.K.; Verhulst, S. Trade-off between growth and immune function: A meta-analysis of selection experiments. Funct. Ecol. 2011, 25, 74–80.
99. Kim, J.E.; Bunderson, B.R.; Croasdell, A.; Reed, K.M.; Coulombe, R.A., Jr. Alpha-class glutathione S-transferases in wild turkeys (Meleagris gallopavo): Characterization and role in resistance to the carcinogenic mycotoxin aflatoxin B1. PLoS ONE 2013, 8, e60662.
100. Bunderson, B.R.; Kim, J.E.; Croasdell, A.; Mendoza, K.M.; Reed, K.M.; Coulombe, R.A., Jr. Heterologous expression and functional characterization of avian mu-class glutathione S-transferases. Comp. Biochem. Physiol. C Toxicol. Pharmacol. 2013, 158, 109–116.
101. Smela, M.E.; Hamm, M.L.; Henderson, P.T.; Harris, C.M.; Harris, T.M.; Essigmann, J.M. The aflatoxin B1formamidopyrimidine adduct plays a major role in causing the types of mutations observed in human hepatocellular carcinoma. Proc. Natl. Acad. Sci. USA 2002, 99, 6655–6660.
102. Corrier, D.E. Mycotoxicosis: Mechanisms of immunosuppression. Vet. Immunol. Immunopathol. 1991, 30, 73–87.
103. Guengerich, F.P.; Arneson, K.O.; Williams, K.M.; Deng, Z.; Harris, T.M. Reaction of aflatoxin B1 oxidation products with lysine. Chem. Res. Toxicol. 2002, 15, 780–792.
104. Iwaki, M.; Kitagawa, T.; Akamatsu, Y.; Aibara, K. Cytotoxic effects of aflatoxin B1 and its association with cellular components in chicken embryo primary cultured cells. Biochim. Biophys. Acta 1990, 1035, 146–153.
105. IARC. Aflatoxins. In A Review of Human Carcinogens. Part. F: Chemical Agents and Related Occupations; International Agency for Research on Cancer Working Group on the Evaluation of Carcinogenic Risks to Humans: Lyon, France, 2012.
106. Aguilar, F.; Hussain, S.P.; Cerutti, P. Aflatoxin B1 induces the transversion of G-->T in codon 249 of the p53 tumor suppressor gene in human hepatocytes. Proc. Natl. Acad. Sci. USA 1993, 90, 8586–8590.
107. Shen, H.M.; Ong, C.N. Mutations of the p53 tumor suppressor gene and ras oncogenes in aflatoxin hepatocarcinogenesis. Mutat. Res. 1996, 366, 23–44.
108. Wild, C.P.; Montesano, R. A model of interaction: Aflatoxins and hepatitis viruses in liver cancer aetiology and prevention. Cancer Lett. 2009, 286, 22–28.
109. Butler, W.H.; Greenblatt, M.; Lijinsky, W. Carcinogenesis in rats by aflatoxins B1, G1, and B2. Cancer Res. 1969, 29, 2206–2211.
110. Wogan, G.N.; Newberne, P.M. Dose-response characteristics of aflatoxin B1 carcinogenesis in the rat. Cancer Res. 1967, 27, 2370–2376.

111. Wogan, G.N.; Paglialunga, S.; Newberne, P.M. Carcinogenic effects of low dietary levels of aflatoxin B1 in rats. Food Cosmet. Toxicol. 1974, 12, 681–685.
112. Cova, L.; Wild, C.P.; Mehrotra, R.; Turusov, V.; Shirai, T.; Lambert, V.; Jacquet, C.; Tomatis, L.; Trépo, C.; Montesano, R. Contribution of aflatoxin B1 and hepatitis B virus infection in the induction of liver tumors in ducks. Cancer Res.1990, 50, 2156–2163.
113. Cullen, J.M.; Marion, P.L.; Sherman, G.J.; Hong, X.; Newbold, J.E. Hepatic neoplasms in aflatoxin B1-treated, congenital duck hepatitis B virus-infected, and virus-free Pekin ducks. Cancer Res. 1990, 50, 4072–4080.
114. Uchida, T.; Suzuki, K.; Esumi, M.; Arii, M.; Shikata, T. Influence of aflatoxin B1 intoxication on duck livers with duck hepatitis B virus infection. Cancer Res. 1988, 48, 1559–1565.
115. Chen, X.; Horn, N.; Applegate, T.J. Efficiency of hydrated sodium calcium aluminosilicate to ameliorate the adverse effects of graded levels of aflatoxin B1 in broiler chicks. Poult. Sci. 2014, 93, 2037–2047.
116. Giambrone, J.J.; Diener, U.L.; Davis, N.D.; Panangala, V.S.; Hoerr, F.J. Effects of purified aflatoxin on broiler chickens. Poult. Sci. 1985, 64, 852–858.
117. Huff, W.E.; Kubena, L.F.; Harvey, R.B.; Corrier, D.E.; Mollenhauer, H.H. Progression of aflatoxicosis in broiler chickens. Poult. Sci. 1986, 65, 1891–1899.
118. Rauber, R.H.; Dilkin, P.; Giacomini, L.Z.; Araújo de Almeida, C.A.; Mallmann, C.A. Performance of turkey poults fed different doses of aflatoxins in the diet. Poult. Sci. 2007, 86, 1620–1624.
119. Richard, J.L.; Pier, A.C.; Cysewski, S.J.; Graham, C.K. Effect of aflatoxin and aspergillosis on turkey poults. Avian Dis.1973, 17, 111–121.
120. Smith, J.W.; Hamilton, P.B. Aflatoxicosis in the broiler chicken. Poult. Sci. 1970, 49, 207–215.
121. Sims, W.M., Jr.; Kelley, D.C.; Sanford, P.E. A study of aflatoxicosis in laying hens. Poult. Sci. 1970, 49, 1082–1084.
122. Verma, J.; Johri, T.S.; Swain, B.K.; Ameena, S. Effect of graded levels of aflatoxin, ochratoxin and their combinations on the performance and immune response of broilers. Br. Poult. Sci. 2004, 45, 512–518.
123. Yarru, L.P.; Settivari, R.S.; Antoniou, E.; Ledoux, D.R.; Rottinghaus, G.E. Toxicological and gene expression analysis of the impact of aflatoxin B1 on hepatic function of male broiler chicks. Poult. Sci. 2009, 88, 360–371.
124. Yarru, L.P.; Settivari, R.S.; Gowda, N.K.; Antoniou, E.; Ledoux, D.R.; Rottinghaus, G.E. Effects of turmeric (Curcuma longa) on the expression of hepatic genes associated with biotransformation, antioxidant, and immune systems in broiler chicks fed aflatoxin. Poult. Sci. 2009, 88, 2620–2627.
125. Lee, J.T.; Jessen, K.A.; Beltran, R.; Starkl, V.; Schatzmayr, G.; Borutova, R.; Caldwell, D.J. Mycotoxin-contaminated diets and deactivating compound in laying hens: 1. Effects on performance characteristics and relative organ weight. Poult. Sci. 2012, 91, 2089–2095.
126. Oliveira, C.A.; Rosmaninho, J.F.; Butkeraitis, P.; Corrêa, B.; Reis, T.A.; Guerra, J.L.; Albuquerque, R.; Moro, M.E. Effect of low levels of dietary aflatoxin B1 on laying Japanese quail. Poult. Sci. 2002, 81, 976–980.
127. Chen, X.; Horn, N.; Cotter, P.F.; Applegate, T.J. Growth, serum biochemistry, complement activity, and liver gene expression responses of Pekin ducklings to graded levels of cultured aflatoxin B1. Poult. Sci. 2014, 93, 2028–2036.
128. Azzam, A.H.; Gabal, M.A. Aflatoxin and immunity in layer hens. Avian Pathol. 1998, 27, 570–577.
129. Garlich, J.D.; Tung, H.T.; Hamilton, P.B. The effects of short term feeding of aflatoxin on egg production and some plasma constituents of the laying hen. Poult. Sci. 1973, 52, 2206–2211.
130. Howarth, B., Jr.; Wyatt, R.D. Effect of dietary aflatoxin on fertility, hatchability, and progeny performance of broiler breeder hens. Appl. Environ. Microbiol. 1976, 31, 680–684.

131. Khan, W.A.; Khan, M.Z.; Khan, A.; Hassan, Z.U.; Rafique, S.; Saleemi, M.K.; Ahad, A. Dietary vitamin E in White Leghorn layer breeder hens: A strategy to combat aflatoxin B1-induced damage. Avian Pathol. 2014, 43, 389–395.
132. Lee, J.T.; Jessen, K.A.; Beltran, R.; Starkl, V.; Schatzmayr, G.; Borutova, R.; Caldwell, D.J. Effects of mycotoxin-contaminated diets and deactivating compound in laying hens: 2. Effects on white shell egg quality and characteristics. Poult. Sci. 2012, 91, 2096–2104.
133. Oliveira, C.A.F.; Rosmaninho, J.F.; Castro, A.L.; Butkeraitis, P.; Alves Reis, T.; Corrêa, B. Aflatoxin residues in eggs of laying Japanese quail after long-term administration of rations containing low levels of aflatoxin B1. Food Addit. Contam. 2003, 20, 648–653.
134. Ortatatli, M.; Oğuz, H. Ameliorative effects of dietary clinoptilolite on pathological changes in broiler chickens during aflatoxicosis. Res. Vet. Sci. 2001, 71, 59–66.
135. Merkley, J.W.; Maxwell, R.J.; Phillips, J.G.; Huff, W.E. Hepatic fatty acid profiles in aflatoxin-exposed broiler chickens. Poult. Sci. 1987, 66, 59–67.
136. Siloto, E.V.; Oliveira, E.F.; Sartori, J.R.; Fascina, V.B.; Martins, B.A.; Ledoux, D.R.; Rottinghaus, G.E.; Sartori, D.R. Lipid metabolism of commercial layers fed diets containing aflatoxin, fumonisin, and a binder. Poult. Sci. 2013, 92, 2077–2083.
137. Klein, P.J.; van Vleet, T.R.; Hall, J.O.; Coulombe, R.A., Jr. Dietary butylated hydroxytoluene protects against aflatoxicosis in turkeys. Toxicol. Appl. Pharmacol. 2002, 182, 11–19.
138. Ortatatli, M.; Oğuz, H.; Hatipoğlu, F.; Karaman, M. Evaluation of pathological changes in broilers during chronic aflatoxin (50 and 100 ppb) and clinoptilolite exposure. Res. Vet. Sci. 2005, 78, 61–68.
139. Chen, K.; Fang, J.; Peng, X.; Cui, H.; Chen, J.; Wang, F.; Chen, Z.; Zuo, Z.; Deng, J.; Lai, W.; Zhou, Y. Effect of selenium supplementation on aflatoxin B1-induced histopathological lesions and apoptosis in bursa of Fabricius in broilers. Food Chem. Toxicol. 2014, 74, 91–97.
140. Kubena, L.F.; Harvey, R.B.; Bailey, R.H.; Buckley, S.A.; Rottinghaus, G.E. Effects of a hydrated sodium calcium aluminosilicate (T-Bind) on mycotoxicosis in young broiler chickens. Poult. Sci. 1998, 77, 1502–1509.
141. Tung, H.T.; Wyatt, R.D.; Thaxton, P.; Hamilton, P.B. Concentrations of serum proteins during aflatoxicosis. Toxicol. Appl. Pharmacol. 1975, 34, 320–326.
142. Fernández, A.; Verde, M.T.; Gomez, J.; Gascon, M.; Ramos, J.J. Changes in the prothrombin time, haematology and serum proteins during experimental aflatoxicosis in hens and broiler chickens. Res. Vet. Sci. 1995, 58, 119–122.
143. Doerr, J.A.; Huff, W.E.; Tung, H.T.; Wyatt, R.D.; Hamilton, P.B. A survey of T-2 toxin, ochratoxin, and aflatoxin for their effects on the coagulation of blood in young broiler chickens. Poult. Sci. 1974, 53, 1728–1734.
144. Doerr, J.A.; Wyatt, R.D.; Hamilton, P.B. Impairment of coagulation function during aflatoxicosis in young chickens. Toxicol. Appl. Pharmacol. 1976, 35, 437–446.
145. Doerr, J.A.; Hamilton, P.B. Aflatoxicosis and intrinsic coagulation function in broiler chickens. Poult. Sci. 1981, 60, 1406–1411.
146. Witlock, D.R.; Wyatt, R.D. Effect of dietary aflatoxin on hemostasis of young turkey poults. Poult. Sci. 1981, 60, 528–531.
147. Donaldson, W.E.; Tung, H.T.; Hamilton, P.B. Depression of fatty acid synthesis in chick liver (Gallus domesticus) by aflatoxin. Comp. Biochem. Physiol. B 1972, 41, 843–847.
148. Hoerr, F.J. Clinical aspects of immunosuppression in poultry. Avian Dis. 2010, 54, 2–15.
149. Chen, K.; Shu, G.; Peng, X.; Fang, J.; Cui, H.; Chen, J.; Wang, F.; Chen, Z.; Zuo, Z.; Deng, J.; Geng, Y.; Lai, W. Protective role of sodium selenite on histopathological lesions decreased T-cell subsets and increased apoptosis of thymus in broilers intoxicated with aflatoxin B1. Food Chem. Toxicol. 2013, 59, 446–454.
150. Chen, K.; Peng, X.; Rang, J.; Cui, H.; Zuo, Z.; Deng, J.; Chen, Z.; Geng, Y.; Lai, W.; Tang, L.; Yang, Q. Effects of dietary selenium on histopathological changes and T cells of

spleen in broilers exposed to aflatoxin B1. Int. J. Environ. Res. Public Health 2014, 11, 1904–1913.
151. Pier, A.C.; Heddleston, K.L.; Cysewski, S.J.; Patteron, J.M. Effect of aflatoxin on immunity in turkeys. II. Reversal of impaired resistance to bacterial infection by passive transfer of plasma. Avian Dis. 1972, 16, 381–387.
152. Thaxton, J.P.; Tung, H.T.; Hamilton, P.B. Immunosuppression in chickens by aflatoxin. Poult. Sci. 1974, 53, 721–725.
153. Edrington, T.S.; Kubena, L.F.; Harvey, R.B.; Rottinghaus, G.E. Influence of a superactivated charcoal on the toxic effects of aflatoxin or T-2 toxin in growing broilers. Poult. Sci. 1997, 76, 1205–1211.
154. Huff, W.E.; Kubena, L.F.; Harvey, R.B.; Hagler, W.M., Jr.; Swanson, S.P.; Phillips, T.D.; Creger, C.R. Individual and combined effects of aflatoxin and deoxynivalenol (DON, vomitoxin) in broiler chickens. Poult. Sci. 1986, 65, 1291–1298.
155. Huff, W.E.; Harvey, R.B.; Kubena, L.F.; Rottinghaus, G.E. Toxic synergism between aflatoxin and T-2 toxin in broiler chickens. Poult. Sci. 1988, 67, 1418–1423.
156. Peng, X.; Zhang, K.; Bai, S.; Ding, X.; Zeng, Q.; Yang, J.; Fang, J.; Chen, K. Histological lesions, cell cycle arrest, apoptosis and T cell subsets changes of spleen in chicken fed aflatoxin-contaminated corn. Int. J. Environ. Res. Public Health 2014, 11, 8567–8580.
157. Chang, C.F.; Hamilton, P.B. Impaired phagocytosis by heterophils from chickens during aflatoxicosis. Toxicol. Appl. Pharmacol. 1979, 48, 459–466.
158. Ibrahim, I.K.; Shareef, A.M.; Al-Joubory, K.M. Ameliorative effects of sodium bentonite on phagocytosis and Newcastle disease antibody formation in broiler chickens during aflatoxicosis. Res. Vet. Sci. 2000, 69, 119–122.
159. Ghosh, R.C.; Chauhan, H.V.; Jha, G.J. Suppression of cell-mediated immunity by purified aflatoxin B1 in broiler chicks. Vet. Immunol. Immunopathol. 1991, 28, 165–172.
160. Neldon-Ortiz, D.L.; Qureshi, M.A. Effects of AFB1 embryonic exposure on chicken mononuclear phagocytic cell functions. Dev. Comp. Immunol. 1992, 16, 187–196.
161. Qureshi, M.A.; Brake, J.; Hamilton, P.B.; Hagler, W.M., Jr.; Nesheim, S. Dietary exposure of broiler breeders to aflatoxin results in immune dysfunction in progeny chicks. Poult. Sci. 1998, 77, 812–819.
162. Chang, C.F.; Hamilton, P.B. Impairment of phagocytosis in chicken monocytes during aflatoxicosis. Poult. Sci. 1979, 58, 562–566.
163. Neldon-Ortiz, D.L.; Qureshi, M.A. Direct and microsomal activated aflatoxin B1 exposure and its effects on turkey peritoneal macrophage functions in vitro. Toxicol. Appl. Pharmacol. 1991, 109, 432–442.
164. Neldon-Ortiz, D.L.; Qureshi, M.A. The effects of direct and microsomal activated aflatoxin B1 on chicken peritoneal macrophages in vitro. Vet. Immunol. Immunopathol. 1992, 31, 61–76.
165. Chang, C.F.; Hamilton, P.B. Refractory phagocytosis by chicken thrombocytes during aflatoxicosis. Poult. Sci. 1979, 58, 559–561.
166. Ghosh, R.C.; Chauhan, H.V.; Roy, S. Immunosuppression in broilers under experimental aflatoxicosis. Br. Vet. J. 1990, 146, 457–462.
167. Giambrone, J.J.; Ewert, D.L.; Wyatt, R.D.; Eidson, C.S. Effect of aflatoxin on the humoral and cell-mediated immune systems of the chicken. Am. J. Vet. Res. 1978, 39, 305–308.
168. Chen, K.; Yuan, S.; Chen, J.; Peng, X.; Wang, F.; Cui, H.; Fang, J. Effects of sodium selenite on the decreased percentage of T cell subsets, contents of serum IL-2 and IFN-γ induced by aflatoxin B1 in broilers. Res. Vet. Sci. 2013, 95, 143–145.
169. Chen, J.; Chen, K.; Yuan, S.; Peng, X.; Fang, J.; Wang, F.; Cui, H.; Chen, Z.; Yuan, J.; Geng, Y. Effects of aflatoxin B1 on oxidative stress markers and apoptosis in spleens in broilers. Toxicol. Ind. Health 2013.
170. Wang, F.; Shu, G.; Peng, X.; Fang, J.; Chen, K.; Cui, H.; Chen, Z.; Zuo, Z.; Deng, J.; Gene, Y.; Lai, W. Protective effects of sodium selenite against aflatoxin-B1 induced

oxidative stress and apoptosis in broiler spleen. Int. J. Environ. Res. Public Health 2013, 10, 2834–2844.

171. He, Y.; Fang, J.; Peng, X.; Cui, H.; Zuo, Z.; Deng, J.; Chen, Z.; Lai, W.; Shu, G.; Tang, L. Effects of sodium selenite on aflatoxin B1-induced decrease of ileac T cell and the mRNA contents of IL-2, IL-6, and TNF-α in broilers. Biol. Trace Elem. Res. 2014, 159, 167–173.

172. Jiang, M.; Peng, X.; Fang, J.; Cui, H.; Yu, Z.; Chen, Z. Effects of aflatoxin B1 on T-cell subsets and mRNA expression of cytokines in the intestine of broilers. Int. J. Mol. Sci. 2015, 16, 6945–6959.

173. Campbell, M.L., Jr.; May, J.D.; Huff, W.E.; Doerr, J.A. Evaluation of immunity of young broiler chickens during simultaneous aflatoxicosis and ochratoxicosis. Poult. Sci. 1983, 62, 2138–2144.

174. Pier, A.C.; Heddleston, K.L.; Boney, W.A.; Lukert, P.K. The effect of aflatoxin on immunity. Proc. XIX Con. Mund. Med. Vet. Zootech. 1971, 1, 216–219.

175. Stewart, R.G.; Skeeles, J.K.; Wyatt, R.D.; Brown, J.; Page, R.K.; Russell, I.D.; Lukert, P.D. The effect of aflatoxin on complement activity in broiler chickens. Poult. Sci. 1985, 64, 616–619.

176. Azzam, A.H.; Gabal, M.A. Interaction of aflatoxin in the feed and immunization against selected infectious diseases. I. Infectious bursal disease. Avian Pathol. 1997, 26, 317–325.

177. Gabal, M.A.; Azzam, A.H. Interaction of aflatoxin in the feed and immunization against selected infectious diseases in poultry. II. Effect on one-day-old layer chicks simultaneously vaccinated against Newcastle disease, infectious bronchitis and infectious bursal disease. Avian Pathol. 1998, 27, 290–295.

178. Boulton, S.L.; Dick, J.W.; Hughes, B.L. Effects of dietary aflatoxin and ammonia-inactivated aflatoxin on Newcastle disease antibody titers in layer-breeders. Avian Dis. 1982, 26, 1–6.

179. Hegazy, S.M.; Azzam, A.; Gabal, M.A. Interaction of naturally occurring aflatoxins in poultry feed and immunization against fowl cholera. Poult. Sci. 1991, 70, 2425–2428.

180. Pier, A.C.; Heddleston, K.L. The effect of aflatoxin on immunity in turkeys. I. Impairment of actively acquired resistance to bacterial challenge. Avian Dis. 1970, 14, 797–809.

181. Pier, A.C. Effects of aflatoxin on immunity. J. Am. Vet. Med. Assoc. 1973, 163, 1268–1269.

182. Boonchuvit, B.; Hamilton, P.B. Interaction of aflatoxin and paratyphoid infection in broiler chickens. Poult. Sci. 1975, 54, 1567–1573.

183. Hegazy, S.M.; Adachi, Y. Comparison of the effects of dietary selenium, zinc, and selenium and zinc supplementation on growth and immune response between chick groups that were inoculated with Salmonella and aflatoxin or Salmonella. Poult. Sci. 2000, 79, 331–335.

184. Wyatt, R.D.; Hamilton, P.B. Interaction between aflatoxicosis and a natural infection of chickens with Salmonella. Appl. Microbiol. 1975, 30, 870–872.

185. Edds, G.T.; Nair, K.P.; Simpson, C.F. Effect of aflatoxin B1 on resistance in poultry against cecal coccidiosis and Marek's disease. Am. J. Vet. Res. 1973, 34, 819–826.

186. Edds, G.T.; Simpson, C.F. Cecal coccidiosis in poultry as affected by prior exposure to aflatoxin B1. Am. J. Vet. Res.1976, 37, 65–68.

187. Witlock, D.R.; Wyatt, R.D.; Anderson, W.I. Relationship between Eimeria adenoeides infection and aflatoxicosis in turkey poults. Poult. Sci. 1982, 61, 1293–1297.

188. Wyatt, R.D.; Ruff, M.D.; Page, R.K. Interaction of aflatoxin with Eimeria tenella infection and monensin in young broiler chickens. Avian Dis. 1975, 19, 730–740.

189. Hamilton, P.B.; Harris, J.R. Interaction of aflatoxicosis with Candida albicans infections and other stresses in chickens. Poult. Sci. 1971, 50, 906–912.

190. Chang, C.F.; Hamilton, P.B. Increased severity and new symptoms of infectious bursal disease during aflatoxicosis in broiler chickens. Poult. Sci. 1982, 61, 1061–1068.

191. Giambrone, J.J.; Partadiredja, M.; Eidson, C.S.; Kleven, S.H.; Wyatt, R.D. Interaction of aflatoxin with infectious bursal disease virus infection in young chickens. Avian Dis. 1978, 22, 431–439.
192. Batra, P.; Pruthi, A.K.; Sadana, J.R. Effect of aflatoxin B1 on the efficacy of turkey herpesvirus vaccine against Marek's disease. Res. Vet. Sci. 1991, 51, 115–119.
193. Tessari, E.N.; Oliveira, C.A.; Cardoso, A.L.; Ledoux, D.R.; Rottinghaus, G.E. Effects of aflatoxin B1 and fumonisin B1on body weight, antibody titres and histology of broiler chicks. Br. Poult. Sci. 2006, 47, 357–364.
194. Yunus, A.W.; Ghareeb, K.; Abd-El-Fattah, A.A.; Twaruzek, M.; Böhm, J. Gross intestinal adaptations in relation to broiler performance during chronic aflatoxin exposure. Poult. Sci. 2011, 90, 1683–1689.
195. Yunus, A.W.; Razzazi-Fazeli, E.; Bohm, J. Aflatoxin B1 in affecting broiler's performance, immunity, and gastrointestinal tract: A review of history and contemporary issues. Toxins 2011, 3, 566–590.
196. Applegate, T.J.; Schatzmayr, G.; Prickel, K.; Troche, C.; Jiang, Z. Effect of aflatoxin culture on intestinal function and nutrient loss in laying hens. Poult. Sci. 2009, 88, 1235–1241.
197. Yang, J.; Bai, F.; Zhang, K.; Lv, X.; Bai, S.; Zhao, L.; Peng, X.; Ding, X.; Li, Y.; Zhang, J. Effects of feeding corn naturally contaminated with AFB1 and AFB2 on performance and aflatoxin residues in broilers. Czech. J. Anim. Sci. 2012, 57, 506–515.
198. Ledoux, D.R.; Rottinghaus, G.E.; Bermudez, A.J.; Alonso-Debolt, M. Efficacy of a hydrated sodium calcium aluminosilicate to ameliorate the toxic effects of aflatoxin in broiler chicks. Poult. Sci. 1999, 78, 204–210.
199. Hamilton, P.B.; Garlich, J.D. Failure of vitamin supplementation to alter the fatty liver syndrome caused by aflatoxin. Poult. Sci. 1972, 51, 688–692.
200. Hamilton, P.B.; Tung, H.T.; Wyatt, R.D.; Donaldson, W.E. Interaction of dietary aflatoxin with some vitamin deficiencies. Poult. Sci. 1974, 53, 871–877.
201. Rogers, S.R.; Pesti, G.M.; Wyatt, R.D. Effect of tryptophan supplementation on aflatoxicosis in laying hens. Poult. Sci.1991, 70, 307–312.
202. Hamilton, P.B.; Tung, H.T.; Harris, J.R.; Gainer, J.H.; Donaldson, W.E. The effect of dietary fat on aflatoxicosis in turkeys. Poult. Sci. 1972, 51, 165–170.
203. Richardson, K.E.; Nelson, L.A.; Hamilton, P.B. Effect of dietary fat level on dose response relationships during aflatoxicosis in young chickens. Poult. Sci. 1987, 66, 1470–1478.
204. Smith, J.W.; Hill, C.H.; Hamilton, P.B. The effect of dietary modifications on aflatoxicosis in the broiler chicken. Poult. Sci. 1971, 50, 768–774.
205. Kermanshahi, H.; Akbari, M.R.; Maleki, M.; Behgar, M. Effect of prolonged low level inclusion of aflatoxin B1 into diet on performance, nutrient digestibility, histopathology and blood enzymes of broiler chickens. J. Anim. Vet. Adv.2007, 6, 686–692.
206. Verma, J.; Swain, B.K.; Johri, T.S. Effect of various levels of aflatoxin and ochratoxin A and combinations thereof on protein and energy utilisation in broilers. J. Sci. Food Agric. 2002, 82, 1412–1417.
207. Verma, J.; Johri, T.S.; Swain, B.K. Effect of aflatoxin, ochratoxin and their combination on protein and energy utilisation in white leghorn laying hens. J. Sci. Food Agric. 2007, 87, 760–764.
208. Ruff, M.D.; Wyatt, R.D. Intestinal absorption of L-methionine and glucose in chickens with aflatoxicosis. Toxicol. Appl. Pharmacol. 1976, 37, 257–262.
209. Oliveira, C.A.; Kobashigawa, E.; Reis, T.A.; Mestieri, L.; Albuquerque, R.; Corrêa, B. Aflatoxin B1 residues in eggs of laying hens fed a diet containing different levels of the mycotoxin. Food Addit. Contam. 2000, 17, 459–462.
210. Celik, I.; Oguz, H.; Demet, O.; Boydak, M.; Donmez, H.H.; Sur, E.; Nizamlioglu, F. Embryotoxicity assay of aflatoxin produced by Aspergillus parasiticus NRRL 2999. Br. Poul. Sci. 2000, 41, 401–409.

211. Dietert, R.R.; Qureshi, M.A.; Nanna, U.C.; Bloom, S.E. Embryonic exposure to aflatoxin-B1: Mutagenicity and influence on development and immunity. Environ. Mutagen. 1985, 7, 715–725.
212. Edrington, T.S.; Harvey, R.B.; Kubena, L.F. Toxic effects of aflatoxin B1 and ochratoxin A, alone and in combination, on chicken embryos. Bull. Environ. Contam. Toxicol. 1995, 54, 331–336.
213. Oznurlu, Y.; Celik, I.; Sur, E.; Ozaydın, T.; Oğuz, H.; Altunbaş, K. Determination of the effects of aflatoxin B1 given in ovo on the proximal tibial growth plate of broiler chickens: Histological, histometric and immunohistochemical findings. Avian Pathol. 2012, 41, 469–477.
214. Sur, E.; Celik, I. Effects of aflatoxin B1 on the development of the bursa of Fabricius and blood lymphocyte acid phosphatase of the chicken. Br. Poult. Sci. 2003, 44, 558–566.
215. Williams, J.G.; Deschl, U.; William, G.M. DNA damage in fetal liver cells of turkey and chicken eggs dosed with aflatoxin B1. Arch. Toxicol. 2011, 85, 1167–1172.
216. Perrone, C.E.; Ahr, H.J.; Duan, J.D.; Jeffrey, A.M.; Schmidt, U.; Williams, G.M.; Enzmann, H.H. Embryonic turkey liver: Activities of biotransformation enzymes and activation of DNA-reactive carcinogens. Arch. Toxicol. 2004, 78, 589–598.
217. Sur, E.; Celik, I.; Oznurlu, Y.; Aydin, M.F.; Oguz, H.; Kurtoglu, V.; Ozaydin, T. Enzyme histochemical and serological investigations on the immune system from chickens treated in ovo with aflatoxin B1 (AFB1). Revue Méd. Vét. 2011, 162, 443–448.
218. Ul-Hassan, Z.; Khan, M.Z.; Khan, A.; Javed, I. Immunological status of the progeny of breeder hens kept on ochratoxin A (OTA)- and aflatoxin B1 (AFB1)-contaminated feeds. J. Immunotoxicol. 2012, 9, 381–391.
219. Lilleberg, S.L.; Cabonce, M.A.; Raju, N.R.; Wagner, L.M.; Kier, L.D. Alterations in the structural gene and the expression of p53 in rat liver tumors induced by aflatoxin B1. Mol. Carcinog. 1992, 6, 159–172.
220. Harris, A.J.; Shaddock, J.G.; Manjanatha, M.G.; Lisenbey, J.A.; Casciano, D.A. Identification of differentially expressed genes in aflatoxin B1-treated cultured primary rat hepatocytes and Fischer 344 rats. Carcinogenesis 1998, 19, 1451–1458.
221. Merrick, B.A.; Phadke, D.P.; Auerbach, S.S.; Mav, D.; Stiegelmeyer, S.M.; Shah, R.R.; Tice, R.R. RNA-Seq profiling reveals novel hepatic gene expression pattern in aflatoxin B1 treated rats. PLoS ONE 2013, 8, e61768.
222. Dugyala, R.R.; Sharma, R.P. The effect of aflatoxin B1 on cytokine mRNA and corresponding protein levels in peritoneal macrophages and splenic lymphocytes. Int. J. Immunopharmacol. 1996, 18, 599–608.
223. Han, S.H.; Jeon, Y.J.; Yea, S.S.; Yang, K.H. Suppression of the interleukin-2 gene expression by aflatoxin B1 is mediated through the down-regulation of the NF-AT and AP-1 transcription factors. Toxicol. Lett. 1999, 108, 1–10.
224. Hinton, D.M.; Myers, M.J.; Raybourne, R.A.; Francke-Carroll, S.; Sotomayor, R.E.; Shaddock, J.; Warbritton, A.; Chou, M.W. Immunotoxicity of aflatoxin B1 in rats: Effects on lymphocytes and the inflammatory response in a chronic intermittent dosing study. Toxicol. Sci. 2003, 73, 362–377.
225. Meissonnier, G.M.; Pinton, P.; Laffitte, J.; Cossalter, A.M.; Gong, Y.Y.; Wild, C.P.; Bertin, G.; Galtier, P.; Oswald, I.P. Immunotoxicity of aflatoxin B1: Impairment of the cell-mediated response to vaccine antigen and modulation of cytokine expression. Toxicol. Appl. Pharmacol. 2008, 231, 142–149.
226. Qian, G.; Tang, L.; Guo, X.; Wang, F.; Massey, M.E.; Su, J.; Guo, T.L.; Williams, J.H.; Phillips, T.D.; Wang, J.S. Aflatoxin B1 modulates the expression of phenotypic markers and cytokines by splenic lymphocytes of male F344 rats. J. Appl. Toxicol. 2014, 34, 241–249.
227. Murarolli, R.A. Effects of Aflatoxin B1 (AFB1) on Hepatic Gene Expression in Pigs and Turkeys. Doctoral Dissertation, University of Missouri, Columbia, MO, USA, 2013.
228. Tilton, S.C.; Gerwick, L.G.; Hendricks, J.D.; Rosato, C.S.; Corley-Smith, G.; Givan, S.A.; Bailey, G.S.; Bayne, C.J.; Williams, D.E. Use of a rainbow trout oligonucleotide

microarray to determine transcriptional patterns in aflatoxin B1-induced hepatocellular carcinoma compared to adjacent liver. Toxicol. Sci. 2005, 88, 319–330.
229. Afanassieff, M.; Goto, R.M.; Ha, J.; Sherman, M.A.; Zhong, L.; Auffray, C.; Coudert, F.; Zoorob, R.; Miller, M.M. At least one class I gene in restriction fragment pattern-Y (Rfp.-Y.), the second MHC gene cluster in the chicken, is transcribed, polymorphic, and shows divergent specialization in antigen binding region. J. Immunol. 2001, 166, 3324–3333.
230. Bauer, M.M.; Reed, K.M. Extended sequence of the turkey MHC B-locus and sequence variation in the highly polymorphic B-G loci. Immunogenetics 2011, 63, 209–221.
231. Chaves, L.D.; Krueth, S.B.; Bauer, M.M.; Reed, K.M. Sequence of a turkey BAC clone identifies MHC class III orthologs and supports ancient origins of immunological gene clusters. Cytogenet. Genome Res. 2011, 132, 55–63.
232. Chaves, L.D.; Krueth, S.B.; Reed, K.M. Characterization of the turkey MHC chromosome through genetic and physical mapping. Cytogenet. Genome Res. 2007, 117, 213–220.
233. Chaves, L.D.; Krueth, S.B.; Reed, K.M. Defining the turkey MHC: Sequence and genes of the B locus. J. Immunol. 2009, 183, 6530–6537.
234. Delany, M.E.; Robinson, C.M.; Goto, R.M.; Miller, M.M. Architecture and organization of chicken microchromosome 16: Order of the NOR, MHC-Y, and MHC-B subregions. J. Hered. 2009, 100, 507–514.
235. Kaufman, J.; Milne, S.; Göbel, T.W.F.; Walker, B.A.; Jacob, J.P.; Auffray, C.; Zoorob, R.; Beck, S. The chicken B locus is a minimal essential major histocompatibility complex. Nature 1999, 401, 923–925.
236. Miller, M.M.; Goto, R.; Bernot, A.; Zoorob, R.; Auffray, C.; Bumstead, N.; Briles, W.E. Two Mhc class I and two Mhc class II genes map to the chicken Rfp.-Y. system outside the B complex. Proc. Natl. Acad. Sci. USA 1994, 91, 4397–4401.
237. Reed, K.M.; Bauer, M.M.; Monson, M.S.; Benoit, B.; Chaves, L.D.; O'Hare, T.H.; Delany, M.E. Defining the turkey MHC: Identification of expressed class I- and class IIB-like genes independent of the MHC-B. Immunogenetics 2011, 63, 753–771.
238. Monson, M.S.; Settlage, R.E.; Mendoza, K.M.; Rawal, S.; El-Nezami, H.S.; Coulombe, R.A.; Reed, K.M. Modulation of the spleen transcriptome in domestic turkey (Meleagris gallopavo) in response to aflatoxin B1 and probiotics. Immunogenetics 2015, 67, 163–178.
239. Rustemeyer, S.M.; Lamberson, W.R.; Ledoux, D.R.; Wells, K.; Austin, K.J.; Cammack, K.M. Effects of dietary aflatoxin on the hepatic expression of apoptosis genes in growing barrows. J. Anim. Sci. 2005, 89, 916–925.
240. Monson, M.S.; Settlage, R.E.; McMahon, K.W.; Mendoza, K.M.; Rawal, S.; el-Nezami, H.S.; Coulombe, R.A.; Reed, K.M. Response of the hepatic transcriptome to aflatoxin B1 in domestic turkey (Meleagris gallopavo). PLoS ONE 2014, 9, e100930.
241. Bothast, R.J.; Nofsinger, G.W.; Lagoda, A.A.; Black, L.T. Integrated process for ammonia inactivation of aflatoxin-contaminated corn and ethanol fermentation. Appl. Environ. Microbiol. 1982, 43, 961–963.
242. Brekke, O.L.; Sinnhuber, R.O.; Peplinski, A.J.; Wales, J.H.; Putnam, G.B.; Lee, D.J.; Ciegler, A. Aflatoxin in corn: Ammonia inactivation and bioassay with rainbow trout. Appl. Environ. Microbiol. 1977, 34, 34–37.
243. Jalili, M.; Jinap, S.; Son, R. The effect of chemical treatment on reduction of aflatoxins and ochratoxin a in black and white pepper during washing. Food Addit. Contam. Part A Chem. Anal. Control. Expo. Risk Assess. 2011, 28, 485–493.
244. Mann, G.E.; Gardner, H.K.; Booth, A.N.; Gumbmann, M.R. Aflatoxin inactivation. Chemical and biological properties of ammonia and methylamine treated cottonseed meal. J. Agric. Food Chem. 1971, 19, 1155–1158.
245. Vesonder, R.F.; Beckwith, A.C.; Ciegler, A.; Dimler, R.J. Ammonium hydroxide treatment of aflatoxin B1. Some chemical characteristics and biological effects. J. Agric. Food Chem. 1975, 23, 242–243.

246. Weng, C.Y.; Martinez, A.J.; Park, D.L. Efficacy and permanency of ammonia treatment in reducing aflatoxin levels in corn. Food Addit. Contam. 1994, 11, 649–658.
247. Codifer, L.P., Jr.; Mann, G.E.; Dollear, F.G. Aflatoxin inactivation: Treatment of peanut meal with formaldehyde and calcium hydroxide. J. Am. Oil Chem. Soc. 1976, 53, 204–206.
248. Patel, U.D.; Govindarajan, P.; Dave, P.J. Inactivation of aflatoxin B1 by using the synergistic effect of hydrogen peroxide and gamma radiation. Appl. Environ. Microbiol. 1989, 55, 465–467.
249. Moerck, K.E.; McElfresh, P.; Wohlman, A.; Hilton, B.W. Aflatoxin destruction in corn using sodium bisulfite, sodium hydroxide and aqueous ammonia. J. Food Prot. 1980, 43, 571–574.
250. Yang, C.Y. Comparative studies on the detoxification of aflatoxins by sodium hypochlorite and commercial bleaches. Appl. Microbiol. 1972, 24, 885–890.
251. Gregory, J.F., 3rd; Edds, G.T. Effect of dietary selenium on the metabolism of aflatoxin B1 in turkeys. Food Chem. Toxicol. 1984, 22, 637–642.
252. Diaz, G.J.; Cortés, A.; Botero, L. Evaluation of the ability of a feed additive to ameliorate the adverse effects of aflatoxins in turkey poults. Br. Poult. Sci. 2009, 50, 240–250.
253. Kubena, L.F.; Harvey, R.B.; Huff, W.E.; Corrier, D.E.; Phillips, T.D.; Rottinghaus, G.E. Efficacy of a hydrated sodium calcium aluminosilicate to reduce the toxicity of aflatoxin and T-2 toxin. Poult. Sci. 1990, 69, 1078–1086.
254. Kubena, L.F.; Harvey, R.B.; Phillips, T.D.; Corrier, D.E.; Huff, W.E. Diminution of aflatoxicosis in growing chickens by the dietary addition of a hydrated, sodium calcium aluminosilicate. Poult. Sci. 1990, 69, 727–735.
255. Kubena, L.F.; Huff, W.E.; Harvey, R.B.; Yersin, A.G.; Elissalde, M.H.; Witzel, D.A.; Giroir, L.E.; Phillips, T.D.; Petersen, H.D. Effects of a hydrated sodium calcium aluminosilicate on growing turkey poults during aflatoxicosis. Poult. Sci.1991, 70, 1823–1830.
256. Kubena, L.F.; Harvey, R.B.; Phillips, T.D.; Clement, B.A. Effect of hydrated sodium calcium aluminosilicates on aflatoxicosis in broiler chicks. Poult. Sci. 1993, 72, 651–657.
257. Zhao, J.; Shirley, R.B.; Dibner, J.D.; Uraizee, F.; Officer, M.; Kitchell, M.; Vazquez-Anon, M.; Knight, C.D. Comparison of hydrated sodium calcium aluminosilicate and yeast cell wall on counteracting aflatoxicosis in broiler chicks. Poult. Sci. 2010, 89, 2147–2156.
258. Magnoli, A.P.; Monge, M.P.; Miazzo, R.D.; Cavaglieri, L.R.; Magnoli, C.E.; Merkis, C.I.; Cristofolini, A.L.; Dalcero, A.M.; Chiacchiera, S.M. Effect of low levels of aflatoxin B1 on performance, biochemical parameters, and aflatoxin B1in broiler liver tissues in the presence of monensin and sodium bentonite. Poult. Sci. 2011, 90, 48–58.
259. Coulombe, R.A.; Guarisco, J.A.; Klein, P.J.; Hall, J.O. Chemoprevention of aflatoxicosis in poultry by dietary butylated hydroxytoluene. Anim. Feed Sci. Technol. 2005, 121, 217–225.
260. Guarisco, J.A.; Hall, J.O.; Coulombe, R.A., Jr. Mechanisms of butylated hydroxytoluene chemoprevention of aflatoxicosis-inhibition of aflatoxin B1 metabolism. Toxicol. Appl. Pharmacol. 2008, 227, 339–346.
261. Guarisco, J.A.; Hall, J.O.; Coulombe, R.A., Jr. Butylated hydroxytoluene chemoprevention of aflatoxicosis—Effects on aflatoxin B1 bioavailability, hepatic DNA adduct formation, and biliary excretion. Food Chem. Toxicol. 2008, 46, 3727–3731.
262. Klein, P.J.; van Vleet, T.R.; Hall, J.O.; Coulombe, R.A., Jr. Effects of dietary butylated hydroxytoluene on aflatoxin B1-relevant metabolic enzymes in turkeys. Food Chem. Toxicol. 2003, 41, 671–678.
263. Larsen, C.; Ehrich, M.; Driscoll, C.; Gross, W.B. Aflatoxin-antioxidant effects on growth of young chicks. Poult. Sci.1985, 64, 2287–2291.

264. Gowda, N.K.; Ledoux, D.R.; Rottinghaus, G.E.; Bermudez, A.J.; Chen, Y.C. Antioxidant efficacy of curcuminoids from turmeric (Curcuma longa L.) powder in broiler chickens fed diets containing aflatoxin B1. Br. J. Nutr. 2009, 102, 1629–1634.
265. Bagherzadeh Kasmani, F.; Karimi Torshizi, M.A.; Allameh, A.; Shariatmadari, F. A novel aflatoxin-binding Bacillus probiotic: Performance, serum biochemistry, and immunological parameters in Japanese quail. Poult. Sci. 2012, 91, 1846–1853.
266. Oatley, J.T.; Rarick, M.D.; Ji, G.E.; Linz, J.E. Binding of aflatoxin B1 to Bifidobacteria in vitro. J. Food Prot. 2000, 63, 1133–1136.
267. Peltonen, K.; El-Nezami, H.; Haskard, C.; Ahokas, J.; Salminen, S. Aflatoxin B1 binding by dairy strains of lactic acid bacteria and Bifidobacteria. J. Dairy Sci. 2001, 84, 2152–2156.
268. Shahin, A.A.M. Removal of aflatoxin B1 from contaminated liquid media by dairy lactic acid bacteria. Int. J. Agric. Biol. 2007, 9, 71–75.
269. Topcu, A.; Bulat, T.; Wishah, R.; Boyac, I.H. Detoxification of aflatoxin B1 and patulin by Enterococcus faecium strains. Int. J. Food Microbiol. 2010, 139, 202–205.
270. Deabes, M.M.; Darwish, H.R.; Abdel-Aziz, K.B.; Farag, I.M.; Nada, S.A.; Tawfek, N.S. Protective effects of Lactobacillus rhamnosus GG on aflatoxins-induced toxicities in male albino mice. J. Environ. Anal. Toxicol. 2012, 2, 132.
271. El-Nezami, H.; Salminen, S.J.; Ahokas, J. Biological control of food carcinogens with use of Lactobacillus GG. Nutr. Today 1996, 21, 41S–42S.
272. El-Nezami, H.; Kankaanpää, P.; Salminen, S.; Ahokas, J. Ability of dairy strains of lactic acid bacteria to bind a common food carcinogen, aflatoxin B1. Food Chem. Toxicol. 1998, 36, 321–326.
273. El-Nezami, H.; Kankaanpää, P.; Salminen, S.; Ahokas, J. Physicochemical alterations enhance the ability of dairy strains of lactic acid bacteria to remove aflatoxin from contaminated media. J. Food Prot. 1998, 61, 466–468.
274. El-Nezami, H.; Mykkänen, H.; Kankaanpää, P.; Salminen, S.; Ahokas, J. Ability of Lactobacillus and Propionibacterium strains to remove aflatoxin B1 from the chicken duodenum. J. Food Prot. 2000, 63, 549–552.
275. Fazell, M.R.; Hajimohammadali, M.; Moshkani, A.; Samadi, N.; Jamalifar, H.; Khoshayand, M.R.; Vaghari, E.; Pouragahi, S. Aflatoxin B1 binding capacity of autochthonous strains of lactic acid bacteria. J. Food Prot. 2009, 72, 189–192.
276. Gratz, S.; Mykkänen, H.; Ouwehand, A.C.; Juvonen, R.; Salminen, S.; el-Nezami, H. Intestinal mucus alters the ability of probiotic bacteria to bind aflatoxin B1 in vitro. Appl. Environ. Microbiol. 2004, 70, 6306–6308.
277. Gratz, S.; Täubel, M.; Juvonen, R.O.; Viluksela, M.; Turner, P.C.; Mykkänen, H.; el-Nezami, H. Lactobacillus rhamnosus strain GG modulates intestinal absorption, fecal excretion and toxicity of aflatoxin B1 in rats. Appl. Environ. Microbiol.2006, 72, 7398–7400.
278. Gratz, S.; Wu, Q.K.; el-Nezami, H.; Juvonen, R.O.; Mykkänen, H.; Turner, P.C. Lactobacillus rhamnosus strain GG reduces aflatoxin B1 transport, metabolism, and toxicity in Caco-2 cells. Appl. Environ. Microbiol. 2007, 73, 3958–3964.
279. Haskard, C.; Binnion, C.; Ahokas, J. Factors affecting the sequestration of aflatoxin by Lactobacillus rhamnosus strain GG. Chem. Biol. Interact. 2000, 128, 39–49.
280. Haskard, C.A.; el-Nezami, H.S.; Kankaanpää, P.E.; Salminen, S.; Ahokas, J.T. Surface binding of aflatoxin B1 by lactic acid bacteria. Appl. Environ. Microbiol. 2001, 67, 3086–3091.
281. Hernandez-Mendoza, A.; Garcia, H.S.; Steele, J.L. Screening of Lactobacillus casei strains for their ability to bind aflatoxin B1. Food Chem. Toxicol. 2009, 47, 1064–1068.
282. Hernandez-Mendoza, A.; Guzman de Peña, D.; Garcia, H.S. Key role of teichoic acids on aflatoxin B1 binding by probiotic bacteria. J. Appl. Microbiol. 2009, 107, 395–403.
283. Hernandez-Mendoza, A.; Rivas-Jimenez, L.; Garcia, H.S. Assessment of aflatoxin B1 binding to Lactobacillus reuteri by microscopy and fluorescence techniques. Food Biotechnol. 2011, 25, 140–150.

284. Kankaanpää, P.; Tuomola, E.; El-Nezami, H.; Ahokas, J.; Salminen, S.J. Binding of aflatoxin B1 alters the adhesion properties of Lactobacillus rhamnosus strain GG in a Caco-2 model. J. Food Prot. 2000, 63, 412–414.
285. Lahtinen, S.J.; Haskard, C.A.; Ouwehand, A.C.; Salminen, S.J.; Ahokas, J.T. Binding of aflatoxin B1 to cell wall components of Lactobacillus rhamnosus strain GG. Food Addit. Contam. 2004, 21, 158–164.
286. Lee, Y.K.; el-Nezami, H.; Haskard, C.A.; Gratz, S.; Puong, K.Y.; Salminen, S.; Mykkänen, H. Kinetics of adsorption and desorption of aflatoxin B1 by viable and nonviable bacteria. J. Food Prot. 2003, 66, 426–430.
287. Peltonen, K.D.; el-Nezami, H.S.; Salminen, S.J.; Ahokas, J.T. Binding of aflatoxin B1 by probiotic bacteria. J. Sci. Food Agric. 2000, 80, 1942–1945.
288. Turbic, A.; Ahokas, J.T.; Haskard, C.A. Selective in vitro binding of dietary mutagens, individually or in combination, by lactic acid bacteria. Food Addit. Contam. 2002, 19, 144–152.
289. El-Nezami, H.; Mykkänen, H.; Kankaanpää, P.; Suomalaine, T.; Salminen, S.; Ahokas, J. Ability of a mixture of Lactobacillus and Propionibacterium to influence the faecal aflatoxin content in healthy Egyptian volunteers: A pilot clinical study. Biosci. Microflora 2000, 19, 41–45.
290. El-Nezami, H.S.; Polychronaki, N.N.; Ma, J.; Zhu, H.; Ling, W.; Salminen, E.K.; Juvonen, R.O.; Salminen, S.J.; Poussa, T.; Mykkänen, H.M. Probiotic supplementation reduces a biomarker for increased risk of liver cancer in young men from Southern China. Am. J. Clin. Nutr. 2006, 83, 1199–1203.
291. Lanza, G.M.; Washburn, K.W.; Wyatt, R.D.; Marks, H.L. Genetic variation of physiological response to aflatoxin in Gallus domesticus. Theor. Appl. Genet. 1982, 63, 207–212.
292. Lanza, G.M.; Washburn, K.W.; Wyatt, R.D.; Marks, H.L. Effect of dietary aflatoxin concentration on the assessment of genetic variability of response in a randombred population of chickens. Genetics 1983, 104, 123–131.
293. Manning, R.O.; Wyatt, R.D.; Marks, H.L.; Fletcher, O.J. Effects of dietary aflatoxin in aflatoxin-resistant and control lines of chickens. Poult. Sci. 1990, 69, 922–928.
294. Scott, T.R.; Rowland, S.M.; Rodgers, R.S.; Bodine, A.B. Genetic selection for aflatoxin B1 resistance influences chicken T-cell and thymocyte proliferation. Dev. Comp. Immunol. 1991, 15, 383–391.
295. Wyatt, R.D.; Marks, H.L.; Manning, R.O. Selection for resistance to aflatoxin in chickens. Poult. Sci. 1987, 66, 1901–1904.
296. Marks, H.L.; Wyatt, R.D. Genetic resistance to aflatoxin in Japanese quail. Science 1979, 206, 1329–1330.
297. Pegram, R.A.; Wyatt, R.D.; Marks, H.L. Comparative responses of genetically resistant and nonselected Japanese quail to dietary aflatoxin. Poult. Sci. 1985, 64, 266–272.
298. Pegram, R.A.; Wyatt, R.D.; Marks, H.L. The relationship of certain blood parameters to aflatoxin resistance in Japanese quail. Poult. Sci. 1986, 65, 1652–1658.
299. Rodeheaver, D.P.; Wyatt, R.D.; Marks, H.L. Relation of serum alpha-amylase to aflatoxin resistance in Japanese quail. Avian Dis. 1986, 30, 568–573.
300. Monson, M.S. Hepatotoxic and Immunomodulatory Transcriptome Responses to Aflatoxin B1 in the Turkey (Meleagris Gallopavo). Ph.D. Thesis, University of Minnesota, Saint Paul, MN, USA, May 2015.
301. Coulombe, R.A. USDA project: Functional Genomics to Enhance Aflatoxin Resistance in Poultry. Utah State University: Logan, UT, USA, 2015.

CHAPTER 10

Individual and Combined Effects of Subchronic Exposure of Three Fusarium Toxins (Fumonisin B, Deoxynivalenol and Zearalenone) in Rabbit Bucks

Judit Szabó-Fodor[1*], Mariam Kachlek[1], Sándor Cseh[3], Bence Somoskői[3], András Szabó[1,2], Zsófia Blochné Bodnár[2], Gábor Tornyos[2], Miklós Mézes[4], Krisztián Balogh[4], Róbert Glávits[5], Dóra Hafner[2] and Melinda Kovács[1,2]

[1]*Faculty of Agricultural and Environmental Sciences, MTA-KE Mycotoxins in the Food Chain Research Group, Kaposvár University, Hungary*

[2]*Faculty of Agricultural and Environmental Sciences, Kaposvár University, Hungary*

[3]*SZIU Faculty of Veterinary Science, István, Hungary*

[4]*SZIU Faculty of Agricultural and Environmental Sciences, Hungary*

ABSTRACT

Objective of the study was to determine reproductive toxicity of Fusarium toxins orally at three subchronic doses on adult Pannon White male rabbits. The four treatments were: control (C, toxin-free diet), F (5 mg/kg FB1), DZ (1 mg/kg DON+0.25 mg/kg ZEA), FDZ (5 mg/kg FB1+1 mg/kg DON+0.25

mg/kg ZEA) for 65 days (n=15/treatment). The doses were pre-determined according the EU limits in finished feed for young pig (in the absence of limits for rabbits' feed; based on the European Commission Recommendation 2006/576/EC and the European Commission Directive 2003/100/EC). The most pronounced effects of the toxins were exerted on the reproductive processes. The ratio of spermatozoa showing progressive forward motility decreased ($P<0.05$) from 80% to 67% in the FDZ group by day 60. Differences were found between the groups DZ (66.3% ± 23.7) and C (80.2% ± 11.2) in spermatozoa morphology. GnRH treated animals produced less testosterone in FDZ animals, compared to the other three groups ($P<0.05$). In the comet assay the individual fumonisin treatment resulted in significantly less 0 comets (intact cells), compared to all others. Based on the prevalence of score, lower ($P<0.0001$) damage was observed in FDZ group, as compared to F and DZ. Among the mycotoxins studied, additive or less than additive effect was found in case of spermatogenesis and sperm cell morphology, synergism in testosterone production, while FB1 acted antagonistically against DON+ZEA in comet assay. All mycotoxins provoked moderate lipid-peroxidation, based on the changes of glutathione concentration, glutathione peroxidase activity and formation of malondialdehyde and conjugated dienes and trienes, and exerted slight genotoxicity based on comet assay, FB1 being antagonistic towards DON+ZEA. In F, DZ and FDZ animals the intensity of spermatogenesis decreased by 43, 31 and 64%, respectively, which was reflected by lack of differentiated spermatozoa, thinning of the germinal epithelium, the appearance of multinuclear giant cells, indicative of the disturbance of meiosis and mitosis of the germinal epithelial cells and in some cases the lack of spermatogonia.

Keywords: Fumonisin; Deoxynivalenol; Zearalenon; Spermatogenesis; Testosterone production; Rabbit

INTRODUCTION

The worldwide contamination of foods and feeds with mycotoxins is a significant problem. Studies have shown extensive mycotoxin contamination in both developing and developed countries [1-3], as effect of global climate change [4].

The toxicity of mycotoxins differs depending on the kind of toxin, and in animals it is related to the species, the dose ingested, the duration of the exposure, and their gender and age. The main classes of Fusarium mycotoxins with respect to animal health and production are the non-oestrogenic trichothecenes such as **DEOXYNIVALENOL** (DON) and T-2 toxin (T-2), the myco-oestrogens including zearalenone (ZEA) and its zearalenol metabolites [5] and fumonisins.

Fumonisins (B1 and B2) are cancer-promoting metabolites of *Fusarium proliferatum* and *Fusarium verticillioides* that have a long-chain hydrocarbon unit (similar to that of sphingosine and sphinganine), which plays a role in their toxicity. **FUMONISIN** B1 (FB1) is the most toxic and has been shown to promote tumour in rats and cause equine leukoencephalomalacia and porcine pulmonary oedema. FB, the most abundant of the numerous fumonisin analogues, was classified by the IARC as a Group 2B carcinogen (possibly carcinogenic in humans) [6].

Although DON is the least toxic type of trichothecene, it can cause significant harmful effects in animals and humans [7]. DON causes a broad variety of toxic effects in animals, and its toxicity is well recognized in mammals. The main effect at the cellular level is the inhibition of protein synthesis via binding to the ribosomal subunit [8]. Chronic oral exposure induces anorexia, decreased weight gain, reduction in feed conversion, gastrointestinal hemorrhaging, inflammation and immune system alterations.

ZEA competes with the naturally produced hormone estradiol-17β for binding sites (estradiol receptors) in various organs in the body of both genders. ZEA can obstruct normal steroid hormone (estradiol, testosterone, progesterone) synthesis in the ovaries and testicles of livestock.

ZEA is found, especially, as a contaminant in corn. ZEA, which is produced mainly by *F. graminearum* and *F. culmorum* and commonly co-occurs with DON and its derivatives. It is among one of the most frequently encountered mycotoxins in grain from FHB (fusarium head blight)-diseased small-grain cereals throughout the Mediterranean countries. It may co-occur with DON in grains such as wheat, barley, oats and corn and fumonisins in corn [9]. Generally, DON is found in higher doses than ZEA when this occurs [10]. Most studies indicate that in artificial environment high moisture favours the production of both classes of mycotoxins, but the optimum temperatures for trichothecene and ZEA production in Fusarium-infected grain appears to be specific to the substrate, species and individual metabolites. The accumulated data revealed definite geographical differences in the level and frequency of DON, NIV and ZEA in wheat and barley. Streit et al. [11] reviewed mycotoxin co-occurrences in animal feed in Europe since 2004. Since Fusarium species are the most frequent fungal pathogens on field crops, it was not surprising that B-trichothecenes (DON), ZEA, and FBs were the major co-contaminants.

Reproductive efficiency is a very important economic factor in animal production. Exposure to several Fusarium mycotoxins have been linked to reproductive disorders as it was reviewed by Cortinovis et al [12]. The effect of these toxins on reproduction has been widely studied in female animals, while knowledge about their effect on reproduction processes in males is limited. As far as it is known they can affect spermatogenesis, and sperm quality, Leydig cell function (testosterone production), gonadotropin secretion, and fertility [13-18].

Humans and animals are generally exposed to multiple mycotoxins in parallel, as some mycotoxins typically co-occur in cereals. However, toxicological data, risk assessments are based on, are provided by studies in which only the individual effects of certain toxins are investigated. The simultaneous exposure of animals to more than one toxin is of concern and requires more study [19]. Synergistic effects may explain why animals sometimes respond negatively to mycotoxin levels much lower than those reported in scientific studies as able to cause mycotoxicoses.

In a survey conducted over a period of 4.5 years in countries of Southern Europe (Portugal, Spain, Italy, Greece and Cyprus) the Fusarium mycotoxins were found to be the major contaminants (fumonisins, type B trichothecenes and ZEA) of feed material and compounded feed samples [20]. There are several studies about mycotoxins' interaction (mainly in vitro), but only a few report the combined effects of three specific Fusarium mycotoxins which are more likely to co-occur; i.e. FB, DON and ZEA. Although there are studies about the three aforementioned mycotoxins very few focus on tertiary mixtures [21-24]. In rabbit bucks, ZEN impairs spermatogenesis and decreases libido, although only at high doses (117.3 mg/kg feed) [25]. Pregnant rabbit does fed on a DON-contaminated diet (0.3 and 0.6 mg/kg) showed marked bodyweight loss, but teratogenic effect was not proven [26]. FB1 was found to be nephron- and hepatotoxic in rabbits [27], and it has also been shown to negatively influence haematopoiesis by impairing bone marrow function [28]. Lung oedema was found only in a small number of rabbits fed a FB1-contaminated diet [29]. The teratogenic effect of FB1 was also described using an oral dose of 300 mg/d for 14 d [18] but a lack of embryotoxicity was found at lower (0.1, 0.5 or 1.00 mg/d) oral doses [30].

Grenier and Oswald [31] performed a meta-analysis of published raw data on mycotoxin interactions in vivo which varies according to the animal species, the dose of toxins, the length of exposure, but also the parameters measured; and classified the interaction into the following categories: synergistic, additive, less than additive, and antagonistic. The authors also differentiated between three types of synergistic effects and two types of antagonisms. Such characterization of mycotoxin interactions is helpful in experimental designs and interpretations of combined toxicity outcomes and should be included into further investigations on mycotoxin interactions.

The aim of this experiment was to study the dietary, low-dose of FB, DON and ZEA mycotoxins' individual and combined effects on breeding rabbit bucks, in particular with regards to reproductive toxicity.

MATERIALS AND METHODS

EXPERIMENTAL ANIMALS, HOUSING AND FEEDING

The experimental animals were Pannon White rabbit breeding bucks (24 weeks of age, 4.0 ± 0.5 kg mean bodyweight, n=60). They were individually housed in wire mesh cages (42 × 50 cm) in a closed building, with 16 light h/day. Average temperature ranged from 16°C to 18°C and the farm had overpressure ventilation.

The animals received a commercial diet containing 10.3 MJ digestible energy/kg, 15.5% crude protein, 4.0% crude fat and 14.7% crude fibre for a total of 65 feeding days. The feedstuffs provided were available ad libitum, and the rabbits also had free access to drinking water.

TOXIN PRODUCTION

Fusarium verticillioides (for FB1) and *Fusarium graminearum* (for DON and ZEA) (NRRL 20960 [MRC 826] and NRRL 5883, respectively) fungal culture (7 days old) was grown on 0.5 strength potato dextrose agar (PDA; Chemika-Biochemica, Basle, Switzerland). Agar discs (5 mm) were prepared with cork borer (Boekel Scientifica, Pennsylvania, USA), which were then stored at 10°C in darkness in test tubes containing sterile distilled water (10 discs/10 ml).

For toxin production, maize (40 g) was soaked in distilled water (40 ml) at room temperature for 1 hour in Erlenmeyer flasks (500 ml), closed with cotton wool plugs. This was followed by the addition of the inoculated agar discs (10 agar discs per flask) to the two-times autoclaved (20 min.) matrix. The cultures were then stored and incubated at 24°C (FB1), 28°C (DON) and 18°C (ZEA) for 3 weeks, respectively. The flasks were shaken twice every day during the first week of incubation. When the incubation time was complete the fungus-infected cereal was dried at room temperature and ground.

The homogenized fungal cultures contained FB, DON and ZEA at concentrations of 3300, 2010 and 1298 mg/kg, respectively. The European Commission has made recommendations (2006/576/EC) regarding the maximum level of several mycotoxins in complete diets (European Commission, 2006) and introduced regulations (2003/100/EC) regarding aflatoxins (European Commission, 2003). However, these only apply to certain cases, particularly with regard to rabbit feed. Thus in this study doses were pre-determined according the EU limits (based on the European Commission Recommendation 2006/576/EC and the European Commission Directive 2003/100/EC) in finished feed for young pig as the most sensitive animal towards these fusarium mycotoxins among livestock. Fungal cultures were mixed into the feed of experimental animals, based on the presented dose in Table 1.

Table 1. Experimental groups.

Group	Mycotoxin concentration (mg/kg) of the feed	n
C (control)	0	15
F (FB1)	5	15
DZ (DON+ZEA)	1+0.25	15
FDZ (FB1+DON+ZEA)	5+1+0.25	15

LC-MS ANALYSIS

LC-MS analysis was performed by a Shimadzu Prominence UFLC separation system equipped with a LC-MS-2020 single quadrupole (ultra-fast) liquid chromatograph mass spectrometer (Shimadzu, Kyoto, Japan) comprised a vacuum degasser, a binary pump (20AD), a column oven (CTO 20A) autosampler (SIL 20ACHT), and mass analyser (MS 2020) with both atmospheric-pressure chemical ionization (APCI) and electrospray ionization (ESI) systems. Optimized mass spectra were obtained with an interface voltage of 4.5 kV, a detector voltage of 1.05 kV in negative mode, 1.25 kV in positive mode. Heat block temperature was 200°C and a desolvation gas temperature was 200°C. For nebulizing and drying gas, nitrogen was used (1.5 L/min and 15 L/min flow rate, respectively). Chromatographic separation was performed at 50°C and achieved on an RP-18 (2.1 × 100 mm, 2.6 μm, Kinetex™ Phenomenex USA) stationary phase applying gradient elution 0.3 mL/min eluent total flow rate for mycotoxins and 0.4 mL/min for silymarin flavonoids, with A: 0.1% AcOH and B: 0.1% AcOH in methanol as eluent. With optimum method performance characteristics analytes were quantified using external calibration.

Rabbit-feed samples were milled and extracted using 1% AcOH containing 75:25=MeOH:H_2O (V/V) for F-2 CH_3CN:water 1:1 (V/V) for FB1 and water for DON and ZEA as extraction solvent. The extracts were shaken at room temperature for an hour then decanted and the supernatant was collected. 1 mL of clean water extract was applied to the immunoaffinity column (IAC; Vicam, DON test) which contains specific antibodies for DON for purification. The IAC was washed with 5 mL water, and DON was slowly eluted in 2 mL methanol. ZEA and FB1 were measured by dilute and shot method. Romer Mix 4 (containing trichotecenes+**ZEARALENON** at 10 mg/L) and Romer MIX 3 (containing FB1-2 at 50 mg/L) primary stock solution used as reference. 1 μL of each samples were analysed with a gradient: (0 min) 5% B, (3 min) 60% B, (8 min) 100% B, followed by a holding-time of 3 min at 100% eluent B and 3 min column re-equilibration at eluent A pumped at a flow rate of 0.3 mL/min. DON is detected as $[M+AcO]^-$ at m/z=355, ZEA at m/z=317 as $[M-H]^-$, FB1 at m/z=722$[M+H]^+$. The limit of detection (LOD) for FB, ZEA and DON was 3.0, 5.0 and 5.0 μg/kg, while the limit of quantification was (LOQ) 10, 1.0 and 2.0 μg/kg, respectively.

EXPERIMENTAL DESIGN

Bucks were divided into 4 experimental groups. One group of the experimental animals served as control, while into the feed of the other three groups fungal culture was mixed in a pre-defined concentration (Table 1).

The daily feed intake was registered by measuring back the left-over feed amount (in the first 3 weeks daily, and weekly thereafter), while bodyweight was recorded once every week.

The health status of the animals was observed throughout the experiment, morbidity and mortality was logged daily.

On days 30 and 60 blood and sperm was sampled (n=15/group), and a gonadotropin-releasing hormone (GnRH) test was performed (blood samplings related to GnRH test: n=6/group). Blood was sampled from the marginal ear vein, while sperm was collected after a training period into artificial vaginas and seminal plasma was separated with centrifugation.

At the end of the study (on day 65) animals were exsanguinated after stunning. The weight of liver, kidneys, testicles and spleen was measured and macroscopic changes were analyzed and recorded. After dissection (n=15/group) samples were taken for histopathological analysis; the testis, liver and kidney was fixed in 10% neutrally buffered formalin. For the analysis of antioxidant status 2-2 g samples were taken from the liver.

The experimental protocol was authorized by the Food Chain Safety and Animal Health Directorate of the Somogy County Agricultural Office, under allowance number SOI/31/1679-11/2014.

CLINICAL CHEMICAL PARAMETERS

The plasma total protein (TP), albumin (ALB), globulin (GLOB), total cholesterol (tCHOL), triglyceride (TG), glucose (GLU), fructosamine (FA), creatinine (CREA) and bilirubin (BIL) concentrations, and the activity of aspartate aminotransferase (AST), alanine aminotransferase (ALT), gamma-glutamyl transferase (GGT), lactate dehydrogenase (LDH) and alkaline phosphatase (ALKP) were determined in a veterinary laboratory (Vet-med Laboratory, Budapest, Hungary), using Roche Hitachi 912 Chemistry Analyzer (Hitachi, Tokyo, Japan) with commercial diagnostic kits (Diagnosticum LTD., Budapest, Hungary).

ANTIOXIDANT STATUS

For the determination of lipid peroxidation, the samples of blood plasma, red blood cell (1:9) hemolysate (RBCH) and liver were stored at -70°C until analysis. Lipid peroxidation was determined by the quantification of malondialdehyde (MDA) levels with 2-thiobarbituric acid method in blood plasma and RBCH [32] and liver homogenate [33], and determination of conjugated dienes (CD) and trienes (CT) according to the AOAC [34] method in the liver. Among the components of the antioxidant system some parameters of the glutathione redox system was determined in blood plasma, RBCH and liver. The amount of reduced glutathione (GSH) measured by the method of Sedlak and Lindsay [35] and the activity of glutathione peroxidase (GPx) according to Lawrence and Burk [36].

GNRH TEST AND DETERMINATION OF TESTOSTERONE CONCENTRATION

Experimental and control bucks were treated i.m. with 0.2 ml GnRH analogue (Receptal; Intervet, Boxmeer, The Netherlands) for the analysis of the toxin effect on the Leydig cell function. The levels of testosterone hormone were determined from blood samples taken just prior to GnRH analogue injection (0 min) and thereafter in the subsequent 2 hours in every 25 minutes (a total of 6 blood samplings).

The testosterone concentration was determined with a direct ^{3}H-radioimmunoassay method [37] adopted and validated for rodents' (chinchilla rabbit and Angora rabbit) plasma as described previously [38].

SPERMATOLOGY

Spermatological analyses covered the following parameters: pH, sperm cell concentration (improved Neubauer cell counting chamber), motility, morphology (native and stained) and acrosomal integrity (Spermac™ staining, Beernem, Belgium) of the spermatozoa. Motility was evaluated with a computer assisted sperm analyzer (Medealab™ CASA System, Erlangen, Germany). Moreover, vital test, hyposmotic and peroxidase tests were carried out. A minimum number of 200 spermatozoa were examined for morphology and 500 for motility evaluation [39].

COMET ASSAY

For comet assay sperm was sampled on day 60 of the experiment (n=15/group). The method was adapted from human spermium examination protocols [40,41], with the following modifications. The semen was washed three times in PBS and re-suspended in PBS to a final of 10^6 cell/ml number. Onto the microscope slides pre-coated with 1% NMP agarose 10 μl cell suspensions and 75 μl 1% LMP agarose were loaded. The decondensation was performed in two steps, first the slides were soaked for 1 hour at 4°C in a lysis buffer with dithiothreitol, and second the slides were soaked for 1 hour at 37°C in a lysis buffer with proteinase K. After lysis the slides were washed in sterile redistilled water to eliminate the salt adhered to the gel. Electrophoresis was performed at 300 mA and 25 V for 30 minutes at 4°C. After washing and drying the slides were stained with ethidium bromide. All chemicals used in this study were obtained from Sigma-Aldrich Ltd. (Budapest, Hungary). Specialized chemicals used were: Histopaque-1077 and RPMI-1640 medium (Sigma-Aldrich Ltd., Budapest, Hungary).

The fluorescence images were generated using an Alpha-Optika B-600TiFL fluorescence microscope (Optika Microscopes, Bergamo, Italy). Scoring was carried out according to Singh et al. [42] and Collins et al. [43], in which comets are classified into scores of '0', '1','2', '3' and '4' according to DNA damage and head/tail migration. Each single comet was scored visually and assigned into an arbitrary unit from 0 to 4, depending on the relative intensity of DNA fluorescence in the tail; 800 cells/group were counted.

HISTOPATHOLOGICAL ANALYSIS

After registering the macroscopic pathological signs on the internal and external organs, testicles, liver, kidneys and spleen were stored in 10% neutrally buffered formalin and were embedded into paraffin. For light microscopic analysis microtome slides of 5 micrometer were prepared and stained with hematoxylin-eosin.

The histopathological analysis was performed according to the Act/2011 (03.30) of the Hungarian Ministry of Agriculture and Rural Development and was in accordance with the ethical guidelines of the OECD Good Laboratory Practice for Chemicals [44].

STATISTICAL ANALYSES

Statistical analyses were performed using IBM SPSS 20.0 [45] software. Data processing and the mathematical-statistical calculations were performed using the compare means (independent-samples-t-test, oneway ANOVA with Tukey post-hoc test), correlate and descriptive statistics modules. In case of comet assay crosstabs options were used for chi-square test.

RESULTS

FEED CONSUMPTION AND BODY WEIGHT

The feed intake of the rabbit bucks was not different among the groups. No significant difference in body weight among groups was detected at any of the 12 time points (data not shown), average body weight of the groups was between 4252 and 4442 g by the end of the experiment.

CLINICO-CHEMICAL PARAMETERS

No inter-group differences ($P \geq 0.05$) were found for ALB, TG, GLU, FA, CREA and BIL concentrations, and AST, ALT, GGT, LDH and ALKP activities.

Differences in TP, GLOB and tCHOL at the two sampling dates (day 30 and 60) are summarized in Table 2. In spite of the slight significant differences between treatments all data were within the reference ranges, which are 54-75 g/L for TP, 15-27 g/L for GLOB and 0.3-3.00 mmol/L for tCHOL [46]. Only GLOB concentration exceeded the upper limit of 27 g/L in the C and FDZ groups on day 30 (29 and 28 g/L, respectively).

Table 2. Clinico-chemical parameters of rabbits on day 30 and 60.

		Day 30		Day 60	
		mean	SD	mean	SD
TP (g/L)	C	69.85^{b}	5.85	62.16^{a}	3.02
	F	68.57^{b}	1.72	64.73^{ab}	3.97
	DZ	61.33^{a}	4.28	65.81^{ab}	7.34
	FDZ	69.57^{b}	1.92	67.74^{b}	3.40
GLOB (g/L)	C	28.65^{b}	3.54	22.38^{a}	2.23
	F	27.13^{b}	1.61	22.96^{a}	2.09
	DZ	20.23^{a}	3.39	24.10^{ab}	5.95
	FDZ	27.98^{b}	1.63	26.67^{b}	3.12
tCHOL (mmol/L)	C	1.45^{ab}	0.12	1.51^{b}	0.35
	F	1.62^{b}	0.19	1.21^{a}	0.09
	DZ	1.33^{a}	0.16	1.29^{a}	0.14
	FDZ	1.58^{ab}	0.17	1.44^{ab}	0.14

[a,b]numbers with different superscripts indicate significant differences ($P \leq 0.05$) between groups (C, F, DZ and FDZ).

ANTIOXIDANT STATUS

On day 30 only MDA concentration in the RBCH showed significant difference between treatments F (35.7 ± 4.5 mmol/l) and DZ (34.2 ± 1.6 mmol/l), as compared to C (43.4 ± 3.5 mmol/l) and FDZ (43.7 ± 7.6 mmol/l), but other antioxidant parameters did not change significantly (data not shown).

Parameters showing significant treatment effect on day 60 are summarised in Table 3. At day 60, DZ treatment resulted in significantly increased GPx activity in the red blood cells and MDA formation both in RBCH and plasma, while less GSH concentration in the blood plasma. As a result of peroxidation of dienoic and trienoic fatty acids, the concentration of conjugated dienes and trienes increased as a result of the DZ exposure, as compared to the combined effect of three mycotoxins (FDZ).

TESTOSTERONE CONCENTRATION

The testosterone concentration was significantly different due to mycotoxin exposure only at day 60 at the sampling minutes of 75, 90 and 115 (Figure 1). Figure 1 clearly demonstrates that Leydig cells of the animals intoxicated with the three mycotoxins in combination (FDZ) for a total of 60 days synthesize significantly less testosterone as a response to exogenous GnRH. In the F group blood testosterone level was slightly lowered, while the two toxin combination (DZ) induced a more expressed decline in the concentration, although the differences were not statistically significant.

Table 3. Antioxidant parameters measured on day 60.

Parameter	Groups			
	C	F	DZ	FDZ
GPx-RBCH[1]	2.7 ± 1.0^{a}	3.5 ± 1.2^{a}	4.9 ± 1.1^{b}	3.1 ± 0.8^{a}
MDA-RBCH[2]	30.3 ± 3.0^{ab}	28.3 ± 3.2^{a}	33.9 ± 7.5^{b}	28.8 ± 3.2^{a}
GSH-plasma[3]	2.2 ± 0.2^{a}	2.2 ± 0.5^{a}	2.0 ± 0.2^{a}	2.6 ± 0.5^{b}
MDA-plasma[2]	18.9 ± 2.8^{a}	18.7 ± 3.0^{a}	24.5 ± 8.0^{b}	16.8 ± 2.9^{a}
CD liver[4]	0.51 ± 0.03^{ab}	0.51 ± 0.02^{ab}	0.53 ± 0.05^{b}	0.48 ± 0.05^{a}
CT liver[4]	0.19 ± 0.01^{a}	0.20 ± 0.01^{ab}	0.21 ± 0.04^{b}	0.18 ± 0.01^{a}

[1]U/g protein,[2]μmol/ml,[3]μmol/g protein,[4]Abs 232 nm

[a,b]numbers with different superscripts indicate significant differences ($P \le 0.05$) between groups (C, F, DZ and FDZ)

SPERMATOLOGICAL ANALYSIS

No effect of toxin treatment was detected on semen pH (ranging from 6.4 to 8.2 in each group), quantity of the semen (on average 1 ml in each group) and concentration of spermatozoa (ranging from 2.4 to 2.6 × 10^7/ml). Comparing the sperm motility between groups, significant effect could also not be proven. The ratio of spermatozoa showing progressive forward motility was around 80% at the beginning of the experiment and decreased from 80 ± 1.7% to 67 ± 4% in the semen of FDZ animals showing significant difference when compared to the other three treatment groups. Between the normal morphology of the spermatozoa of the DZ and C animals, statistically significant differences were found (Table 4).

Table 4. The ratio of spermatozoa with normal morphology in the semen after 60 days of mycotoxin exposure (%, mean ± SD).

Group	Spermatozoa with normal morphology
C	80.2 ± 11.2[b]
F	76.0 ± 9.0[ab]
DZ	66.3 ± 23.7[a]
FDZ	68.9 ± 14.1[ab]
[a,b]different superscripts in the column indicate significant between-group differences (P ≤0.05)	

The most frequent morphological abnormalities were: abnormality of the tail, retention of cytoplasmic drop, absence of the acrosome and altered head. A cell was considered altered if at least one defect was present.

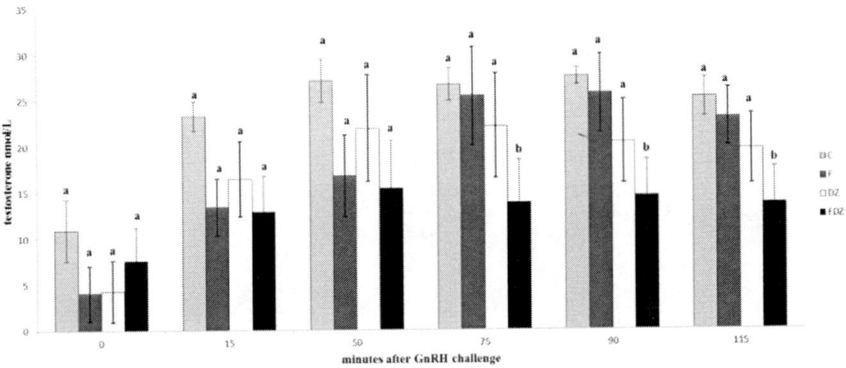

Figure 1. Gonadotropin-releasing hormone (GnRH; mean and standard deviation) induced testosterone production at day 60. a,bnumbers with different superscripts indicate significant differences (P ≤ 0.05) between groups (C, F, DZ and FDZ).

COMET ASSAY

All treatments caused DNA damage and 98.6, 91.6 and 91.8% of the treated cells could be categorized as having 1 to 3 scores in the F, DZ and FDZ group, respectively (Figure 2). Cells with 0 values showed the highest prevalence in group C, while cells with the slightest damage (score 1) were dominant in all toxin treated groups. FB1 resulted significantly less 0 comets (intact cells) compared to the other treatments.

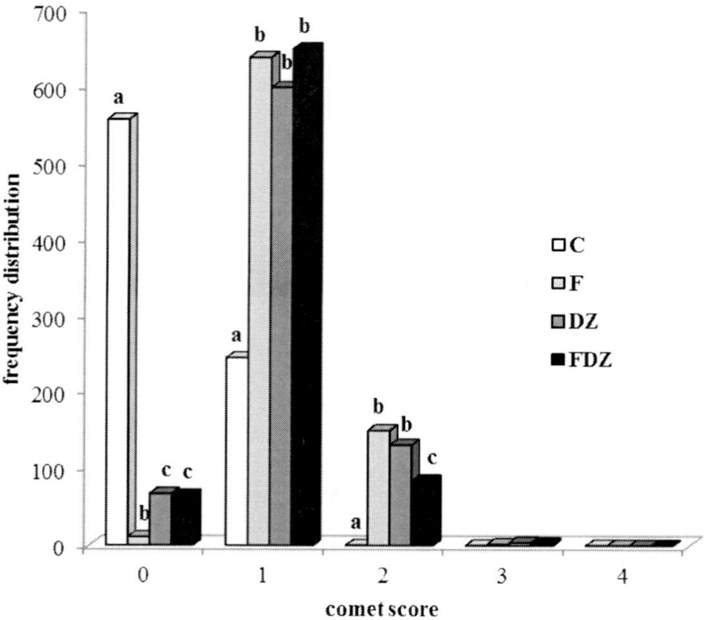

Figure 2. Number of cells with the respective comet values (values of comet scores: 0 to 4). a,bcolumns with different superscripts indicate significant differences ($P<0.0001$)

Cells, with score 1 (Figure 3a) occurred in nearly similar proportions in case of the three toxin treatments. The comet value 2 (Figure 3b) showed the highest prevalence in samples of F and DZ animals, FDZ treatment resulted in significantly less cells of this type. The number of cells with score 3 (Figure 3c) occurred only in 0.6% of the categorized cells. No cells were found with a comet assay score of 4 in the experiment.

HISTOPATHOLOGY

No sign of mycotoxicosis was detected. In animals consumed FB1, DZ and FDZ contaminated diet the intensity of spermatogensis decreased by 43, 31 and 64%, respectively (Table 5), which was reflected by the lack of differentiated spermatozoa, thinning of the germinal epithelium, the appearance of multinuclear giant cells indicative of the disturbance of meiosis and mitosis of the germinal epithelial cells and in some cases the lack of spermatogonia.

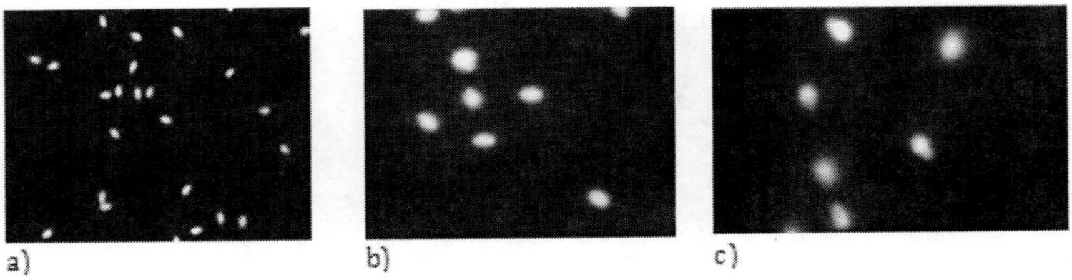

Figure 3: Comet assay (DNA damage) scoring: images of comets from bucks' spermatozoa stained with EtBr. Classes 0 (a), 1 (b) and 2 (c) were used for visual scoring.

Group	NO	Number of animals showing alteration				
		1	2	3	All together	%
C	15	0	0	0	0	0
F	8	6	0	0	6	43
DZ	9	4	0	0	4	31
FDZ	5	7	1	1	9	64

NO=number of animals without alteration, 1=slight/small area/low occurrence frequency, 2=intermediate severity/intermediate area/intermediate occurrence frequency, 3=marked/extensive/high occurrence frequency

Table 5: Decreased intensity of spermatogenesis as revealed by the histopathology of the testicles.

These histological findings were observed to different severities in the seminiferous tubules. In case of slight alteration (grade 1) disturbance of sperm cell formation was observed in 20-30% of the seminiferous tubules, while in case of grade 2 and 3, 30-70 or 100 % of the tubules, respectively, showed decreased **SPERMATOGENESIS**. Decreased spermatogenesis was manifested in lack of well differentiated sperm cells and less primary, secondary spermatocytes and spermatids, and thinning of the seminiferous epithelium as a consequence (grade 1). In case of intermediate severity (grade 2) these alterations were observed in higher ratio of the tubules, the disturbance of meiosis and mitosis was indicated by the appearance of multinuclear cells. Cases, when all tubules were attacked and moreover, in some tubules even the initial spermiogenetic cells, spermatogonia were also absent, were classified as marked damage (grade 3) (Figures 4 and 5).

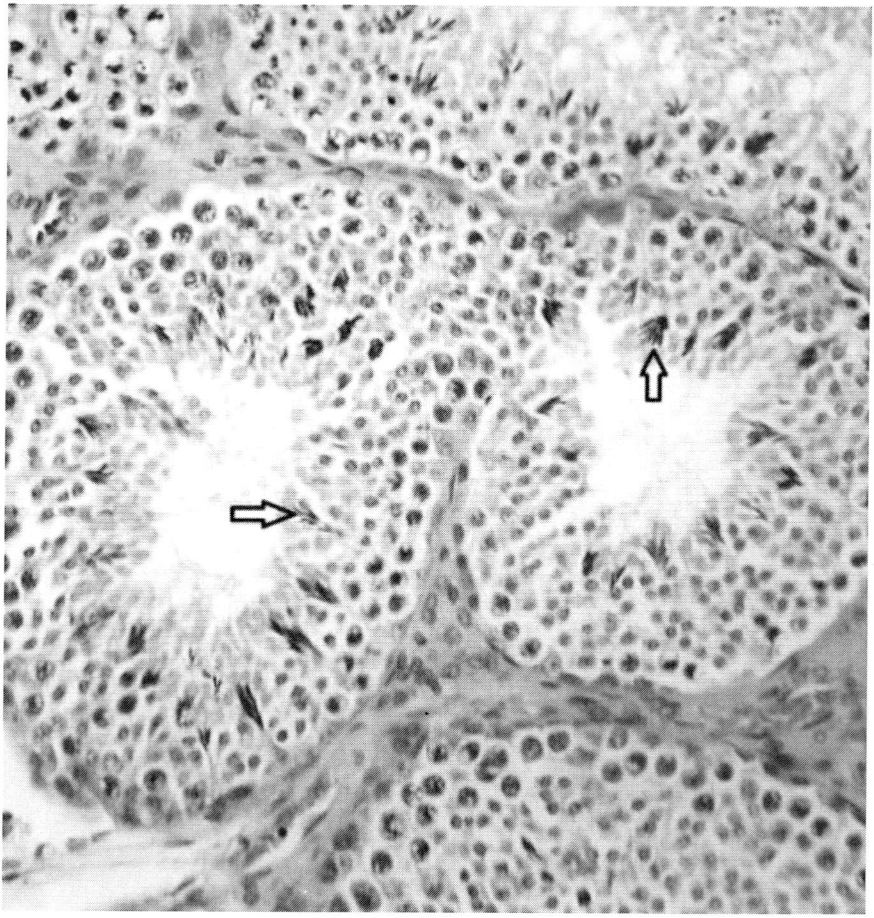

Figure 4: Active spermatogenesis in a healthy control animal's seminiferous tubules. The cells at different differentiation levels in the epithelium germinativum (spermatogonia, stage I and II spermatocytes, spermatides and mature spermia (↑) are as well visible. Haematoxylin-eosin staining, 400X.

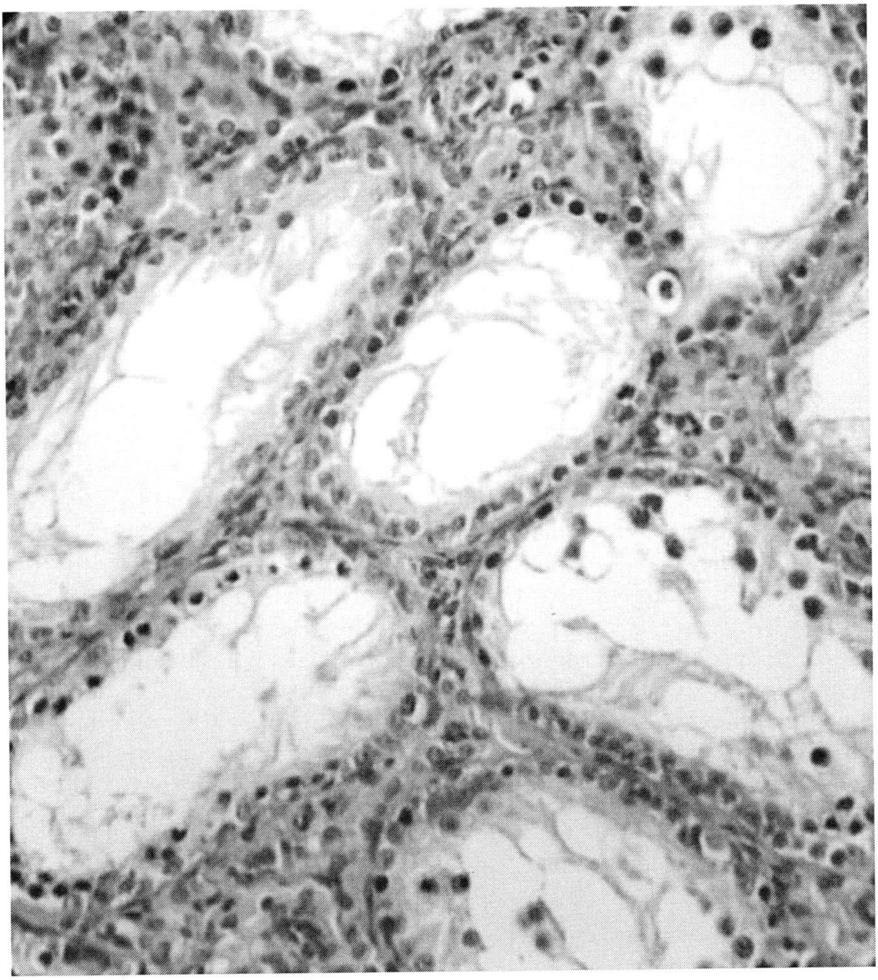

Figure 5: A part of the seminiferous tubule after FDZ treatment. The marked lack of germinal epithelialcells is visible. Haematoxylin-eosin staining, 400X.

In the testicles of the toxin treated animals the cytomorphology and the proportion of the Leydig interstitial cells was not significantly different from the controls. In the Malpighian bodies of the spleen of every DZ and FDZ animal, slight lymphocyte depletion (slight thinning of the T and B dependent zones of the Malpighian tubules) was observed. However, the cytomorphology of lymphoblast and lymphocyte cells was not different from the control. In the livers and kidneys no detectable alteration attributable to mycotoxins was found.

ORGAN MASSES

Among all organs which were measured for weight (testicles, liver, kidneys, spleen) only the weight of the spleen was significantly different between groups. It was the most (1.84 ± 0.49^{b}) in DZ animals, and least (1.46 ± 0.38^{a}) in C, while 1.65 ± 0.30^{ab} and 1.65 ± 0.37^{ab} in F and FDZ, respectively. However the relative organ mass values (data not shown) of the different groups were not significantly different.

DISCUSSION

The present study intended to determine the negative effect of a combined, 65-day oral Fusarium mycotoxin administration on the reproductive traits of 24-week old breeding Pannon White rabbit bucks. The low toxin exposure levels were chosen taking the limit values given by the 2006/576/EC recommendation into consideration. The in feed concentrations of FB, DON and ZEA (5, 1 and 0.25 mg/kg feed, respectively) revealed 169-193 μg FB1/kg BW, 33.7-38.7 μg DON/kg BW, and 8.5-9.7 μg ZEA/kg bw exposure. Results indicate that a prolonged low-dose mycotoxin exposure may adversely affect male reproduction as it has been previously demonstrated in case of T-2 toxin. The combined toxin effect indicates highly complex interactions among the mycotoxins, appearing differently in the divergent physiological processes. Among the mycotoxins studied here additive, synergistic, as well as antagonistic effects were detected (Table 6).

Table 6: Summary of the interactions between FB, ZEA and DON experienced in the experiment (classified according to Grenier and Oswald, 2011).

Parameter examined	Type of interaction
feed intake	antagonism (not significant)
lipid peroxidation (MDA, GPx, CD, CT)	antagonism
testosterone production	synergism
spermatogenesis by histology	additive
sperm cells with normal morphology	less than additive
weight of the spleen	antagonism
genotoxicity	antagonism

FEED CONSUMPTION AND BODY WEIGHT

The only observation to mention concerning feed consumption and body weight (BW) was that the mean feed intake (FI) of DZ and FDZ animals was somewhat lower, as compared to F and C (significance was not proven). This result may mostly be attributed to DON, since this is the most well-known characteristic effect of trichothecene mycotoxins [47]. In the study of Hewit et al. [48] using naturally contaminated (4.2 mg/kg DON) corn for Fusarium mycotoxin exposure in growing rabbits, no feed refusal was observed and the toxin treated animals had slightly increased BW gain, suggesting that rabbits are less sensitive to this mycotoxin. Deep discussion of a mere tendency is void, but observing data at weeks 4 and 5 again indicate the "likewise positive effect" of low-dose FB, since the lowest FI was found for DZ, but FDZ and F were less suppressed. This is supported by the result on non-human primates of Gelderblom et al. [49], where long-term feeding of FB1 alone had no negative effect on FI and BW, even at higher doses. In that study apparent no-effect threshold in terms of kidney and liver damage was estimated at 8.21 and 13.25 mg total fumonisin/kg diet, being about the double exposure, as compared to the present study. Also no effect on final live weight and weight gain was observed when male rabbits were fed diets containing 12.3 and 24.5 mg/kg FB1 for 5 weeks [50]. Summarized, our study indicates that DON provides its characteristic feed-intake lowering effect in a slight, but insignificant rate at the dose according to Commission Recommendation (2006/576/EC), while fumonisin shows no determinative effect on BW and FI at these concentrations, in breeding rabbit bucks.

CLINICAL CHEMICAL PARAMETERS AND ANTIOXIDANT STATUS

The health status of the animals can be monitored by regular determination of certain clinical chemical parameters. Despite that the liver and the kidney is considered to be the primary target organs of many mycotoxins (like trichothecenes and FB1) in all species examined, gross hepatic lesions and disorder was not induced by any of the treatments, as underscored by the unaltered ALT, AST, GGT and CREA values, the identical liver and kidney mass values and the lack of histopathological alterations in all groups. These organs were likewise tolerant towards the exposure of low dose of these mycotoxins, without providing degenerative processes. Sprando et al. [14] described an analogous observation, namely few lesions of various tissues in control and treated animals, either incidental or related to experimental techniques (gavage) and unrelated to the test substance. The effects of low (10 mg/kgBW) and high (100 mg/kgBW) oral doses of purified ZEN were studied by Čonková et al [47] in rabbits. Low ZEN doses resulted in a significant increase in ALP activity, while high ZEN doses showed significant increases of AST, ALT, AP, GGT, and LDH activity, indicating possible liver toxicity due to the chronic effects of the toxin.

The alteration found in the TP was attributed to the concentration changes of globulin, since albumin concentration was constant. According to Kim et al. [51], DON stress lowers immunoglobulin expression in mice, but ZEA was found to have no influence on it. This could be the reason of the slight temporary decrease in GLOB and TP of DZ animals at day 30. The results at day 60 are difficult to explain, because difference between groups was rather attributable to the lowered GLOB level in control animals than toxin treatments. However fumonisin treatment decreased GLOB concentration at day 60, compared to day 30. Results of the FDZ group on day 60 are in accordance with those of Tessari et al. [52], describing that FB1 (added to an aflatoxin B1 containing diet) has primary additive effect on the immunological response in birds. It is interesting to highlight that FB1 alone and in higher doses (8 mg/kg feed) had a sex-dependent immunosuppressive effect, male pigs being more prone [53]. The FDZ treatment evokes the question, whether the low dose of FB1 or the interaction between these toxins augmented the immune response. Some kind of anomaly in the immune system was supported by histology, in that in both DZ and FDZ animals, slight lymphocyte depletion (slight thinning of the T and B dependent zones of the Malpighian tubules) was observed in the spleen, without any change in the cytomorphology of lymphoblasts and lymphocytes. In the experiment Sprando et al. [14] similar finding was described when rats orally exposed with 5 mg/kg DON.

The pathophysiological significance of the changes in tCHOL level, which was around 5-20%, is questionable, taking the broad physiological ranges (0.3-3.00 mmol/L) into consideration.

The increased GPx activity in the red blood cells of DZ animals by day 60 reflects more lipid hydroperoxide production, as the role of the enzyme is to reduce lipid hydroperoxides to their corresponding alcohols and this way to reduce free hydrogen peroxide to water. This is supported by the highest MDA concentrations both in the red blood cells and the blood plasma, referring to augmented in vivo lipid peroxidation [54], and also by the decreased GSH, as co-substrate of GPx, level in plasma. One of the mechanisms by which DON exerts its toxicity is inducing oxidative stress within the cells [55]. ZEA is also known to induce generation of reactive oxygen species (ROS) and consequently lipidperoxidation, while fumonisin was characterised as moderate oxidant mycotoxin [56]. Our results show general agreement with those of Minervini et al. [57], reporting that in vitro FB1 dose at the level of the EC Regulation is not inducing oxidative stress in intestinal cells. On the other hand FB1 has proven to induce ROS production in broiler chicks, but only at extreme high (100 mg/kg feed) dose [58]. Our findings suggest that FB1 in low dose may act antagonistically towards Z and D, which has not been reported yet. Though there is no literature on the mitigating property of FB1 on lipid peroxidation. We thus summarize DZ combination was the most harmful in terms of initial (CD and CT) and terminal (MDA) phases of lipid peroxidation after long-term exposure (60 days); meanwhile FB1 at the EC Regulation level did not cause significant oxidative stress, even more relieved those effects triggered by ZEA and DON.

GN-RH INDUCED TESTOSTERONE PRODUCTION

The testosterone concentration was not significantly influenced by one (F) and two toxins (DZ), but the FDZ treatment dramatically lowered its synthesis and/or release from the Leydig cells by the end of the sampling intervals (minutes 75, 90 and 115), suggesting synergistic effect. As far as the authors are aware, there exists only a single study on male reproduction traits as affected by orally administered DON, which was performed in rats [14]. In this study 0.5, 1, 2.5 and 5 mg/kg DON was applied, and the authors found a dose-dependent down-regulation of testosterone synthesis after 28 days of exposure. Trichothecene toxins may decrease the **TESTOSTERONE PRODUCTION** of the Leydig cells by inhibiting early steps of the steroidogenic pathway, i.e. the conversion of pregnenolone to progesterone [25], as it was also observed in our previous study in case of T-2 exposure [38]. These Fusarium toxins may act indirectly as well, on the pituitary gland, or affect Sertoli cell inhibin production, as shown in case of DON [14]. In present study a direct effect of the mycotoxin on Leydig cells was not supported by the histopathological findings.

In fact not DON, but zearalenone is the most potent factor, directly inhibiting testosterone biosynthesis in the Leydig cells, adding that this inhibition is concentration- and time-dependent, happening via changing the nuclear estrogen receptor signaling [59], however the exact molecular mechanism is not known yet. For FB1 direct toxicological studies are lacking, but according to Lu et al. [60], elevated ceramide levels are directly apoptotic towards Leydig cells, which is inhibited by FB, a ceramide-synthase blocking substance. Thus FB1 had no deleterious effect on the rat testicular Leydig cell function in vitro. In summary the main, testosterone synthesis blocking effect in our study was primarily attributed to ZEA. Some kind of synergistic effect occurred between FB1 and DON+ZEA. Taking the common cellular modes of actions of these toxins into consideration, this might be explained with respect to the DNA-synthesis inhibiting effect. FB1 affected the DNA synthesis together with ZEA in a more than additive, rather synergistic way in a DNA inhibition assay, although this effect was not fully confirmed when it was further analysed in a second stage [61].

SPERMATOLOGY, HISTOPATHOLOGY AND COMET ASSAY

Toxin effect and interaction between the Fusarium toxins applied, manifested in lower sperm motility (FDZ), less spermatozoa with normal morphology (DZ), additive effect in the decreased intensity of spermatogenesis as shown by the histology of the testes, while antagonism concerning genotoxicity measured by comet assay.

Concerning the individual effects of these mycotoxins on sperm motility, only a few data are available. FB1 fed with male rabbits at 7.5 and 10.0 mg/kg dietary levels decreased sperm motility by approximately 18%, the possible explanation was the inhibition of cyclic 3'5'AMP activity and calcium ion transport, or, the decreased formation of the acidic epididymidal glycoprotein synthesis required to maintain motility [50]. Changes, after increasing oral exposure of DON were not consequent, and considered to be random [14], while in an in vitro study no significant effect of DON on motility was found using **RABBIT** spermatozoa (Medvedová et al., 2012). Boeira et al. [62] reported compromised spermatozoa motility after feeding mice with ZEA, while 12 mg daily ZEA exposure for 8 weeks did not alter sperm motility in adult rams [63]. Further data from in vitro studies support no effect of ZEA on boar spermatozoa [64], while on the contrary Rajkovic et al. [65] and Tsakmakidis et al. [66] described impairment of progressive motility. Interestingly in an in vitro study with stallion sperm ZEA caused hyperactivation in motility [67].

As the cellular level effects by which the above described mycotoxins may adversely influence motility are not well described yet, the combined effect is also difficult to explain. Decreased sperm motility caused by the exposure of the three mycotoxins (FDZ) by day 60 might be attributable also to the decreased testosterone concentration as a result of synergistic effect between FB, DON and ZEA observed also by day 60. The male reproductive organs are strongly androgen-dependent in respect of structure, as well as function. Testosterone supports spermatogenesis, sperm maturation, seminal plasma production and sexual functions [68]. So the decreased testosterone secretion may give an explanation also for the more morphologically abnormal cells, and the decreased intensity of spermatogenesis shown by histology. However, direct effect of the particular toxins is also presumable. Interestingly FB1 had only a slight adverse effect on the cell morphology and spermatogenesis. The characteristic toxic effect of FB1 is the inhibition of the sphingolipid synthesis, thus causing cellular ceramide depletion and elevation of shinganine, which is cytotoxic, and has growth inhibitory and pre-apoptotic effects [69]. This results in disturbances of cell growth, differentiation and morphology. In rabbits orally administered FB1 caused similar morphological and structural abnormalities in sperm morphology and spermatogenesis, and 7.5 mg/kg in the diet for FB1 as LOAEL (lowest observed adverse effect level) was suggested [50]. In our study even lower dose, 5 mg/kg had exerted adverse effect; however sensitivity to toxic substance is influenced by several factors. Compared to F group, DZ was similarly compromised. In case F and DZ, disturbance of sperm cell formation was observed only in 20-30% of the seminiferous tubules, however the combined treatment of DON and ZEA resulted in more cells of abnormal morphology. After exposure of the three toxin (FDZ) in 1 animal 30-70, while in another 100% of the seminiferous tubules were concerned, indicating additive effect. The effect of ZEA on spermatogenesis is more thoroughly studied compared to FB1 and DON. In orally exposed mice Zatecka et al. [70] detected decreased sperm concentration, and increased number of morphologically abnormal and also apoptotic spermatozoa. No pathological patterns in the seminiferous tubules were observed and the spermatogenesis was not interrupted, only the number of spermatogonia and spermatocytes were lower (not significantly) in exposed animals (25 μg/kg BW). Kim et al. [71] highlighted that apoptosis is the main effect by which ZEA causes germ cell depletion and atrophy of the testes. Germ cell apoptosis and thus impaired spermatogenesis was described also by Cho et al [72].

The in vitro exposure to ZEA induced cytotoxicity in boar sperm cells [64]. DON in 2.5 and 5 mg/kg concentration caused germ cell degeneration, decreased sperm cell release and abnormal cell development [14]. As the authors reported, DON may inhibit protein, RNA and DNA synthesis, and be cytotoxic to certain cells, which could cause the harmful effect in the most sensitive pre-leptotene spermatocytes.

Genotoxicity of Fusarium mycotoxins on sperm cells has been reported previously in case of FB1 [17] and ZEA [17,64,66,67], using sperm chromatin structure assay (SCSA), while no data on genotoxicity of DON related to sperm cells are available. A recent study showed that DON decreased cell viability, caused damage to the membrane, the chromosomes and the DNA in human lymphocytes, and potentially induced genotoxicity by the depletion of the antioxidant capacity. On the other hand according to Golli-Bennour and Bacha [56] DON is a non-oxidant mycotoxin, and its genotoxicity is the result of a direct effect on DNA fragmentation and caspade-dependent apoptosis.

In case of FB, Minervini et al. [17] found high degree of individual variability, when chromatin structure stability was checked in vitro, using equine spermatozoa. Interestingly no increase in ROS production was observed even in those cases when genotoxicity was detected, so the damage was attributed to mitochondrial dysfunction related to the altered ceramide metabolism. This mode of action might influence motility as well, knowing that spermatozoa are used as biosensors when testing mitochondrial toxicity, because of the strong influence of mitochondrial function on motility [17].

When we examined the single and combined genotoxic effect of FB1 and DON+ZEA on rabbit sperm cells, an antagonistic effect was observed, highly in accordance with the antioxidant parameters of the same experiment. It is also worth to mention, that none of the toxins had strong genotoxic effect as shown by the total lack of highly damaged cells (score 4). To our knowledge this is the first study analyzing the DNA damage on sperm cells by comet assay in orally mycotoxin exposed animals. There is only one in vitro study using sperm cells for comet assay in mycotoxin genotoxicity studies [73-75], in which human spermatozoa and lymphocytes were assessed for their sensitivity to the genotoxic effect of DON. It was described that spermatozoa are more sensitive towards DON (in terms of DNA damage) as compared to lymphocytes. The reason may be that spermatozoa are a highly specialized cell types with no cytoplasm and extremely condensed chromatin, which seems to be prone towards toxic stimuli. Thus the moderate genotoxic effect of DON has been proven and spermatozoa comet assay as a rapid test for genotoxicity recommended.

CONCLUSIONS

In this study three orally administered fusariotoxins (FB, DON and ZEA) were tested in adult male rabbits, focusing primarily on their reproduction endpoints. The dietary levels were relatively low according to the EU limits in finished feed for young pig, in the absence of limits for rabbits' feed (based on the European Commission Recommendation 2006/576/EC and the European Commission Directive 2003/100/EC). Results indicate that a prolonged low-dose mycotoxin exposure may adversely affect male reproduction. Reproductive processes show higher sensitivity to toxic effects, shown by the different parameters examined. Among the mycotoxins studied additive or less than additive effect was found in case of spermatogenesis and sperm cell morphology, synergism in testosterone production, while FB1 acted antagonistic against DON+ZEA on feed intake, lipid-peroxidation, and genotoxicity. All mycotoxins provoked moderate lipid-peroxidation and exerted slight genotoxicity.

ACKNOWLEDGEMENTS

The research was supported by the Hungarian Academy of Sciences (within the frame of the MTA-KE "Mycotoxins in the food chain" Research Group), by the National Scientific Research Fund (OTKA PD 104823) to KB and János Bolyai Research Grants of the Hungarian Academy of Sciences (BO/261/13 and BO/499/13) to KB and JSzF.

REFERENCES

1. Jelinek CF, Pohland AE, Wood GE (1989) Worldwide occurrence of mycotoxins in foods and feeds-an update. J Assoc Off Anal Chem 72: 223-230.
2. Placinta CM, D'Mello JF, Macdonald AC (1999) A review of worldwide contamination of cereal grains and animal feed with Fusariummycotoxins. Animal Feed Science and Technology 78: 21-37.
3. Bennett JW, Klich M (2003) Mycotoxins. ClinMicrobiol Rev 16: 497-516.
4. Marroquín-Cardona AG, Johnson NM, Phillips TD, Hayes AW (2014) Mycotoxins in a changing global environment-a review. Food ChemToxicol 69: 220-230.
5. Glenn AE (2007) MycotoxigenicFusarium species in animal feed. Animal Feed Science and Technology 137: 213-24.
6. IARC Working Group on the Evaluation of Carcinogenic Risks to Humans (2002) Some traditional herbal medicines, some mycotoxins, naphthalene and styrene. IARC MonogrEvalCarcinog Risks Hum 82: 1-556.
7. Wegulo SN1 (2012) Factors influencing deoxynivalenol accumulation in small grain cereals. Toxins (Basel) 4: 1157-1180.
8. Pestka JJ, Zhou HR, Moon Y, Chung YJ (2004) Cellular and molecular mechanisms for immune modulation by deoxynivalenol and other trichothecenes: unraveling a paradox. ToxicolLett 153: 61-73.
9. González HH, Martínez EJ, Pacin AM, Resnik SL, Sydenham EW (1999) Natural co-occurrence of fumonisins, deoxynivalenol, zearalenone and aflatoxins in field trial corn in Argentina. Food AdditContam 16: 565-569.
10. Osweiler GD (1986) Diagnosis of mycotoxicoses Occurrence and clinical manifestations of trichothecenetoxicoses and zearalenonetoxicoses. Richard JL, Thurston JR, editors. National Animal Disease Center; Ames, Iowa, USA.
11. Streit E, Schwab C, Sulyok M, Naehrer K, Krska R, et al. (2013) Multi-mycotoxin screening reveals the occurrence of 139 different secondary metabolites in feed and feed ingredients. Toxins (Basel) 5: 504-523.
12. Cortinovis C, Pizzo F, Spicer LJ, Caloni F (2013) Fusariummycotoxins: effects on reproductive function in domestic animals-a review. Theriogenology 80: 557-564.
13. Young LG, King GJ (1983) Prolonged feeding of low levels of zearalenone to young boars. Journal of Animal Science 57: 313-314.
14. Sprando RL, Collins TF, Black TN, Olejnik N, Rorie JI, et al. (2005) Characterization of the effect of deoxynivalenol on selected male reproductive endpoints. Food ChemToxicol 43: 623-635.
15. Gbore FA, Egbunike GN (2008) Testicular and epididymal sperm reserves and sperm production of pubertal boars fed dietary fumonisin B(1). AnimReprodSci 105: 392-397.
16. Gbore FA1 (2009) Reproductive organ weights and semen quality of pubertal boars fed dietary fumonisin B1. Animal 3: 1133-1137.
17. Minervini F, Lacalandra GM, Filannino A, Nicassio M, Visconti A, et al. (2010) Effects of in vitro exposure to natural levels of zearalenone and its derivatives on chromatin structure stability in equine spermatozoa. Theriogenology 73: 392-403.

18. Kovács M, Tornyos G, Matics Z, Mézes M, Balogh K, et al. (2013) Effect of chronic T-2 toxin exposure in rabbit bucks, determination of the No Observed Adverse Effect Level (NOAEL). AnimReprodSci 137: 245-252.
19. Speijers GJ, Speijers MH (2004) Combined toxic effects of mycotoxins. ToxicolLett 153: 91-98.
20. Griessler K, Rodrigues I, Handl J, Hofstetter U (2010) Occurrence of mycotoxins in Southern Europe. World Mycotoxin Journal 3: 301-309.
21. Kouadio JH, Dano SD, Moukha S, Mobio TA, Creppy EE (2007) Effects of combinations of Fusariummycotoxins on the inhibition of macromolecular synthesis, malondialdehyde levels, DNA methylation and fragmentation, and viability in Caco-2 cells. Toxicon 49: 306-317.
22. Lei M, Zhang N, Qi D (2013) In vitro investigation of individual and combined cytotoxic effects of aflatoxin B1 and other selected mycotoxins on the cell line porcine kidney 15. ExpToxicolPathol 65: 1149-1157.
23. Wan LY, Turner PC, El-Nezami H (2013) Individual and combined cytotoxic effects of Fusarium toxins (deoxynivalenol, nivalenol, zearalenone and fumonisins B1) on swine jejunal epithelial cells. Food ChemToxicol 57: 276-283.
24. Wan LY, Woo CS, Turner PC, Wan JM, El-Nezami H (2013) Individual and combined effects of Fusarium toxins on the mRNA expression of pro-inflammatory cytokines in swine jejunal epithelial cells. ToxicolLett 220: 238-246.
25. Fenske M, Fink-Gremmels J (1990) Effects of fungal metabolites on testosterone secretion in vitro. Arch Toxicol 64: 72-75.
26. Khera KS, Whalen C, Angers G (1986) A teratology study on vomitoxin (4-deoxynivalenol) in rabbits. Food ChemToxicol 24: 421-424.
27. Gumprecht LA, Marcucci A, Weigel RM, Vesonder RF, Riley RT, et al. (1995) Effects of intravenous fumonisin B1 in rabbits: nephrotoxicity and sphingolipid alterations. Nat Toxins 3: 395-403.
28. Mariscal-Quintanar MG, Garcia-Escamilla RM, Garcia-Escamilla N, Torres-Lopez J, Bautista-Ordonez JA, et al.(1997) Effect of ingesting Aspergillusflavus and Fusariummoniliforme on the cytology of bone marrow and blood and on serum albumin and globulin concentrations. Veterinaria Mexico 28: 75-81.
29. Orova Z (2003) Investigations on the teratogenic effects of fumonisin B1 in swine and rabbit. Thesis, University of Kaposvár, Kaposvár.
30. LaBorde JB, Terry KK, Howard PC, Chen JJ, Collins TF, et al. (1997) Lack of embryotoxicity of fumonisin B1 in New Zealand white rabbits. FundamApplToxicol 40: 120-128.
31. Grenier B, Oswald I (2011) Mycotoxin co-contamination of food and feed: meta-analysis of publications describing toxicological interactions. World Mycotoxin Journal 4: 285-313.
32. Placer ZA, Cushman LL, Johnson BC (1966) Estimation of product of lipid peroxidation (malonyldialdehyde) in biochemical systems. Anal Biochem 16: 359-364.
33. Botsoglou NA, Fletouris DJ, Papageorgiou GE, Vassilopoulos VN, Mantis AJ, et al. (1994) Rapid, sensitive and specific thiobarbituric acid method for measuring lipid peroxidation in animal tissue, food and feedstuff samples. Journal Agricultural and Food Chemistry 42: 1931-1937.
34. AOAC (1984) Official Methods of Analysis. (14thedn), Association of Official Analytical Chemists, Arlington, VA, USA.
35. Sedlak J, Lindsay RH (1968) Estimation of total, protein-bound, and nonprotein sulfhydryl groups in tissue with Ellman's reagent. Anal Biochem 25: 192-205.
36. Lawrence RA, Burk RF (1978) Species, tissue and subcellular distribution of non Se-dependent glutathione peroxidase activity. J Nutr 108: 211-215.
37. Csernus V (1982) Antibodies of high affinity and specificity for RIA determination of progesterone, testosterone, estradiol-17ß and cortisol. In: Görög S (eds) Advances in steroid analysis. Academic Press, Budapest, Hungary.

38. Kovács M, Tornyos G, Matics Z, Kametler L, Rajli V, et al. (2011) Subsequent effect of subacute T-2 toxicosis on spermatozoa, seminal plasma and testosterone production in rabbits. Animal 5: 1563-1569.
39. World Health Organisation (1999) WHO laboratory manual for the examination of human semen and sperm-cervical mucus interaction. (4thedn), Cambridge University Press, UK.
40. Simon L, Carrell DT (2013) Sperm DNA damage measured by comet assay. Methods MolBiol 927: 137-146.
41. Gopalan RC, Emerce E, Wright CW, Karahalil B, Karakaya AE, et al. (2011) Effects of the anti-malarial compound cryptolepine and its analogues in human lymphocytes and sperm in the Comet assay. ToxicolLett 207: 322-325.
42. Singh NP, McCoy MT, Tice RR, Schneider EL (1988) A simple technique for quantitation of low levels of DNA damage in individual cells. Exp Cell Res 175: 184-191.
43. Collins A, Dusinská M, Franklin M, Somorovská M, Petrovská H, Duthie S, et al. (1997) Comet assay in human biomonitoring studies: reliability, validation, and applications. Environ Mol Mutagen 30: 139-146.
44. OECD Principles of Good Laboratory Practice (as revised in 1997). OECD Environmental Health and Safety Publications (OECD).
45. IBM SPSS 20 for Windows. SPSS Inc. Chicago, IL. USA.
46. Rotter BA, Prelusky DB, Pestka JJ (1996) Toxicology of deoxynivalenol (vomitoxin). J Toxicol Environ Health 48: 1-34.
47. Harcourt-Brown F (2002) Textbook of rabbit medicine. (1stedn), Elsevier Science. New York, USA.
48. Hewitt MA, Girgis GN, Brash M, Smith TK (2012) Effects of feed-borne Fusariummycotoxins on performance, serum chemistry, and intestinal histology of New Zealand White fryer rabbits. J AnimSci 90: 4833-4838.
49. Gelderblom WC, Seier JV, Snijman PW, Van Schalkwyk DJ, Shephard GS, et al. (2001) Toxicity of culture material of Fusariumverticillioides strain MRC 826 to nonhuman primates. Environ Health Perspect 109 Suppl 2: 267-276.
50. Ewuola EO, Egbunike GN (2010) Effects of dietary fumonisin B1 on the onset of puberty, semen quality, fertility rates and testicular morphology in male rabbits. Reproduction 139: 439-445.
51. Kim EJ, Jeong SH, Cho JH, Ku HO, Pyo HM, et al. (2008) Plasma haptoglobin and immunoglobulins as diagnostic indicators of deoxynivalenol intoxication. J Vet Sci 9: 257-266.
52. Tessari EN, Oliveira CA, Cardoso AL, Ledoux DR, Rottinghaus GE (2006) Effects of aflatoxin B1 and fumonisin B1 on body weight, antibody titres and histology of broiler chicks. Br PoultSci 47: 357-364.
53. Marin DE, Taranu I, Pascale F, Lionide A, Burlacu R, et al. (2006) Sex-related differences in the immune response of weanling piglets exposed to low doses of fumonisin extract. Br J Nutr 95: 1185-1192.
54. Mead JF, Alfin-Slater RB, Howton DR, Popják G (1986) Lipids: Chemistry, biochemistry and nutrition. Plenum Press, NY, USA.
55. Mishra S, Tripathi A, Chaudhari BP, Dwivedi PD, Pandey HP, et al. (2014) Deoxynivalenol induced mouse skin cell proliferation and inflammation via MAPK pathway. ToxicolApplPharmacol 279: 186-197.
56. El Golli-Bennour E, Bacha H (2011) Hsp70 expression as biomarkers of oxidative stress: mycotoxins' exploration. Toxicology 287: 1-7.
57. Minervini F, Garbetta A, D'Antuono I, Cardinali A, Martino NA, et al. (2014) Toxic mechanisms induced by fumonisin b1 mycotoxin on human intestinal cell line. Arch Environ ContamToxicol 67: 115-123.
58. Poersch AB, Trombetta F, Braga AC, Boeira SP, Oliveira MS, et al. (2014) Involvement of oxidative stress in subacute toxicity induced by fumonisin B1 in broiler chicks. Vet Microbiol 174: 180-185.

59. Liu Q, Wang Y, Gu J, Yuan Y, Liu X, et al. (2014) Zearalenone inhibits testosterone biosynthesis in mouse Leydig cells via the crosstalk of estrogen receptor signaling and orphan nuclear receptor Nur77 expression. Toxicology In Vitro 28: 647-656.
60. Lu ZH, Mu YM, Wang BA, Li XL, Lu JM, et al. (2003) Saturated free fatty acids, palmitic acid and stearic acid, induce apoptosis by stimulation of ceramide generation in rat testicular Leydig cell. BiochemBiophys Res Commun 303: 1002-1007.
61. Tajima O, Schoen ED, Feron VJ, Groten JP (2002) Statistically designed experiments in a tiered approach to screen mixtures of Fusariummycotoxins for possible interactions. Food ChemToxicol 40: 685-695.
62. Boeira SP, Funck VR, Borges Filho C, Del'Fabbro L, de Gomes MG, et al . (2015) Lycopene protects against acute zearalenone-induced oxidative, endocrine, inflammatory and reproductive damages in male mice. Chemico-Biological Interaction 25: 50-57.
63. Milano GD, Odriozola E, Lopez TA (1991) Lack of effect of a diet containing zearalenone on spermatogenesis in rams. Vet Rec 129: 33-35.
64. Benzoni E, Minervini F, Giannoccaro A, Fornelli F, Vigo D, et al. (2008) Influence of in vitro exposure to mycotoxinzearalenone and its derivatives on swine sperm quality. ReprodToxicol 25: 461-467.
65. Rajkovic A, Uyttendaele M, Debevere J (2007) Computer aided boar semen motility analysis for cereulide detection in different food matrices. Int J Food Microbiol 114: 92-99.
66. Tsakmakidis IA, Lymberopoulos AG, Alexopoulos C, Boscos CM, Kyriakis SC (2006) In vitro effect of zearalenone and alpha-zearalenol on boar sperm characteristics and acrosome reaction. ReprodDomestAnim 41: 394-401.
67. Filannino A, Stout TA, Gadella BM, Sostaric E, Pizzi F, et al. (2011) Dose-response effects of estrogenic mycotoxins (zearalenone, alpha- and beta-zearalenol) on motility, hyperactivation and the acrosome reaction of stallion sperm. Reproductive Biology and Endocrinology 9: 134.
68. de Kretser DM, Kerr JB (1994) The cytology of testis. In: E Knobil and JD Neill (eds) The physiology of reproduction. Raven Press Ltd, New York, USA.
69. Müller S, Dekant W, Mally A (2012) Fumonisin B1 and the kidney: modes of action for renal tumor formation by fumonisin B1 in rodents. Food ChemToxicol 50: 3833-3846.
70. Zatecka E, Ded L, Elzeinova F, Kubatova A, Dorosh A, et al. (2014) Effect of zearalenone on reproductive parameters and expression of selected testicular genes in mice. Reproductive Toxicology 45: 20-30.
71. Kim IH, Son HY, Cho SW, Ha CS, Kang BH (2003) Zearalenone induces male germ cell apoptosis in rats. ToxicolLett 138: 185-192.
72. Cho ES, Ryu SY, Jung JY, Park BK, Son HY (2011) Effects of red ginseng extract on zearalenone induced spermatogenesis impairment in rat. J Ginseng Res 35: 294-300.
73. Baumgartner A, Kurzawa-Zegota M, Laubenthal J, Cemeli E, Anderson D (2012) Comet-assay parameters as rapid biomarkers of exposure to dietary/environmental compounds - an in vitro feasibility study on spermatozoa and lymphocytes. Mutation Research 743: 25-35.
74. Conková E, Laciaková A, Pástorová B, Seidel H, Kovác G (2001) The effect of zearalenone on some enzymatic parameters in rabbits. ToxicolLett 121: 145-149.
75. Marína MA, Kalafová M, Schneidgenová N, Maruniaková N (2012) The effect of deoxynivalenol on rabbit spermatozoa motility in vitro. Journal of Microbiology, Biotechnology and Food Sciences 2: 368-377.

CHAPTER 11

Pharmacokinetic/Pharmacodynamic Profiles of Tiamulin in an Experimental Intratracheal Infection Model of *Mycoplasma Gallisepticum*

Xia Xiao[1,2†], Jian Sun[1†], Tao Yang[1], Xi Fang[1], Jie Cheng[1], Yan Q. Xiong[3,4] and Ya-Hong Liu[1,2*]

[1]*National Reference Laboratory of Veterinary Drug Residues (SCAU), College of Veterinary Medicine, South China Agricultural University, Guangzhou, China*

[2]*Jiangsu Co-Innovation Centre for Prevention and Control of Important Animal Infectious Diseases and Zoonoses, Yangzhou, Jiangsu, China*

[3]*Los Angeles Biomedical Research Institute, Harbor-UCLA Medical Center, Torrance, CA, USA*

[4]*David Geffen School of Medicine, UCLA, Los Angeles, CA, USA*

ABSTRACT

Mycoplasma gallisepticum is the most important pathogen in poultry among four pathogenic *Mycoplasma* species. Tiamulin is a pleuromutilin antibiotic that shows a great activity against *M. gallisepticum* and has been approved for use in veterinary medicine particularly for poultry. However, the pharmacokinetic/pharmacodynamics (PK/PD) profiles of tiamulin against *M. gallisepticum* are not well understood. Therefore, in the current studies, we investigated the *in vivo* PK/PD profiles of tiamulin using a well-established experimental intratracheal infection model of *M. gallisepticum*.

The efficacy of tiamulin against *M. gallisepticum* was studied in 8-day-old chickens after intramuscular (i.m.) administration at 10 doses between 0–80 mg/kg. Liquid chromatography-tandem mass spectrometry (LC-MS/MS) was used to evaluate the PK parameters of tiamulin following i.m. administration at doses of 5, 40, and 80 mg/kg in *Mycoplasma gallisepticum*-infected neutropenic chickens. Real-time PCR (RT-PCR) was used for quantitative detection of *M. gallisepticum*. The MIC of tiamulin against *M. gallisepticum* strain S6 was 0.03 μg/mL. The PK/PD index, AUC_{24h}/MIC, correlated well with the *in vivo* antibacterial efficacy. The *in vivo* data suggest that animal dosage regimens should supply AUC_{24h}/MIC of tiamulin of 382.68 h for 2 $\log_{10}$ ccu equivalents *M. gallisepticum* reduction. To attain that goal, the administered dose is expected to be 45 mg/kg b.w. for treatment of *M. gallisepticum* infection with an MIC_{90} of 0.03 μg/mL.

Keywords: tiamulin, *M. gallisepticum*, *in vivo* PK/PD, chicken

INTRODUCTION

Mycoplasma gallisepticum (*M. gallisepticum*), a multi-host pathogen, is the most pathogenic agent of chronic respiratory disease (CRD) in poultry and turkey (1, 2). It can also be transmitted from poultry to house finches as well as other similar species and cause outbreaks of upper respiratory disease (3). Because it produces vast losses in the poultry industry, *M. gallisepticum* is considered as the most economically important pathogen of the four pathogenic *Mycoplasma* species (4).

Current means of controlling *M. gallisepticum* infections among avian species include biosurveillance practices, vaccines, and medication (5). Despite control measures, *M. gallisepticum* may be present in chick flocks that are maintained for long growing periods with minimal biosecurity. Some live attenuated *M. gallisepticum* vaccines are approved for use only within the commercial egg layer industry (6). The effectiveness of other subunit vaccines may be limited due to efficacy and associated costs (7). Although some vaccines are effective, *M. gallisepticum* outbreaks within vaccinated flocks has been observed (7), suggesting better medication control are required.

Antibiotics have been extensively used in the areas with intensive and varied population of poultry flocks for controlling *M. gallisepticum* or other pathogen infections (8). The use of antimicrobial agents has been considered as an economic method for controlling *M. gallisepticum* infections (8). Several kinds of antibiotics (macrolides, tetracyclines, fluoroquinolones, and pleuromutilin) have displayed good activity against *M. gallisepticum* (9, 10). However, it has also been noticed that antibacterial usage over time can induce resistance in this organism (11, 12).

Tiamulin is a semisynthetic derivative of the diterpene antibiotic pleuromutilin used in swine and poultry for treatment and prophylaxis of dysentery, pneumonia, and mycoplasmal infections (13). Although it has been used over 30 years, no significant changes of susceptibility on this pathogen were observed previously (11, 14). However, one recent investigation indicated that tiamulin resistant isolates were seen after tiamulin treatment (15). Therefore, it is important to study the optimal tiamulin regimen in order to maximize antibacterial activity and to prevent emergence of resistance. It is well-known that pharmacokinetic (PK) and pharmacodynamics (PDs) profiles of antibiotics provide useful information in the establishment of optimal dose regimens for better clinical setting management and prohibit resistance emergence (16, 17). To our best knowledge, PK/PD profiles of tiamulin against *M. gallisepticum* are very limited. Therefore, in the current experiment, we performed *in vivo* PK/PD studies of tiamulin against *M. gallisepticum* after intramuscular (i.m.) administration. Though the i.m. administration has not been approved for tiamulin in chicks, this study would provide foundation for tiamulin injectable formulation for chicks in future. The goals of the present investigations were to: (1) evaluate the PK profiles and dose proportionality of tiamulin in an intratracheal infection animal model; (2) provide the magnitude of the PK/PD index AUC_{24h}/MIC for different extent of efficacies; and (3) establish a rational dosage regimen that optimizes tiamulin efficacy with respect to bacteriological and clinical outcomes.

MATERIALS AND METHODS

Bacteria, Chemicals, Susceptibility Assay, and Animals

A well-characterized *M. gallisepticum* standard virulent strain S6 was purchased from the Chinese Veterinary Microorganism Culture Collection Center (Beijing, China). Tiamulin Fumarate (>99%) was kindly supplied by the Hebei Yuanzheng Pharmaceutical Company (Hebei, China). The MIC of tiamulin on the strain S6 was determined by a standard micro-dilution method according to recommended protocols (18, 19). Three hundred fifty Sanhuang chickens of 1-day-old weighting 35~45 g supplied by Guangdong Academy of Agricultural Sciences (Guangzhou, Guangdong, China) were used in this experiment. Birds were free of *M. gallisepticum* and fed with clean water and antibacterial-free fodder.

In Vitro Standard DNA Preparing

An *in vitro* DNA standard curve was established according to our previous report (19). Briefly, 36 h incubated *M. gallisepticum* medium was centrifuged for 10 min at 1,500 rpm and then resuspending the pellet in 0.6 mL fresh *M. gallisepticum* medium. 0.1 mL of the sample was serial diluted for bacteria counting by culture method [color change unit per millimeter (ccu/mL)]. Meanwhile, DNA was isolated from the sample and serial 10-fold dilutions (10^0–10^{-6}) prepared from the 0.6 mL sample with a bacteria DNA kit (Omega Bio-tek, Inc., Norcross, GA, USA). The DNA copies of *M. gallisepticum* were determined by real-time PCR (RT-PCR) (20). The DNA standard curve was plotted by the number of *M. gallisepticum* calculated from the culture method and cycle threshold (C_t) values obtained using RT-PCR results.

Neutropenia Model

Two days post arrival, the chicken neutropenia model published by our laboratory (19) was established in this study in order to eliminate the immunity variance of different chickens and study the efficacy of tiamulin solely, *via* intramuscular administration of cyclophosphamide at 60 mg/kg for 3 days (21). Birds were severely granulocytopenic (absolute leukocyte count <1,000/mm^3) and remained so for 8 days after the third injection of cyclophosphamide. The South China Agriculture University Animal ethics committee approved all *in vivo* experiments with an approved number of 2014-08. In addition, all husbandry practices and experimental operations were performed with full consideration of animal welfare.

M. gallisepticum Intratracheal Infection Model

M. gallisepticum mainly invades the respiratory system in chickens. Thus, an *M. gallisepticum* intratracheal infection model was utilized in this study according to Xia's report (19). Briefly, 24 h post the last dose of cyclophosphamide, 0.2 mL of solution containing approximately 10^8 color change unit (ccu) of the *M. gallisepticum* strain was inoculated intratracheally to neutropenic chickens for 3 days [95% infective dose (ID_{95}) for the studied strain]. Initial pathogen loading was quantified according to our established method described in our previous report (19). Briefly, at 24 h after the last infection dose, chickens were euthanized and trachea, air sac, and lungs were collected, homogenized in 2 mL PBS and centrifuged at 500 rpm for 5 min. An aliquot of 0.5 mL supernatant was used for DNA extraction with a bacterial DNA kit (Omega Bio-tek, Inc., Norcross, GA, USA) as described above. DNA copies of *M. gallisepticum* in these samples were measured by RT-PCR as described above. The amount of *M. gallisepticum* was calculated using the DNA copies *via* the *in vitro* standard curve.

Determination of DNA Copies of *M. gallisepticum* Using RT-PCR

The method determinating DNA copies of *M. gallisepticum* in different samples by RT-PCR were identical with our previous report (19). All RT-PCR reactions were performed on a BIO-RAD iQ 5 (Bio-Rad Laboratories, Inc., USA) using the SYBR premix Ex Taq™ (TaKaRa, Shiga, Japan). C_t values were defined as the cycle number yielding a maximum value of the second derivative of the amplification curve of the sample. Samples were defined as positive when both a measurable C_t and the expected T_m (±0.5°C) were seen. The standard samples and a negative control (elution buffer) were included in each run.

Tiamulin Pharmacokinetics in Neutropenic Intratracheal Infection Model

The infected neutropenic chickens were administrated with tiamulin intramuscularly at single doses of 5, 40, or 80 mg/kg b.w. 1.5 mL blood was sampled from the neck vein at 5, 10, 30 min, 1, 2, 4, 6, 8, 12, and 24 h after drug administration (10 chickens/time point). Blood samples were incubated immediately at room temperature for 1 h and then placed in 4°C for 2 h to enable clot retraction. Serum was obtained by centrifuging at 3,000 rpm for 10 min and frozen at −20°C immediately until analysis within 2 weeks. Tiamulin was extracted from serum by acetonitrile with a proportion of 2:1. Concentrations of tiamulin in serum were determined *via* a high-performance liquid chromatography-tandem mass spectrometry (LC-MS/MS) method, which was developed by our group and reported previously (22, 23). The recovery and precision were calculated by analysis of spiked samples at three concentration levels (5 replicates of each concentration). Mean recoveries of tiamulin that spiked at three concentration levels were in the range of 86.0–92.7%. The limit of quantitation (24) was confirmed at 2.5 ng/mL. The coefficient of correlation (r^2) was 0.9995 for the linear range of 2.5–500 ng/mL. The intra-day and inter-day coefficients of variation were determined to be 5.6 and 10.7%.

Efficacy of Tiamulin in Neutropenic Chicken Intratracheal Infection Model

To evaluate the efficacy of tiamulin at 24 h post-three day infections, either 0.85% NaCl (controls) or tiamulin at 5, 10, 20, 30, 40, 50, 60, 70, or 80 mg/kg were administrated to infected neutropenic chickens intramuscularly once daily for 3 days (five chickens/dose). At 24 h after the last drug administration, the amounts of *M. gallisepticum* in each chicken were calculated using the method described above.

Pharmacokinetics and Pharmacodynamics Analysis

The PK profiles of tiamulin were analyzed by the non-compartmental model with uniform weighting using the WinNonlin software (version 6.1; Pharsight, CA, USA). The surrogate marker of antibacterial activity, AUC_{24h}/MIC were calculated using *in vitro* MIC value and PK parameters obtained from three doses of i.m. administrations of tiamulin. The bacteria loading for each animal was calculated according to $C_t$values and the *in vitro* standard DNA curve. The efficacy of tiamulin was evaluated by the reduction of *M. gallisepticum* compared to the initial bacteria count before drug treatment. The *in vivo* PK/PD relationship of tiamulin against *M. gallisepticum* was studied using the sigmoid E_{max} model WINNONLIN software (version 6.1; Pharsight, CA, USA) with the equation as follows:

E=E0+Emax×CNeECN50+CNeE=E0+Emax×CeNEC50N+CeN

where E_0 is the change in $\log_{10}$ ccu equivalents/mL in the control sample (absence of tiamulin), E_{max} is the difference in $\log_{10}$ ccu equivalents/mL of the greatest amount of kill, C_e is the AUC_{24h}/MIC in the effect compartment, EC_{50} is the AUC_{24h}/MIC value producing a 50% reduction in bacterial counts, and N is the Hill coefficient that describes the steepness of the curve (25).

Dosage Calculation

In order to deduce a more rational regimen, the general formula was employed to estimate dosages for different magnitudes of efficiency (26).

$$\text{Dose} = \frac{CL_{\text{per hour}} \times (AUC_{24} / MIC) \times MIC_{90}}{F \times f_u}$$

where dose is the optimal dose (milligram/kilogram/day), CL is the body clearance (liter/kilogram/day), AUC/MIC is the breakpoint marker for the desired effect (hour), MIC_{90} is the MIC inhibiting 90% of strains (milligram/liter), F is the bioavailability, and f_u is the free drug fraction.

RESULTS

Susceptibility Testing

The MIC of tiamulin against the studied strain was 0.03 μg/mL.

In Vitro Standard DNA Curve

The correlation between the C_t values and $\log_{10}$ ccu/mL reached statistic significance with the equation of $y = -0.3087x + 10.44$ and R^2 of 0.9988 (Figure 1). The limitation of detection was 3×10^2 equivalents ccu/mL. The recovery rates at different dilutions were 56.5 ± 5.1%.

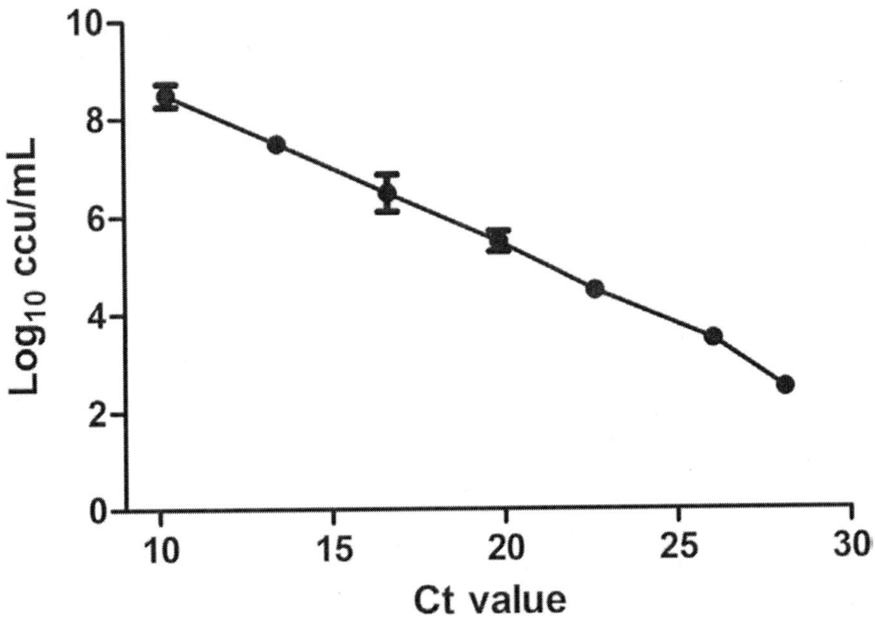

FIGURE 1. ***IN VITRO*** **DNA STANDARD CURVE**. Relationship of DNA standards between C_t value and culture results (Log_{10} ccu/mL). Mean of three different RT-PCR runs.

M. gallisepticum Intratracheal Infection Model

Major clinical symptoms, including depression, mouth breathing, eye closures, were observed from the infected animals. In addition, histopathology studies showed that airsacculitis as the cardinal symptom of *M. gallisepticum* infection was observed in 95% infected chickens. The mean *M. gallisepticum* load was 2.2 × 10^6 ccu equivalents/mL for all inoculated chickens. The morbidity and mortality rates were 95 and 17% at 5 days post-infection, respectively. Neither clinical symptoms nor airsacculitis were observed in the control animals. Bacteriological assays were also negative in the control group.

Tiamulin Pharmacokinetics Profiles in Neutropenic Intratracheal Infection Model

The main PK parameters are presented in Table 1 and Figure 2. The C_{max} were 2.05, 8.8, and 14 μg/mL for 5, 40, and 80 mg/kg doses, respectively, which were observed at 0.167 h after administration. The half-life ($T_{1/2\beta}$) was about 1.24 h for all three different doses. A second peak was observed for all the doses administered at 8–12 h. Importantly, a significant correlation between doses and AUC_{24h} was observed (R^2 = 0.999, Figure 3). As the AUC_{24h} were increasing in a dose-dependent manner from 5 to 80 mg/kg, the AUC_{24h} of other dose regimens were calculated according to the linear relationship. The parameters of CL were parameterized as CL/F because of the extravascular administration while MRT_{last} was the ratio of $AUMC_{last}/AUC_{last}$.

Table 1. Pharmacokinetic parameters of tiamulin in serum following intramuscular administration at a single dose of 5, 40, or 80 mg/kg in *m. Gallisepticum*-infected chickens (n = 10/group).

Parameters (units)	5 mg/kg	40 mg/kg	80 mg/kg
C_{max}(μg/mL)	2.05	8.8	14
T_{max}(h)	0.167	0.167	0.167
$T_{1/2\beta}$ (h)	1.15	1.39	1.17
AUC_{24h} (μg h/mL)	1.07	8.61	18.21
CL/F (L/h/kg)	4.53	4.6	4.37
MRT_{last} (h)	8.15	3.46	2.33

C_{max}, maximum serum concentration; T_{max}, time of maximum serum concentration; $T_{1/2\beta}$, elimination half-life; AUC_{24h}, 24 h area under serum concentration-time curve; CL/F, clearance divided by bioavailability; MRT_{last}, residence time.

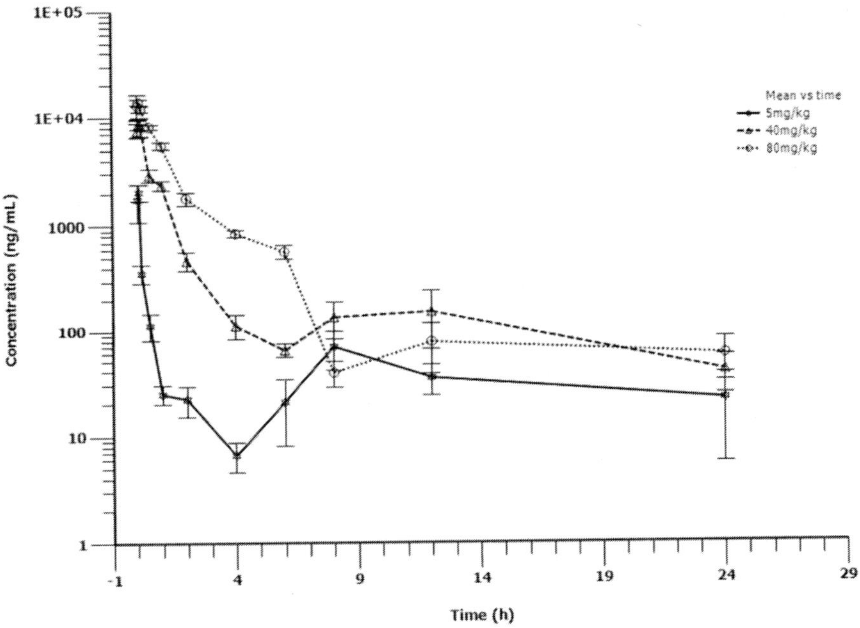

Figure 2. Serum concentrations of tiamulin following i.m. Administration at a single dose of 5, 40, or 80 mg/kg in *m. Gallisepticum* intratracheal infection model (n = 10/time point).

In Vivo Efficacy of Tiamulin in Neutropenic Chicken Intratracheal Infection Model

The C_t values of respiratory tissues of chickens administrated with doses from 5 to 80 mg/kg increased with the increasing of doses, implying that the bacteria loading decreased with the dose increasing. The $C_t$value increased sharply between dosages of 10 and 50 mg/kg while gently from 50 to 80 mg/kg (Figure 4). Limbs Twitch was observed in chickens treated with doses of 60, 70, 80 mg/kg for about 10 min; however, no death emerged.

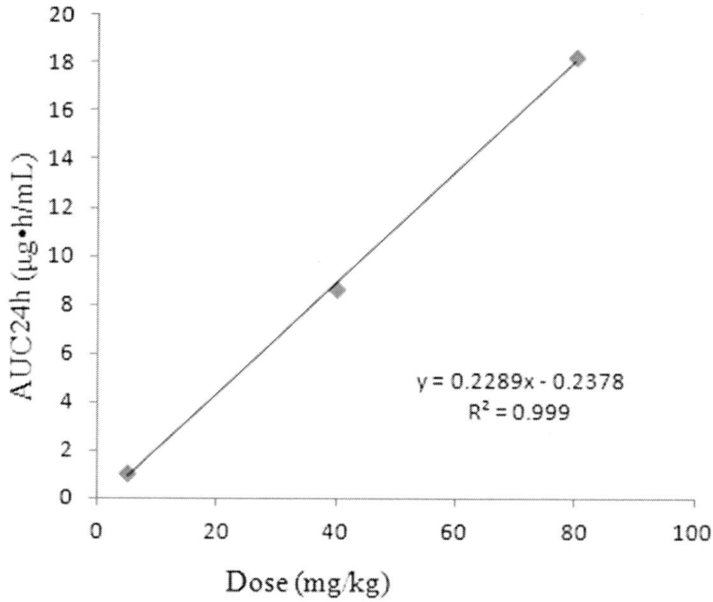

Figure 3. Linear regression plots between administered dose and auc24h values.

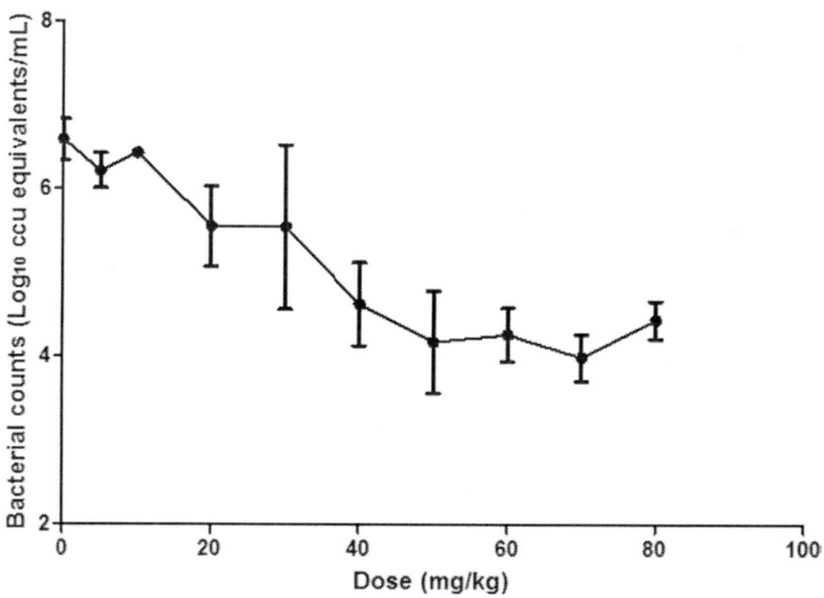

Figure 4. The calculated *in vivo m. Gallisepticum* counts after tiamulin treatment. **(*n* = 4–7/dose).**

Tiamulin PK/PD Profiles

The PK/PD indices AUC_{24h}/MIC was integrated using the PK parameters, dose proportionality, and MIC data. The effect (*E*) was calculated as the reduction of *M. gallisepticum* using the unit of Log_{10} ccu equivalent/mL. The relationship between the ratio of AUC_{24h}/MIC and antibiotic efficacy was described using the Sigmoid E_{max} model (Figure 5). The AUC_{24h}/MIC corrected well with the effects ($R^2 = 0.8979$). The values of AUC_{24h}/MIC for mycoplasmastasis (0 $\log_{10}$ ccu/mL reduction), for 1 $\log_{10}$ ccu reduction and 2 $\log_{10}$ ccu reduction are 98.98, 206.56, 382.58, respectively (Table 2). The EC_{50} was 211.19 h, and the slope of the graph (*N*) was 3.68.

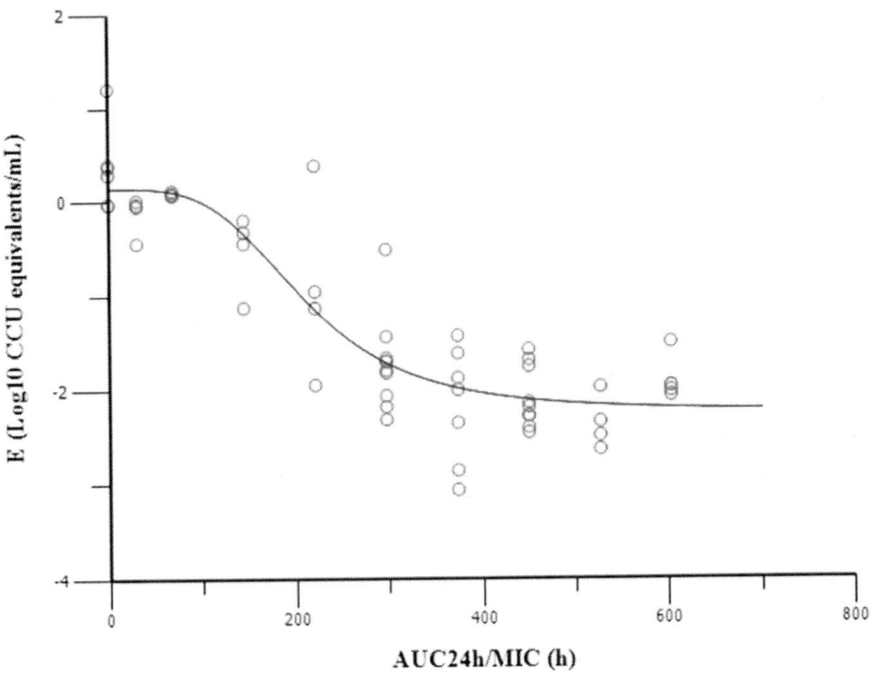

Figure 5. Sigmoid E_{max} Relationships Of Tiamulin Between Antimycoplasmal Effect (*E*, Log_{10} Ccu Equivalents/Ml) And *In Vivo* Auc_{24h}/ Mic Ratio Against Mg In Serum Of Chickens.

Table 2. Tiamulin pk/pd analysis with the parameter of auc_{24h}/mic in *m. Gallisepticum* intratracheal infection model.

Parameter	Value	SD
E_{max} (Log_{10} ccu equivalents/mL)	2.38	0.24
E_0 (Log_{10} ccu equivalents/mL)	0.14	0.13
EC_{50} (h)	211.19	20.3
AUC_{24h}/MIC for 0 log_{10} ccu equivalents/mL reduction (h)	98.98	
AUC_{24h}/MIC for 1 log_{10} ccu equivalents/mL reduction (h)	206.56	
AUC_{24h}/MIC for 2 log_{10} ccu equivalents/mL reduction (h)	382.58	
Slope (*N*)	3.68	1.2

E_0 is the change in Log_{10} ccu equivalents/mL after 24 h incubation in the control sample (absence of drug) compared with the initial inoculum. E_{max} is the difference in effect of the greatest amount of kill. EC_{50} is the AUC_{24h}/MIC value producing a 50% reduction in bacterial counts from the initial inoculum. AUC_{24h}/MIC is the 24 h area under concentration-time curve/minimum inhibitory concentration ratios. N is the Hill coefficient that describes the steepness of the AUC_{24h}/MIC-effect curve.

Dosage Calculation

To calculate dosages, the bioavailability was taken into account owing to the extravascular route of administration, and the free drug fraction was not required for using PD data generated *in vivo*. In order to provide a dosage regimen attaining maximum effect for chickens infected with *M. gallisepticum* with an MIC_{90} of 0.03 μg/mL, a dose of 45 mg/kg for 3 days was recommended.

DISCUSSION

M. gallisepticum, a highly transmissible, persistent pathogen in chickens, turkeys, and some wild birds, causes considerable economical losses to the poultry industry all over the world (27). Burch and Valks (28) had clarified that tiamulin possessed excellent activity against 241 *M. gallisepticum* strains by comparing the MIC of tiamulin against *M. gallisepticum* and the C_{max} or concentration at steady state (C_{ss}) in chicken blood in 2002 (28). However, its chemotherapeutic properties are not well described. In addition, the use of PK/PD model to identify the PD activity by integrating the PK characters, MIC, and pathogen loading outcome has proven helpful in designing rational dosage regimens in humans and animals (29, 30). Burch and Alvarez (31) evaluated the relationship between C_{max}/MIC or AUC/MIC and prevention or treatment results of tiamulin in *M. gallisepticum* infection in 3-week-old chickens (31). The PK/PD parameters in his study were with some rough as he used the reported PKs of tiamulin in 7-week-old health chickens to predict that in 3-week-old *M. gallisepticum*-infected chickens. And the PDs of that study was evaluated by scores of lesions, and whether *M. gallisepticum* was isolated or not from air sacs. It is known that isolation of *M. gallisepticum* is easy to influence by overgrowth of faster growing *Mycoplasma* species or impeded by other organisms or no growth in subculture, so it may not be a good method to judge the PDs. What's more, the study did not calculate the PDs from the aspect of bacteria loading reduction, which is important in PK/PD study. Thus, in this study, *in vivo* PK/PD profiles of tiamulin were established using PK characters in infected chickens, MIC, and pathogen-loading outcome.

This was the first report about the PKs of tiamulin in *M. gallisepticum* intratracheal-infected chicks. Tiamulin was rapidly absorbed with the peak concentrations for 5, 40, 80 mg/kg dose of 2.05, 8.8, 14.0 μg/mL achieved at 0.167 h. The half-life was in the range of 1.15–1.39 h, which was lower than the value reported in dogs (4.7 h) (32), which indicated that the elimination of tiamulin was of significant difference among species. The rate and extent of absorption was constant among the tested doses with the values of clearance divided by bioavailability (CL/F) of 4.53, 4.6, 4.37 L/h/kg, respectively. That is to say, tiamulin showed dose-dependent PKs when given as a single i.m. dose of 5, 45, or 80 mg/kg b.w. This was in accordance with the report that tiamulin showed dose proportionality in the range of 10–25 mg/kg b.w. in dogs (32). This phenomenon was also confirmed in our previously research with valnemulin, which is also a semisynthetic derivative of the diterpene antibiotic pleuromutilin (19). A second peak was observed for all the doses administration at 8–12 h. Tiamulin mainly accumulates and metabolizes *via* liver. It was speculated that, according to the chemical structure, a relatively high liposolubility of tiamulin would result in enterohepatic recycling, which is possibly the main reason that the second peak was observed (33).

The parameters of tiamulin (AUC_{24h}) that showed dose proportionality in the range of 5–80 mg/kg following i.m. administration allowed us to calculate the AUC_{24h} for other dosage administrations. As reported for other pleuromutilin derivatives (34, 35), the PK/PD surrogate of pleuromutilin derivatives is AUC_{24h}/MIC. The data from the present multiple dosage studies of tiamulin confirmed the conclusion that the AUC_{24h}/MIC was the PK/PD index of tiamulin. The values of AUC_{24h}/MIC for mycoplasmastasis (0

$\log_{10}$ ccu equivalents reduction), activity of 1 $\log_{10}$ ccu equivalents reduction, and 2 $\log_{10}$ ccu equivalents reduction were 124, 205, and 327 h, respectively, which were much lower than those obtained from an *in vivo* study of valnemulin (28,820, 38,030, and 56,256 h, respectively) (36). One possible reason for the differences was that the concentration in serum was not that real one act on *M. gallisepticum* as the infection site is air sac or respiratory system, and antibiotic concentrations in air sac or lung are usually different from those in serum (14). Another reason could be that the distribution volume of timulin was much wider than that of valnemulin in chickens, which means that the tissue concentration of tiamulin maybe higher than that of valnemulin (36). Tiamulin did not reach a mycoplasmacidal effect (3 $\log_{10}$ ccu equivalents reduction), which was in accordance with Burch and Alvarez's (31) results that even tiamulin concentration far exceeded the MBC could not eliminate the pathogen in treatment procedure (31).

As was observed in this investigation, doses over 60 mg/kg would trigger neuro-toxicity. The calculated dose of 45 mg/kg for 3 days was sufficient for clinical treatment of *M. gallisepticum* infection with an MIC_{90} of 0.03 μg/mL. This regimen was in the range of recommended doses for drinking water (30–60 mg/kg) in poultry but more than twice of the recommended dose range for i.m. administration (10–20 mg/kg) in pigs (13). It was also lower than the neuro-toxicity dose. Though the i.m. administration has not been approved for tiamulin in chicks, this result would provide foundation for tiamulin injectable formulation for chicks in future.

The conventional method for evaluating the efficacy of an antibacterial agent is bacteria counting by culture method. However, the cultivation technique is expensive and laborious and time consuming (37). Also, this gold standard for *M. gallisepticum* was usually influenced by overgrowth of faster growing *Mycoplasma* species or impeded by other organisms or no growth in subculture. The RT-PCR had been confirmed in a previous study for qualitative and quantitative detection of *M. gallisepticum* (36). The specificity, sensitivity, and reproducibility were sufficient for quantitative detection of *M. gallisepticum* in clinical samples (20).

Neutropenic chickens were used in this study to evade host immunity factors that may play important roles in the *M. gallisepticum* infections as well as the efficacy of antimicrobial therapy. A previous study suggested that the PK/PD index magnitude necessary for successful therapy is reduced in animal models in the presence of neutrophils (38). Therefore, further studies on elucidation of host immunity–pathogen interactions are essential for the drug. Furthermore, a population PK approach should be conducted in the future to derive a population clearance of tiamulin for use in dose calculation. In addition, MIC distribution should be evaluated to take the variation of susceptibility of *M. gallisepticum* to tiamulin into account.

The present study characterized the *in vivo* activity of tiamulin against *M. gallisepticum* in a neutropenic chicken model. The PK/PD surrogate AUC_{24h}/MIC correlated well with the *in vivo* antibacterial efficacy. The *in vivo* data suggest that animal dosage regimens should supply AUC_{24h}/MIC of tiamulin of 382.68 h for 2 $\log_{10}$ ccu equivalents *M. gallisepticum* reduction. These studies suggest that tiamulin, if used for treatment of *M. gallisepticum* infection with an MIC_{90} of 0.03 μg/mL, would benefit from 45 mg/kg i.m. once daily for 3 days.

AUTHOR CONTRIBUTIONS

XX, JS, and Y-HL designed this study; TY, XF, JC, and XX carried out the whole experiments; YX helped to analysis the data and revised this manuscript; XX and JS analyzed the data and write this manuscript and Y-HL revised it.

CONFLICT OF INTEREST STATEMENT

The authors declare that the research was conducted in the absence of any commercial or financial relationships that could be construed as a potential conflict of interest.

FUNDING

This study was supported by a research program for Changjiang Scholars, Innovative Research Team in University of Ministry of Education of China (Grant IRT13063), Science and Technology Planning Project of Guangdong Province China (Grant 2012A 020800004). The funders had no role in study design, data collection, and analysis, decision to publish, or preparation of the manuscript.

REFERENCES

1. Adelman JS, Kirkpatrick L, Grodio JL, Hawley DM. House finch populations differ in early inflammatory signaling and pathogen tolerance at the peak of *Mycoplasma gallisepticum* infection. *Am Nat* (2013) 181(5):674–89.
2. Abolnik C, Gouws J. Extended survival times of *Mycoplasma gallisepticum* and *Mycoplasma synoviae* on kanekalon synthetic hair fibres. *Poult Sci* (2014) 93(1):8–11.

3. Dhondt AA, DeCoste JC, Ley DH, Hochachka WM. Diverse wild bird host range of *Mycoplasma gallisepticum* in eastern North America. *PLoS One* (2014) 9(7):e103553.
4. Mohammed HO, Carpenter TE, Yamamoto R. Economic impact of *Mycoplasma gallisepticum* and *M. synoviae* in commercial layer flocks. *Avian Dis* (1987) 31(3):477–82.
5. Kleven SH. Control of avian mycoplasma infections in commercial poultry. *Avian Dis* (2008) 52(3):367–74.
6. Peebles ED, Jacob R, Branton SL, Evans JD, Leigh SA, Gerard PD. Effects of different vaccine combinations against *Mycoplasma gallisepticum* on the digestive and reproductive organ characteristics of commercial egg-laying hens. *Poult Sci* (2015) 94(12):2898–904.
7. Lin H, Tian Y, Zhang Z, Wu L, Chen Y. Quantification of piperazine phosphate in human plasma by high-performance liquid chromatography-electrospray ionization tandem mass spectrometry employing precolumn derivatization with dansyl chloride. *Anal Chim Acta* (2010) 664(1):40–8.
8. Stipkovits L, Burch DGS. Antibiotic resistance of mycoplasmas of chickens and turkey origin. *Page 179 in Proceedings Xth World Veterinary Poultry Association Congress, Sydney, Australia*. Sydney, Australia: Australian Veterinary Poultry Association (1993). Abstr. 121.
9. Bradbury JM, Yavari CA, Giles CJ. In vitro evaluation of various antimicrobials against *Mycoplasma gallisepticum* and *Mycoplasma synoviae* by the micro-broth method, and comparison with a commercially-prepared test system. *Avian Pathol* (1994) 23(1):105–15.
10. Hannan PC, Windsor GD, de Jong A, Schmeer N, Stegemann M. Comparative susceptibilities of various animal-pathogenic mycoplasmas to fluoroquinolones. *Antimicrob Agents Chemother* (1997) 41(9):2037–40.
11. Lin MY. In vitro comparison of the activity of various antibiotics and drugs against new Taiwan isolates and standard strains of avian mycoplasma. *Avian Dis* (1987) 31(4):705–12.
12. Gerchman I, Lysnyansky I, Perk S, Levisohn S. In vitro susceptibilities to fluoroquinolones in current and archived *Mycoplasma gallisepticum* and *Mycoplasma synoviae* isolates from meat-type turkeys. *Vet Microbiol* (2008) 131(3–4):266–76.
13. EMA. (1999). Available at: http://www.ema.europa.eu/docs/en_GB/document_library/Maximum_Residue_Limits_-_Report/2009/11/WC500015563.pdf
14. Pakpinyo S, Sasipreeyajan J. Molecular characterization and determination of antimicrobial resistance of *Mycoplasma gallisepticum* isolated from chickens. *Vet Microbiol* (2007) 125(1–2):59–65.
15. Pridmore A. *Antibacterial Activity of Tiamulin, Valnemulin, Tylosin and Lincomycin against Brachyspira and Mycoplasma Isolates: Determination of Minimum Inhibitory Concentration (MIC)*. Report to Novartis (2008).
16. Andes D, Craig WA. Animal model pharmacokinetics and pharmacodynamics: a critical review. *Int J Antimicrob Agents* (2002) 19(4):261–8.
17. Rogers GB, Hoffman LR, Doring G. Novel concepts in evaluating antimicrobial therapy for bacterial lung infections in patients with cystic fibrosis. *J Cyst Fibros* (2011) 10(6):387–400.
18. Hannan PC. Guidelines and recommendations for antimicrobial minimum inhibitory concentration (MIC) testing against veterinary *Mycoplasma* species. International Research Programme on Comparative Mycoplasmology. *Vet Res* (2000) 31(4):373–95.
19. Xiao X, Sun J, Yang T, Fang X, Wu D, Xiong YQ, et al. In vivo pharmacokinetic/pharmacodynamic profiles of valnemulin in an experimental intratracheal *Mycoplasma gallisepticum* infection model. *Antimicrob Agents Chemother* (2015) 59(7):3754–60.
20. Mekkes DR, Feberwee A. Real-time polymerase chain reaction for the qualitative and quantitative detection of *Mycoplasma gallisepticum*. *Avian Pathol* (2005) 34(4):348–54.
21. Craig WA, Andes DR. In vivo pharmacodynamics of ceftobiprole against multiple bacterial pathogens in murine thigh and lung infection models. *Antimicrob Agents Chemother* (2008) 52(10):3492–6.
22. Hu F, Bian K, Liu Y, Su Y, Zhou T, Song X, et al. Development of a modified QUick, Easy, CHeap, Effective, Rugged and Safe method for the determination of multi-class

antimicrobials in vegetables by liquid chromatography tandem mass spectrometry. *J Chromatogr A* (2014) 1368:52–63.

23. Hu FY, He LM, Yang JW, Bian K, Wang ZN, Yang HC, et al. Determination of 26 veterinary antibiotics residues in water matrices by lyophilization in combination with LC-MS/MS. *J Chromatogr B Analyt Technol Biomed Life Sci* (2014) 94(9–950):79–86.
24. Patel N, Pai MP, Rodvold KA, Lomaestro B, Drusano GL, Lodise TP. Vancomycin: we can't get there from here. *Clin Infect Dis* (2011) 52(8):969–74.
25. Burton ME, Shaw LM, Schentag JJ, Evans WE. *Applied Pharmacokinetics & Pharmacodynamics: Principles of Therapeutic Drug Monitoring*. Philadelphia: Lippincott Williams and Wilkins Press (2005). p. 61–5.
26. Toutain PL, Bousquet-Melou A, Martinez M. AUC/MIC: a PK/PD index for antibiotics with a time dimension or simply a dimensionless scoring factor? *J Antimicrob Chemother* (2007) 60(6):1185–8.
27. Evans JD, Leigh SA, Branton SL, Collier SD, Pharr GT, Bearson SMD. *Mycoplasma gallisepticum*: current and developing means to control the avian pathogen. *J Appl Poult Res* (2005) 14:757–63.
28. Burch DGS, Valks M. Comparison of minimal inhibitory concentrations (Mic) against chicken mycoplasma of tiamulin and other antimicrobials and their concentrations in the blood. *Proceedings of the 12th World Veterinary Poultry Congress*. Cairo, Egypt: (2002). 322 p.
29. Andes D, Craig WA. In vivo activities of amoxicillin and amoxicillin-clavulanate against *Streptococcus pneumoniae*: application to breakpoint determinations. *Antimicrob Agents Chemother* (1998) 42(9):2375–9.
30. Craig WA. Pharmacokinetic/pharmacodynamic parameters: rationale for antibacterial dosing of mice and men. *Clin Infect Dis* (1998) 26(1): 1–10; quiz 11–12.
31. Burch DGS, Alvarez R. Cmax & AUC/MIC (AUIC) relationships for tiamulin prevention and treatment of *Mycoplasma gallisepticum* infections in chickens. *The 2nd International Conference on Antimicrobial Agents in Veterinary Medicine*. Ottawa, Canada: (2004). 29 p.
32. Laber G. Investigation of pharmacokinetic parameters of tiamulin after intramuscular and subcutaneous administration in normal dogs. *J Vet Pharmacol Ther* (1988) 11(1):45–9.
33. Mostafavi SA, Foster AT. A double-peak phenomenon in the pharmacokinetics of acebutolol enantiomers after oral administration: discontinuous absorption of acebutolol. *Daru J Faculty Pharm* (2002) 10(4):141–7.
34. Novak R. Are pleuromutilin antibiotics finally fit for human use? *Ann N Y Acad Sci* (2011) 1241:71–81.
35. Zhao DH, Zhou YF, Yu Y, Shi W, Yang X, Xiao X, et al. Integration of pharmacokinetic and pharmacodynamic indices of valnemulin in broiler chickens after a single intravenous and intramuscular administration. *Vet J* (2014) 201(1):109–15.
36. Xiao X, Zhao DH, Yang X, Shi W, Deng H, Ma J, et al. *Mycoplasma gallisepticum* and *Escherichia coli* mixed infection model in broiler chickens for studying valnemulin pharmacokinetics. *J Vet Pharmacol Ther* (2014) 37(1):99–102.
37. Garcia M, Jackwood MW, Levisohn S, Kleven SH. Detection of *Mycoplasma gallisepticum, M. synoviae,* and *M. iowae* by multi-species polymerase chain reaction and restriction fragment length polymorphism. *Avian Dis* (1995) 39(3):606–16.
38. Pechere M, Letarte R, Pechere JC. Efficacy of different dosing schedules of tobramycin for treating a murine *Klebsiella pneumoniae* bronchopneumonia. *J Antimicrob Chemother* (1987) 19(4):487–91.

CHAPTER 12

Evaluation of Bioequivalence of Two Long-Acting 20% Oxytetracycline Formulations in Pigs

Zhixin Lei[1], Qianying Liu[1], Bing Yang[1], Saeed Ahmed[1], Jincheng Xiong[1], Tingting Song[1], Pin Chen[1], Jiyue Cao[1]* and Qigai He[2]*

[1]*National Reference Laboratory of Veterinary Drug Residues, MAO Key Laboratory for Detection of Veterinary Drug Residues, Huazhong Agriculture University, Wuhan, China*

[2]*State Key Laboratory of Agriculture Microbiology, College of Veterinary Medicine, Huazhong Agriculture University, Wuhan, China*

ABSTRACT

The aim of this study was to explore the bioequivalence of long-acting oxytetracycline in two formulations, a reference formulation (Terramycin 20% LA, Pfizer) and a test one (Kangtekang 20% LA, Huishen). Both formulations were administered intramuscularly at 20 mg/kg body weight at each of 24 healthy animals during a two-period crossover parallel experimental design. The oxytetracycline (OTC) concentrations in plasma were measured by high-performance liquid chromatography, and the limit of quantification was 0.05 μg/ml with a recovery ratio of above 90%. Moreover, the descriptive pharmacokinetics parameters (C_{max}, AUC_{0-144h}, and $AUC_{0-\infty}$) were calculated and compared under analysis of variance, and 90% confidence interval (CI) were compared, except for T_{max} analyzed by non-parametric tests based on Wilcoxons's signed rank test. The comparison results of C_{max}, AUC_{0-144h}, $AUC_{0-\infty}$, and T_{max}were 5.066 ± 0.486, 5.071 ± 0.877 μg/ml, 118.926 ± 13.259, 126.179 ± 17.390 μg h/ml, 123.087 ± 13.906, 130.732 ± 18.562 μg h/ml, 0.740 ± 0.278, 0.650 ± 0.258 h, respectively, and did not reveal any significant differences. In addition, 90% CIs of these ratios for reference and test product were within an interval of 80–125%, and the relative bioavailability of test one was (94.291 ± 15.287)%. Therefore, it has been concluded that test OTC was bioequivalent to the reference formulation in pigs.

Keywords: oxytetracycline, pharmacokinetics, bioequivalence, formulations, confidence interval

INTRODUCTION

Oxytetracycline (OTC), a broad-spectrum antibiotic, is widely used to treat Gram-positive (*Streptococcus* sp., *Staphylococcus* sp.) and Gram-negative (*Escherichia coli*) infections in animals (1–3). OTC is a tetracycline derivative produced by *Streptomyces rimosus* (4, 5). Long-acting oxytetracycline (LOTC) formulations are used to treat disease by maintaining an effective concentration in animals for 2 or 3 days. It is manufactured easily and used widely in developing countries as veterinary medicine (6). Long-acting injection can be performed to increase OTC and keep high concentrations in plasma above minimum inhibitory concentrations for several days, decreasing the number of administration per treatment (7, 8). As first LOTC receiving a market authorization, Terramycin LA (Pfizer) is the reference formulation for bioequivalence studies (6, 9–11).

Bioequivalence studies play an important role in new animal drug implementation and to support supplemental application in case of modification of dosage forms, administration routes, or manufacturing process that may have a significant effect on bioavailability (12, 13). Bioequivalence guidelines are established by respective organizations, while the most well-known guidelines are those of Europe and America. As a guideline to bioequivalence in the US, two products are considered for bioequivalence studies, when they are as follow bioequivalent, those in therapeutic ingredient or active ingredients, and available at the site of drug action, affected drug in assimilation rate and extent (14–18). In addition, a guideline in Europe states that bioequivalence studies should be performed within acceptable limits for the active ingredients between two products or administration routes under appropriate conditions (19–21).

According to the European medicines agency guidelines, a pharmacokinetic parameters comparison between two formulations is the best way for a bioequivalence examination of veterinary drugs, in which the area under the plasma concentration time curve to last concentration ($AUC_{0\text{-}t}$), the area under the plasma concentration time curve extrapolated to infinity ($AUC_{0\text{-}\infty}$), the peak maximum plasma concentration (C_{max}), and the time to maximum concentration (T_{max}) are used for bioequivalent analysis (21). Once the bioequivalence has been demonstrated between two formulations, the clinical efficacy of test formulation is equivalent to those observed during the clinical trials of the reference formulation (22–24).

For OTC bioequivalence and pharmacokinetics studies, most investigations have been conducted in cattle and chickens, but fewer in pigs (6, 10, 25–28).

MATERIALS AND METHODS

Drugs and Reagents

Oxytetracycline standard (96.5%), lot number: 15820000, was used for calibration of the analytical method. Two kinds of commercial products of OTC injection containing 20% OTC—a test one (Kangtekang 20% LA, Huishen, lot number: 10453036AA, produced from China) and a reference one (Terramycin 20% LA, Pfizer, lot number: 20050802, produced from America)—were selected for bioequivalence study. The other reagent used in this study was analar, purchased from Sinopharm Group Shanghai Chemical Reagent Co., Ltd.

Animals

Twenty-four landrace healthy pigs (12 males, 12 females), 6 weeks old and weighing 25–30 kg were purchased from the Livestock and Poultry Breeding Center of Hubei Province (Wuhan, China). Pigs were raised with water and antibiotic-free feed to acclimatize for 1 week prior to first drug administration. The environment temperature and relative humidity of the pig enclosure were kept at 20–26°C and 45–65%, respectively.

Bioequivalence Study Design and Sample Collection

The bioequivalence study was performed according to a crossover design with a washout period of 8 days. Pigs were divided randomly into groups A and B of 12 pigs by groups. Pigs in group A received the test product

(Kangtekang 20% LA, Huishen) in the first period and the reference product (Terramycin 20% LA, Pfizer) in the second period and group B received drugs in the opposite order. The two products were administered intramuscularly at a dose of 20 mg/kg body weight. All animals were bred with water and antibiotic-free feed.

Animals were sampled during the following 6 days after i.m. administrating OTC products during each period. Plasma samples (3 ml) were collected at 0.083, 0.25, 0.5, 1, 2, 3, 6, 8, 12, 24, 36, 48, 72, 96, 120, and 144 h into tubes containing heparin after i.m. administration at a single dose of 20 mg/kg body weight. Samples were centrifuged at 3,000 rpm/s for 10 min, and then blood serum was obtained. Samples were stored at −20°C until analyzed by high-performance liquid chromatography (HPLC).

HPLC Determination of OTC Concentration

A C18 reverse-phase column (250 mm × 4.6 mm, 5 μm i.d.; Agilent, USA) was used for HPLC, which was performed with a 355-nm detection wavelength at 30°C. Phases A, B, and C were 0.01 mol/l oxalic acid, acetonitrile, and methanol, respectively, which constituted the mobile phase of HPLC with a volume ratio (A:B:C, 83:10:7) for gradient elution. Plasma samples of 0.5 ml were mixed with 100 μl 10% perchloric acid with vortex oscillation at 3,000 rpm/s, then extracted to obtain supernatant liquid. Supernatants were evaporated to dryness under nitrogen at 45°C, followed by resuspension in the mobile phase at the initial volume, and then 20 μl of resuspension was injected directly into the HPLC system with 0.22 μm organic membranes.

Pharmacokinetic and Bioequivalence Analysis

Descriptive PK parameters were obtained with WinNonlin Professional software (Version 5.2.1) (Certara, USA). The maximum plasma concentrations (C_{max}) and time to C_{max} (T_{max}) were obtained from the plasma concentration versus time data. AUC were calculated by the linear trapezoidal rule until the last sampling time (AUC_{0-144h}) and with extrapolation to infinity ($AUC_{0-\infty}$). $T_{1/2}$ was calculated from the terminal slope (β) estimated by log-linear regression according VICH guideline.

Analysis of variance (ANOVA) was used to compare and assess the effect of the formulations on the test formulation compared with the classical ones, with logarithmically transformed parameters of AUC_{0-144h}, $AUC_{0-\infty}$ (Bioequivalence, August 2015) as recommended by technical guidelines for veterinary drug in China (29, 30). However, T_{max} comparison was performed with a Wilcoxon signed rank test. Parametric 90% confidence intervals (CIs) of the mean of test/reference ratios of AUC_{0-144h}, $AUC_{0-\infty}$, and C_{max} were calculated using the residual variance of ANOVA with the assumption of a multiplicative model. CIs were calculated with SPSS analysis (IBM, USA).

Statistical Analysis

The p values of <0.05 and <0.01 were considered statistically significant and extremely significant, respectively ($*p < 0.05$ and $**p < 0.01$).

RESULTS

OTC HPLC Analysis in Plasma

The proposed method was suitable for OTC quantification in plasma, showing good selectivity above 89% recovery and a good linear relationship from 0.05 to 10 μg/ml. The chromatogram in Figures 1A–D shows the blank, the lower limit of quantification (LLOQ), and measured samples at 5 min after i.m. of two formulation products in plasma, which indicate the proposed method for OTC detection are specific and accurate. The typical regression equation was $y = 0.0507A - 0.0238$, $R^2 = 0.9999$. The LLOQ was 0.05 μg/ml in plasma presented in the Table 1. The coefficient of determination (R^2) of standard curves from 0.05 to 10 μg/ml was 0.9999, and the inter-day and intra-day coefficient variation were below 5% in plasma. In addition, the recovery ratios were in the range of 89.13 ± 2.54 to 92.19 ± 3.75% in plasma (Table 2).

Table 1. Sensibility of oxytertracycline in plasma.

Samples	LLOD (μg/ml)	Lower limit of quantification (μg/ml)
Plasma	0.025	0.05

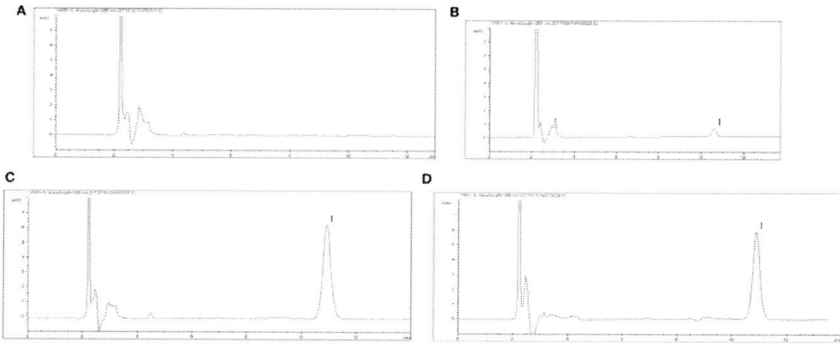

FIGURE 1. CHROMATOGRAMS OF OXYTETRACYCLINE (OTC) IN PLASMA. **(A)** Control blank group in plasma, **(B)** lower limit of quantification (0.05 μg/ml) of OTC in plasma, **(C)** chromatograms of plasma at 1 h after i.m. 20% OTC reference injection, **(D)** chromatograms of plasma at 1 h after i.m. 20% OTC test one, *1*, the peak time of OTC.

Table 2. The result (mean ± sd) of method validation of oxytertracycline in plasma.

Concentration (μg/ml)	Recovery (%)	Inter-day CV (%)	Intra-day CV (%)	Accuracy (RE%)
0.1	92.19 ± 3.75	3.17 ± 0.75	3.61 ± 0.65	7.96
1.0	90.80 ± 2.40	2.04 ± 0.76	4.07 ± 0.86	9.23
10	89.13 ± 2.54	2.08 ± 0.77	2.43 ± 0.78	10.86

CV, coefficient of variability, RE, the accuracy.

Pharmacokinetic Analysis

The mean ± SD of OTC concentration time profile are presented in Figure 2 after i.m. of two formulations, and the main descriptive PK parameters are reported in Table 3.

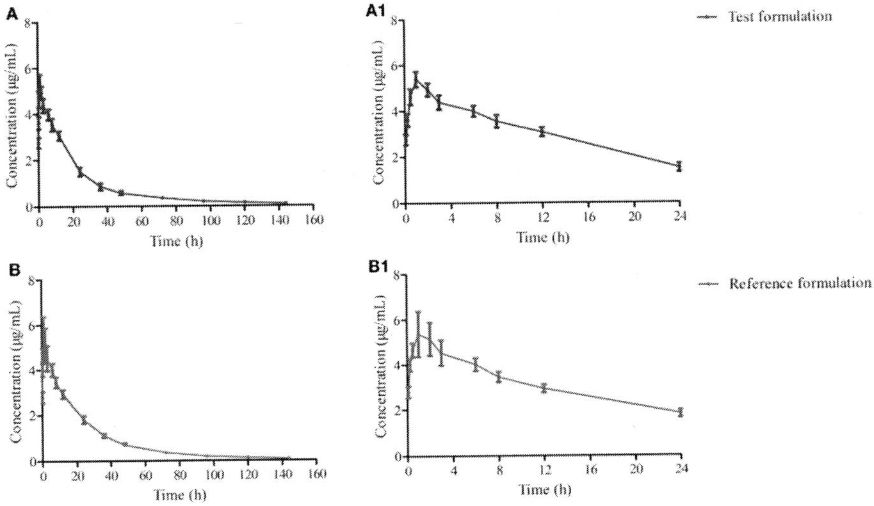

Figure 2. The mean concentration time curves of test and reference formulations in plasma after i.m. (a) present the test one, (a_1) present amplification of panel (a) at 0–24 h, (b) present the reference one, (b_1) present amplification of panel (b) at 0–24 h.

Table 3. Pharmacokinetic parameters for the test formulation and reference formulation, *p* value and relative fraction.

Parameters	Unit	Test	Reference	Analysis of variance	*F* (%)
AUC_{0-144h}	μg h/ml	118.976 ± 13.259	126.179 ± 17.390	0.976	
$AUC_{0-\infty}$	μg h/ml	123.087 ± 13.906	130.732 ± 18.562	0.313	94.291 ± 15.287
C_{max}	μg/ml	5.066 ± 0.486	5.071 ± 0.677	0.874	
T_{max}	h	0.740 ± 0.278	0.650 ± 0.258	>0.05[a]	

F represents relative bioavailability.
[a]Wilcoxon test.

Bioequivalence Analysis

Log-transformed C_{max}, AUC_{0-144h}, $AUC_{0-\infty}$ and untransformed T_{max} of the test formulation (Kangtekang 20% LA, Huishen) were compared with the reference one (Terramycin 20% LA, Pfizer) for a bioavailability study with ANOVA analysis and 90% CI. No statistically significant differences were observed for C_{max}, AUC_{0-144h}, $AUC_{0-\infty}$ in Table 3. The relative bioavailability of the test product compared to the reference one was 94.291 ± 15.287% (Table 3).

The two one-sided *T*-tests and 90% CI analysis displayed the ratios mean of log-transformed C_{max}, AUC_{0-144h}, $AUC_{0-\infty}$, and 90% CI on test to reference formulations were 99.5, 98.8, 99.1% and 91.9–107.2, 99.6–100.7, 92.0–102.1%, which were all in the range of 75–143% within the bioequivalence acceptance range (Table 4).

Table 4. Two one-sided *t*-test and 90% of confidence interval (ci).

Parameters	t_1	t_2	90% CI	Ratio (*T*/*R*) (%)	Acceptable range (%)
AUC_{0-144}	4.928*	9.812	99.6–100.7	98.8	80–125
$AUC_{0-\infty}$	6.335*	8.405	92.0–102.1	99.1	80–125
C_{max}	11.486*	12.108	91.9–107.2	99.5	75–143

*Presents significant difference $p < 0.05$.

DISCUSSION

As long-acting OTC belongs to the tetracycline group, it is a broad-spectrum antimicrobial agent, presenting persistent actions and sustaining absorption from the depot site after i.m. injection (26). Most studies have reported PK and bioequivalence studies in cattle, sheep, goats, and rabbits, but only a few have investigated bioequivalence in pigs (6, 11, 31, 32). In the current study, the analytical methods for OTC quantification detection in plasma were specific, sensitive, accurate and were performed for bioequivalence study between the test product and the reference one in pigs. LOQ (0.05 μg/ml) was available for OTC detection in plasma, which was similar to the one previously described (0.05 μg/ml) by Nora Mestorino et al. and Brentnall et al. (6, 27).

AUC was a useful index for biological availability of the active moiety of a drug formulation regarding the absorption extent. Both AUC_{0-144h} and $AUC_{0-\infty}$ were 118.976 ± 13.259, 126.179 ± 17.390 and 123.087 ± 13.906, 130.732 ± 18.562 μg h/ml for test and reference products, which are higher than previous reports in pigs (33), but the ANOVA analysis revealed these two formulations had no significant differences with $p > 0.05$ (Table 3). C_{max} of these two products were similar to other reports in pigs—about 5.066 ± 0.486 and 5.071 ± 0.877 μg/ml—which presented a high drug concentration in plasma (Table 3) (34–36). Both formulations displayed similar plasma profiles and demonstrated that they were absorbed progressively, reaching C_{max} approximately 0.75 h after administration, keeping 1/20 C_{max} concentration in plasma after 120 h. T_{max}, C_{max}, and $AUC_{0-\infty}$ were 2.01 h, 3.98 μg/ml, 74.87 μg h/ml, and 0.609 h, 4.956 μg/ml,

112.483 μg h/ml in the research of Jie Bao and Xue Ling at a dose of 20 mg/kg, respectively (36, 37). In this study, a lower peak is reached later in comparison with our results. It might be related to the OTC concentration in the drug product, which was 30% higher than the 20% in our test and reference product studied. These results suggest a slower absorption of OTC with 30% formulation in comparison with 20% ones. These parameters of OTC in this study were similar to a previous study (35). The relative bioavailability of the test product compared to the reference one was 94.291 ± 15.287% (Table 3), and was included in the range 0.8–1.25, which satisfied the requirements of the Food and Drug Administration (21).

For a bioequivalence study of these two OTC injection formulations, and to make a statistical comparison, $AUC_{0\text{-last}}$ and $AUC_{0-\infty}$, C_{max}, and T_{max} were selected. When these parameters were not significantly different between the test and reference product, they will be bioequivalent if the CI were comparable. This was demonstrated through the ANOVA of log-transformed value and the two one-sided *T*-tests and 90% CI. In our study, the three parameters revealed there were no significant differences between the test and reference products (t_1 and t_2) in $AUC_{0\text{-last}}$ and $AUC_{0-\infty}$, C_{max}, which were all above 1.717 and had significant differences ($p < 0.05$) with each other (Table 4). The $AUC_{0\text{-}144h}$ and $AUC_{0-\infty}$, C_{max} results presenting 90% of CI were inside the CIs (75–143) set by the European Union. T_{max} presented no significant difference by non-parametric analysis (Table 4). Thus, these results demonstrated that the test OTC injection (Kangtekang 20% LA, Huishen) was bioequivalent to the reference one (Terramycin 20% LA, Pfizer).

ETHICS STATEMENT

The animals that were used in this study were conducted according to relevant guidelines and regulations of Animal Care Center, Hubei Science and Technology Agency in China (SYXK 2013-0044). Animal housing care and experimental protocol were conducted according to the regulation of experimental animal usage in Hubei province of China, and the protocol was approved by the Ethics Committee of Huazhong Agricultural University.

AUTHOR CONTRIBUTIONS

Experiment designation: ZL, QL, JC, and QH; carrying out the experiment: ZL, BY, and JX; manuscript writing: ZL; and manuscript review and modification: SA, TS, and PC.

CONFLICT OF INTEREST STATEMENT

The authors declare that the research was conducted in the absence of any commercial or financial relationships that could be construed as a potential conflict of interest.

ACKNOWLEDGMENTS

This work was granted by Natural Science Foundation of China (no. 31272572).

REFERENCES

1. Nouws JF, Vree TB. Effect of injection site on the bioavailability of an oxytetracycline formulation in ruminant calves. *Vet Q* (1983) 83:165–70.
2. Nouws JF, Vree TB, Termond E, Lohuis J, Van LP, Binkhorst GJ, et al. Pharmacokinetics and renal clearance of oxytetracycline after intravenous and intramuscular administration to dairy cows. *Vet Q* (1985) 7:296.
3. Mevius DJ, Nouws JF, Breukink HJ, Vree TB, Driessens F, Verkaik R. Comparative pharmacokinetics, bioavailability and renal clearance of five parenteral oxytetracycline-20% formulations in dairy cows. *Vet Q* (1986) 8:285.
4. Coetzeea JF, Apleya MD, Kocanb KM, Rurangirwac FR, Donkersgoedd JV. Comparison of three oxytetracycline regimens for the treatment of persistent *Anaplasma marginale* infections in beef cattle. *Vet Parasitol* (2005) 127:61–73.
5. Li RQ, Ren YW, Li J, Huang C, Shao JH, Chen XX, et al. Comparative pharmacokinetics of oxytetracycline in blunt-snout bream (*Megalobrama amblycephala*) with single and multiple-dose oral administration. *Fish Physiol Biochem* (2015) 41:803–9.
6. Mestorino N, Marchetti ML, Lucas MF, Modamio P, Zeinsteger P, Lastra CF, et al. Bioequivalence study of two long-acting formulations of oxytetracycline following intramuscular administration in bovines. *Front Vet Sci* (2016) 3:50.
7. Banting ADL, Baggot JD. Comparison of the pharmacokinetics and local tolerance of three injectable oxytetracycline formulations in pigs. *J Vet Pharmacol Ther* (1996) 19:50–5.
8. Posner LP, Burns P. Injectable anesthetic agents. *Can Vet J* (2009) 32:314–6.
9. Giguère S, Prescott JF, Dowling PM. *Antimicrobial Therapy in Veterinary Medicine*. 5th ed. New York, NY: John Wiley & Sons, Inc (2013).
10. Mestorino N, Mariño HE, Marchetti L, Errecalde JO. Pharmacokinetics and tissue residues of an oxytetracycline/diclofenac combination in cattle. *Rev Sci Tech* (2007) 26:679–90.
11. Escudero E, Carceles CM, Serrano JM. Pharmacokinetics of oxytetracycline in goats: modifications induced by a long-acting formulation. *Vet Rec* (1995) 135:548–52.
12. Ozdemir N, Yildirim M. Bioequivalence study of two long-acting oxytetracycline formulations in sheep. *Vet Res Commun* (2006) 30:929.
13. Zaid AN, Mousa A, Jaradat N, Bustami R. Lornoxicam immediate-release tablets: formulation and bioequivalence study in healthy Mediterranean volunteers using a validated LC-MS/MS method. *Clin Pharmacol Drug Dev* (2017).

14. Bioavailability and bioequivalence requirements; abbreviated applications; proposed revisions – FDA. Proposed rule. *Fed Regist* (1998) 63:64222–8.
15. Food and Drug Administration. *Guidance for Industry Bioavailability and Bioequivalence Studies for Orally Administered Drug Products – General Considerations*. Rockville (2000).
16. Chen ML, Shah V, Patnaik R, Adams W, Hussain A, Conner D, et al. Bioavailability and bioequivalence: an FDA regulatory overview. *Pharm Res* (2001) 18:1645–50.
17. Food and Drug Administration. *Bioavailability and Bioequivalence Studies for Orally Administered Drug Products – General Considerations*. Rockville (2003).
18. Davit BM, Kanfer I, Yu CT, Cardot JM. BCS biowaivers: similarities and differences among EMA, FDA, and WHO requirements. *AAPS J* (2016) 18:612–8.
19. Martinez MN, Hunter RP. Current challenges facing the determination of product bioequivalence in veterinary medicine. *J Vet Pharmacol Ther* (2010) 33:418–33.
20. Galgatte U. Study on requirements of bioequivalence for registration of pharmaceutical products in India, South Africa and Australia. *Saudi Pharm J* (2014) 22:391–402.
21. Kaushal N, Singh SK, Gulati M, Vaidya Y, Kaushik M. Study of regulatory requirements for the conduct of bioequivalence studies in US, Europe, Canada, India, ASEAN and SADC countries: impact on generic drug substitution. *J Appl Pharm Sci* (2016) 6:206–22.
22. Chow SC, Liu JP. *Design and Analysis of Bioavailability and Bioequivalence Studies*. 2nd ed. New Jersey: Berlex Laboratories, Inc (2000).
23. Chow SC, Liu J. *Design and Analysis of Bioavailability and Bioequivalence Studies*. 3rd ed. UK: CRC Press, Taylor and Francis Group (2008).
24. Chow SC, Liu JP. Statistical assessment of biosimilar products. *J Biopharm Stat* (2010) 20:10–30.
25. Gberindyer AF, Okpeh ER, Semaka AA. Pharmacokinetics of short- and long-acting formulations of oxytetracycline after intramuscular administration in chickens. *J Avian Med Surg* (2001) 29:298.
26. Brentnall C, Cheng Z, Mckellar QA, Lees P. Pharmacokinetic-pharmacodynamic integration and modelling of oxytetracycline administered alone and in combination with carprofen in calves. *Res Vet Sci* (2013) 94:687.
27. Brentnall C, Cheng Z, Mckellar QA, Lees P. Influence of oxytetracycline on carprofen pharmacodynamics and pharmacokinetics in calves. *J Vet Pharmacol Ther* (2013) 36:320.
28. Dedonder KD, Gehring R, Tell LA, Riviere JE. Protocol for diversion of confirmed positive bulk raw milk tankers to calf ranches – a review of the pharmacokinetics of tetracyclines and sulfonamides in veal calves. *Anim Health Res Rev* (2016) 17(2):127–36.
29. H. C. Government of Canada, Health Products, V. D. D. Food Branch. *VICH guideline 52: Bioequivalence – Blood Level Bioequivalence Study (Step 4) – 2014 Health Canada Consultation Notice*. Brussels (2014).
30. Pfaller MA, Andes D, Diekema DJ, Espinel-Ingroff A, Sheehan D, CLSI Subcommittee for Antifungal Susceptibility Testing. Wild-type MIC distributions, epidemiological cutoff values and species-specific clinical breakpoints for fluconazole and Candida: time for harmonization of CLSI and EUCAST broth microdilution methods. *Drug Resist Updat* (2010) 13:180–95.
31. Xia W, Nielsen P, Gyrdhansen N. Oxytetracyclines in cattle. A comparison between a conventional and a long-acting preparation. *Acta Vet Scand* (1983) 24:120.
32. Chong W, Kim YJ, Kim SD, Han SK, Ryu PD. Lack of bioequivalence of two oxytetracycline formulations in the rabbit. *J Vet Sci* (2002) 3:25–30.
33. Archimbault P, Navetat H, Vidal R, Douin MJ, Mignot A. Plasma bioavailability of 2 long-acting oxytetracycline formulations in the pig. *Vet Res* (1994) 25:399–404.
34. Cai JH, Xin GU, Liu YN. Study on pharmacokinetics of long-action oxytetracycline injection in swine. *Chin J Vet Drug* (2006) 40(7):17–21.

35. Hall WF, Kniffen TS, Bane DP, Bevill RF, Koritz GD. Plasma concentrations of oxytetracycline in swine after administration of the drug intramuscularly and orally in feed. *J Am Vet Med Assoc* (1989) 194:1265.
36. Liao XL, Zhang BX, Ding HZ. Pharmacokinetics and relatively bioavailability of oxytetracycline long acting injection in swine. *Chin J Vet Drug* (2012) 46(12):19–21,39.
37. Bao J. *Bioequivalence and Pharmacokinetics Studies of Hydrochloride Oxytetracycline Injection in Pigs*. Wuhan: China Agricultural University (2007).

CHAPTER 13

The Effect of Supplementing Air-Dried *Moringa stenopetala* Leaf to Natural Grass Hay on Feed Intake and Growth Performances of Arsi-Bale Goats

Aberra Melesse [1,*], Degnet H/Meskel [2], Sandip Banerjee [1], Aster Abebe [1] and Amsalu Sisay [1]

[1]*School of Animal and Range Sciences, Hawassa University, P.O. Box 05, Hawassa, Ethiopia*

[2]*Wachamo University, Hossana, Ethiopia*

ABSTRACT

The most constraining factor in goat production in the tropics is underfeeding mainly attributed to limitations of feed both in quantity and quality. This study was conducted to assess the effect of supplementing different levels of air-dried *Moringa stenopetala* leaf (MSL) as a protein source on nutrient intake and growth performances of Arsi-Bale male goats. A total of 24 yearling goats with average initial body weight of 13.6 ± 0.25 kg were used in the study. The goats were blocked by live weight into four groups (*n* = 6 per group) and the groups were then randomly allocated into four supplemented treatments. All goats received a basal diet of natural grass hay *ad libitum* and 340 g/head/day concentrate. The treatments were the control diet with no supplementation (Treatment 1, T1) and diets supplemented with MSL at a rate of 120 g/head/day (Treatment 2, T2), 170 g/head/day (Treatment 3, T3) and 220 g/head/day (Treatment 4, T4). The duration of the experiment was 75 days. The results indicated that the average daily feed intake was ($p < 0.001$) higher in goats supplemented with T3 and T4 diets. The total dry matter, organic matter, and crude protein intakes of goats fed with T3 and T4 supplementations were ($p < 0.001$) also higher than those reared in T1 and T2 diets. Goats reared in T3 and T4 diets had lower ($p < 0.05$) feed conversion ratio than those fed with T1 and T2 diets. The final body weight in goats reared in T3 and T4 diets was 18.2 kg and 18.5 kg, respectively, being ($p < 0.05$) higher than those of T1 (15.8 kg) and T2 (16.3 kg). The average daily weight gain in goats fed with T3 and T4 diets was 111 and 114 g/goat/day, respectively, which was ($p < 0.05$) higher than those reared in the control (T1) (54.0 g/goat/day) and T2 (58.1 g/goat/day) diets. It can thus be concluded that goats reared at high level of MSL supplementation (T3 and T4) had better nutrient intake, feed conversion efficiency and growth performances, suggesting its potential as a good protein supplement to natural grass hay at the farmer's management level.

Keywords: Arsi-Bale goats; natural grass hay; *Moringa stenopetala* leaf; nutrient intake; supplementation; weight gain

1. INTRODUCTION

Ethiopia has the largest livestock population among African countries and has different agro-ecological zones ranging from semi-arid to arid environments which are suitable for goat production. There are approximately 570 breeds and types of goats in the world of which 89 are found in Africa [1]. The present estimated population of goats in Ethiopia is 29.3 million [2]. Goats play important roles in communities by improving the livelihood of resource-challenged farmers by creating alternative employment opportunities, enhancing family income by sale of live animals, skin, manure *etc.* Apart from contributing to human nutrition in form of meat and milk, they also act as an income buffer to the risks associated with erratic climatic changes [3]. In addition, demand for goat's meat is on the rise throughout the world, particularly in Ethiopia, due to increased human population coupled with income growth particularly in the cities. Regardless of their good attributes, production of goats remains low due to diseases, poor genotype, management, and nutrition; nevertheless, there is ample scope for improving the productivity of goats [4].

One of the most constraining factors in goat production in the tropics is underfeeding, which is mainly attributed to limitations of feed in both quantity and quality. The situation is exacerbated during the dry season when natural pastures usually dry out and are overgrazed, resulting in low contents of protein and energy in the fodder. As a result, large flocks of productive livestock cannot be maintained on such low quality feeds to meet the basic maintenance requirements of the animals. It is therefore important to supplement the meager available fodder with some amount of concentrates in order to improve intake and digestibility of such poor quality feed resources. However, the use of such supplements is usually limited under smallholder livestock production systems due to their inaccessibility and high cost.

As a result, in order to mitigate the problems associated with the lack of dietary protein, there is a need to look for some alternative but cheap sources of protein. Previous works have reported that leaves of multipurpose trees growing in the backyards of the farmers can be used as a cheap source protein for supplementing livestock [5]. In this regard, leaves from the Moringa tree may serve as a multipurpose protein source for livestock. The uniqueness of this tree is that it does not shed its leaves during the dry season when other sources of fodder become scarce [6]. It is a multipurpose tree cultivated both for human food and animal feed and is found in abundance in certain parts of Southern Ethiopia. Recent studies conducted by [7,8] and [9] have indicated that the leaves of *Moringa stenopetala* are rich in protein (28.2%–36.1%) and contain substantial amounts of essential amino acids as well as macro and trace minerals [10].

There are several studies reporting on the use of the fresh foliage of *Moringa* spp. as a source of livestock feed. According to the report of [11], supplementation of fresh *Moringa oleifera* leaves has resulted in the improvement of milk yield in crossbred cattle. Promising results have been obtained from inclusion of air-dried *M. stenopetala* leaf supplementation to the basal diet of sheep [12] and growing chicken [8,13]. However, information on the feeding value of *M. stenopetala* leaf on the performance of goats in the tropical countries, particularly in Ethiopian, is lacking. Thus, this research was conducted to assess the effect of supplementing air-dried *M. stenopetla* leaf meal to natural grass hay on feed intake and body weight performances of Arsi-Bale goats.

2. MATERIALS AND METHODS

2.1. Experimental Site

The experiment was carried out at the sheep and goat farm of School of Animal and Range Sciences, Hawassa University (Ethiopia), which lies geographically between 7°5′ N latitude and 38°29′ E longitude at an altitude of 1700 m above sea level. The average rainfall ranges from 800 to 1100 mm. The mean minimum and maximum temperatures in the study area are 13.5 °C and 27.6 °C, respectively [14].

2.2. Preparation of Experimental Feeds

Fresh *Moringa stenopetala* leaves were collected from the trees owned by the farmers in Mirab Abaya (district town) which is situated near Arbaminch city. The district is situated between 6°4′ N latitude and 37°34′ E longitude with the mean altitude of 1220 m above the sea level. The fresh leaves were harvested from the available trees regardless of the age of the tree. The leaves were trimmed from its twigs and then shade dried on a plastic sheet. During the drying process, regular turning of leaves was done to ensure uniform drying. The air-dried Moringa leaves were finally transported to the experimental site and ground into coarse powder which hereafter is referred to as air-dried *Moringa stenopetala* leaf (MSL). The powder was packed in bags of 100 kg and stored until used. Natural grass hay was bought from a nearby private farm and hand chopped into the size of 3 to 5 cm for ease of feeding. Samples of MSL and various ingredients were subjected to chemical analysis before being used in the formulation of experimental diets.

2.3. Experimental Animals and Their Management

In this experiment, 24, about one-year-old (age determined by dentition) Arsi-Bale intact male goats were used with an average initial body weight of 13.6 ± 0.25 kg. Goats were purchased from local market and quarantined for a fortnight, during which they were treated with 250 mg of albendazole (Chengdu Qiankun Veterinary Pharmaceuticals Co. Ltd., Chengdu, China), administered through drenching gun for de-worming the animals. Moreover, 1.5 mL Oxytetracycline (Chengdu Qiankun Veterinary Pharmaceuticals Co. Ltd., Chengdu, China) was provided intravenously for three days to treat the animals from coughing caused due to transportation from the market to the experimental station. The barns were also sprayed with Diazinone (Chengdu Qiankun Veterinary pharmaceuticals col. Ltd., Chengdu, China) to take care of any ecto-parasite infestation.

2.4. Experimental Design and Treatment Diets

The feeding trial was a completely randomized design (CRD) consisting of one control and three supplemental treatment diets with six goats within each treatment. The trial was conducted for 75 days exclusive of 15 days of adaptation period. All the experimental goats had *ad libitum* access to natural grass hay and water. They were also provided with a concentrate at a rate of 2.5% of their body weight; this was done according to the recommendation of [15]. Consequently, as the average initial body weight of goat was 13.6 kg, the calculated concentrate supplied per goat per day was 340 g. The diets therefore contained 340 g concentrate only (treatment, T1), 340 g concentrate plus 35% MSL supplementation (120 g) (treatment 2, T2), 340 g concentrate plus 50% MSL supplementation (170 g) (treatment 3, T3), and 340 g concentrate plus 65% MSL supplementation (220 g) (treatment 4, T4). Accordingly, the total concentrate/MSL mixture offered to T1, T2, T3, and T4 treatment groups were 340, 460, 510, and 560 g/head day, respectively. The concentrate was a mixture of 50% wheat bran, 35% maize, 14% Noug (*Guizotia abyssinica*) seed cake, and 1% salt. Natural grass hay and the concentrate/MSL mixture were offered separately.

2.5. Data Collection

At the end of the quarantine period, the goats were weighed early in the morning (prior to being offered any feed) for two consecutive days and the body weight was averaged. Then the goats were blocked into control and supplemented groups (n = 6 per group) using initial body weight and the groups were then randomly allocated into four supplemented treatments. Animals were ear tagged and housed in individual pens and each animal was provided with feeder and watering trough.

Animals were offered the concentrate/MSL mixture twice a day at 8:00 a.m. and 1:00 p.m. in equal portions throughout the experimental period. Samples of daily feeds offered and refused were collected, measured and pooled over experimental period for each feed and animal and stored in plastic bags. Representative samples of the offered and refused feeds for the different treatments were then taken for chemical analysis. The daily average feed intake of hay and concentrate/ MSL mixture was determined separately by the difference between the amounts of feed offered and refused and then averaged. Body weight of each goat was recorded every fortnight after overnight fasting to determine body weight change. Average individual daily weight gain was calculated as the difference between final body weight and initial body weight of the goat divided by the experimental days.

2.6. Chemical Analysis of the Feed

The analysis of ash and moisture contents was conducted using the method of [16]. The dry matter content of the feed was determined by drying the samples at 105 °C overnight in a force air oven. Ash was determined by combusting the samples at 550 °C for 5 h in a muffle furnace. Nitrogen (N) was determined using the Keldjhal method and then the crude protein (CP) was calculated as N × 6.25 [16]. The acid detergent fiber (ADF) and neutral detergent fiber (NDF) contents were analyzed using the method of [17] in an ANKOM® 200 Fiber Analyzer (ANKOM Technology Corp., Fairport, NY, USA). All samples were analyzed in duplicates at Animal Nutrition Laboratory of Hawassa University.

2.7. Statistical Analysis

The collected data were subjected to the General Linear Models of SAS (SAS, 2010, version 9.3, SAS Institute Inc., Cary, NC, USA). When significant differences were observed among treatment means, they were separated by Duncan Multiple Range Test. Comparisons with $p < 0.05$ were considered significant and all statements of statistical differences were based on this level unless noted otherwise. The following linear model summarizes the statistics employed to analyze the data:

$$Yij = \mu + Ai + Aj + eij$$

where *Yij* is the observed *j* variables in the *i*th supplement, μ is the overall mean of the observed variable, *Ai* is the effect due to *i*th supplement levels (i = 0, 120, 170, 220 g/head/day), and *eij* is the random residual error

3. RESULTS

3.1. Chemical Composition of Experimental Feed

Chemical composition of the experimental feeds is presented in Table 1. The crude protein (CP) content of the *M. stenopetala* leaf was the highest followed by that of the Noug seed cake; however the lowest value was observed for the hay offered. The dry matter (DM) content of the feed was more or less similar across the feed ingredients. The organic matter (OM) content was lowest for the *M. stenopetala* leaf while it was highest for maize. The results also indicated that the ether extract (EE) values were highest for Noug seed cake and lowest for hay. The neutral detergent fiber (NDF) values were highest for hay and the lowest for *M. stenopetala* leaf.

Table 1. Proximate composition of ingredients of the concentrate, natural grass hay, and air-dried *Moringa stenopetala* leaf.

Nutrients (%, DM Basis)	Ingredients of the Concentrate Mix			Natural Grass Hay	MSL
	Maize	Wheat Bran	Noug Seed Cake		
Dry matter	93.3	93.6	93.9	95.6	94.4
Organic matter	98.1	88.5	90.1	94.0	88.0
Crude protein	6.50	16.3	25.0	5.10	29.5
Ether extract	9.60	5.42	11.5	2.86	5.90
NDF	10.0	36.5	32.5	61.2	17.8
ADF	4.90	17.5	21.5	25.7	16.5

MSL: air-dried *Moringa stenopetala* leaf meal; NDF: neutral detergent fiber; ADF: acid detergent fiber; DM: Dry matter.

3.2. Nutrient Contents of the Experimental Treatments

The results from the Table 2 indicate that the CP content was highest in T4 diet supplemented with 220 g/head/day MSL and lowest in non-supplemented control diet (T1). The DM content of the feed was similar across the treatments while the organic matter (OM) and ether extract (EE) content was lowest in the control (T1) and the highest values were observed in the goats supplemented with 220 g/head/day MSL diet (T4) while the reverse was true for ash, acid detergent fiber (ADF), and NDF contents.

Table 2. The average proximate compositions (%) of treatment diets composed of natural grass hay, concentrate supplemented with different levels of air-dried *Moringa stenopetala* leaf.

Treatments	DM	Ash	OM	CP	EE	NDF	ADF
T1	92.3	6.20	88.0	16.0	5.7	69.3	56.1
T2	92.7	7.10	92.0	16.7	6.10	45.5	38.3
T3	93.6	8.01	92.9	17.2	8.70	36.5	23.8
T4	94.8	8.03	93.8	18.7	9.50	33.6	22.5

DM, dray matter; OM, organic matter; CP, crude protein; EE, ether extract; NDF, neutral detergent fiber; ADF, acid detergent fiber; T1, no supplementation; T2, 120 g/head/day supplementation with air-dried *Moringa stenopetala* leaf; T3, 170 g/head/day supplementation with air-dried *Moringa stenopetala* leaf; T4, 220 g/head/day supplementation with air-dried *Moringa stenopetala* leaf.

3.3. Nutrient Intakes

As presented in Table 3, goats reared in T3 and T4 diets had ($p < 0.01$) higher intakes of CP, DM, OM, NDF, and ADF for natural grass hay than those fed with T1 and T2. For concentrate/MSL mixtures, the highest and lowest DM, OM, and CP intakes were observed in goats reared in T4 and T1, respectively. Conversely, the NDF and ADF intakes for concentrate/MSL was highest in goats reared in the control diet (T1), but lowest in those fed with the T3 diet. The total intakes of DM, OM, and CP were ($p < 0.001$) higher in goats fed with T4 than those reared in other treatment diets. These parameters were also ($p < 0.001$) higher in goats reared in T3 than those in T2 and T1 diets and goats fed with T2 had significantly higher DM, OM, and CP intakes than those of T1. On the other hand, goats reared in the control diet (T1) had ($p < 0.001$) higher ADF intake than those supplemented with different levels of MSL. The goats fed with T2 diet had ($p < 0.001$) higher NDF intake than other treatments.

3.4. Body Weight Change and Feed Conversion Efficiency

While there were no significant differences between the supplemented groups at the beginning of the experiment, significant differences in body weight were recorded across the different treatment diets at the later stages of the experiment (Table 4). Accordingly, the non-supplemented goats (T1) had the lowest final body weight while those reared on T4 supplementation had the highest. Goats supplemented with T3 and T4 diets had significantly higher values for the final body weight, total and daily weight gain than those reared in T1 and T2 diets. The values also indicated that there was no significant difference between those receiving T3 and T4 supplemented diets. Moreover, growth parameters of goats reared in T1 did not differ from those of T2, indicating two distinctly truncated categories.

As shown in Table 4, the feed intake values were ($p < 0.001$) higher in goats supplemented with increasing levels of MSL (T3 and T4) than those of the non-supplemented (T1) and less supplemented (T2) groups. No significant differences were noted between goats supplemented with T3 and T4 levels and those fed with the T1 and T2 diets. The FCR value was lower in goats reared in T3 and T4, ($p < 0.05$) than those fed with T1 and T2 diets. However, FCR was similar ($p > 0.05$) between goats reared in T1 and T2 diets; and between those in T3 and T4 diets.

Table 3. The effects of hay supplemented with different levels of air-dried *Moringa stenopetala* leaf on nutrient intakes of Arsi-Bale goats.

Parameters (g/head/day)	T1	T2	T3	T4	SE	P
Hay intake						
Dry matter	291 ^{b}	272 ^{b}	372 ^{a}	404 ^{a}	22.3	**
Organic matter	287 ^{b}	268 ^{b}	366 ^{a}	397 ^{a}	21.9	**
Crude protein	15.5 ^{b}	14.7 ^{b}	19.8 ^{a}	21.3 ^{a}	1.21	**
NDF	187 ^{b}	174 ^{b}	239 ^{a}	259 ^{a}	14.3	**
ADF	78.3 ^{b}	73.3 ^{b}	100 ^{a}	108 ^{a}	6.00	**
Supplemental intake						
Dry matter	281 ^{d}	321 ^{c}	370 ^{b}	420 ^{a}	3.73	***
Organic matter	267 ^{d}	319 ^{c}	367 ^{b}	417 ^{a}	3.64	***
Crude protein	48.7 ^{d}	57.8 ^{c}	68.0 ^{b}	83.2 ^{a}	0.689	***
NDF	211 ^{a}	158 ^{b}	144 ^{d}	150 ^{c}	1.604	***
ADF	170 ^{a}	133 ^{b}	94.5 ^{d}	100 ^{c}	1.	***
Total intake						
Dry matter	572 ^{c}	594 ^{c}	742 ^{b}	825 ^{a}	23.7	***
Organic matter	544 ^{c}	587 ^{c}	733 ^{b}	814 ^{a}	23.4	***
Crude protein	64.3 ^{d}	72.4 ^{c}	87.8 ^{b}	105 ^{a}	1.55	***
NDF	397 ^{a}	332 ^{b}	383 ^{a}	408 ^{a}	14.9	**
ADF	249 ^{a}	206 ^{b}	194 ^{b}	208 ^{b}	6.355	***

^{a,b,c,d} Means within the same row with different superscript letters are significantly different ($p < 0.05$); ** $p < 0.01$; *** $p < 0.001$; SE: standard error of the mean; NDF: neural detergent fiber; ADF: acid detergent fiber; T1: no supplementation; T2: 120 g/head/day supplementation with air-dried *Moringa stenopetala* leaf; T3: 170 g/head/day supplementation with air-dried *Moringa stenopetala* leaf; T4: 220 g/head/day supplementation with air-dried *Moringa stenopetala* leaf.

Table 4. The effects of supplementation of different levels of air-dried *Moringa stenopetala* leaf on feed intake and growth performances of Arsi-Bale goats.

Growth Performance	T1	T2	T3	T4	SE	P
Initial body weight (kg/head)	13.5	13.8	13.6	13.7	0.481	NS
Final body weight (kg/head)	15.8 [b]	16.3 [b]	18.2 [a]	18.5 [a]	0.544	**
Total weight gain (kg/head)	2.27 [b]	2.44 [b]	4.65 [a]	4.80 [a]	0.439	***
Average daily gain (g/head)	54.0 [b]	58.1 [b]	111 [a]	114 [a]	10.4	***
Average daily feed intake (g/head)	609 [b]	632 [b]	785 [a]	841 [a]	24.8	***
FCR (g feed/g gain)	12.6 [a]	11.5 [a]	7.69 [b]	7.41 [b]	1.49	*

[a,b] Means within the same row with different superscript letters are significantly different ($p < 0.05$); * $p < 0.05$; ** $p < 0.01$; *** $p < 0.001$; SE = standard error of the mean; FCR: feed conversion ratio; NS: Not significantly different ($p > 0.05$); T1: no supplementation; T2: 120 g/head/day supplementation with air-dried *Moringa stenopetala* leaf; T3: 170 g/head/day supplementation with air-dried *Moringa stenopetala* leaf; T4: 220 g/head/day supplementation with air-dried *Moringa stenopetala* leaf.

4. DISCUSSION

4.1. Nutrient Composition of M. stenopetala and Experimental Feed

The CP values of MSL as assessed in the current study are similar to those reported by [7,8,10,13]. The CP values are also similar to the findings of [18] for *M. oleifera* leaves and [10] for *M. stenopetala* leaf meal. However, the CP content in *M. stenopetala* and *M. oleifera* leaves as reported by [19,20] are lower than those obtained from the present study while high values were reported by [9], which may be attributed to the maturity stage of leaves as the young leaves usually have higher protein values than the older ones. The CP content in the Moringa leaves in the present study indicated that they can be considered good protein supplements to low quality forages in the smallholder livestock production system. The EE values in the present study are similar to those reported by [21,22] for the leaves of *M. stenopetala* and *M. oleifera*, respectively. However, lower EE values were reported by [8] for *M. stenopetala* leaf. This could be due to differences in Moringa species, maturity stage of leaves, and type of soils in which the plans were grown.

The contents of NDF and ADF in MSL in the present study are comparable to the observations of [8] for the same Moringa species; but were higher than those of *M. oleifera* leaf meal reported by [22,23] The ash content in MSL as observed in this study is similar to the observations of [8,19] for *M. stenopetala* leaves. The ash content was also similar to those reported by [10,24] for *M. oleifera* leaves. However, the ash content of MSL in the present study was higher than those reported by [7,21,25]; and lower than those of [22] for *M. oleifera* leaves. Similarly, [18,26] have reported high contents of ash (14.1% and 14.8%, respectively) in the leaves of *M. oleifera.*

The comparatively high NDF and ADF contents of the natural grass hay in the present study indicate that it is of a poor nutritional quality and is not adequate to support maintenance requirements of goats as the CP content is well below the minimum level (7%) required for the microbial function [27]. The CP, DM, and OM contents of the hay used in the current study were consistent with the reports of [28]. The CP, NDF, and ADF values for wheat bran and Noug seed cake were found to be similar to those reported by [29]. The average contents of DM and ADF in Noug seed cake were comparable with the reports of [30]. However, the values for CP and ash in Noug seed cake reported by [30] were lower than observed in the present findings which might be due to the maturity stage of the Noug seed and type of soil in which the plants were grown. The CP, EE, and NDF values of wheat bran as assessed in the current study were comparable with the observations of [31].

4.2. Nutrient Intake

Feed intake is an important parameter in determining the nutritive value of animal's feed as it account for nutrient supply. In the present study, there is significant difference in intakes of DM, OM and CP and these parameters increased with increasing supplementations level of MSL which is consistent with the findings of [11]. Sánchez [11] observed increased intake in cows supplemented with *M. oleifera* leaves to a basal diet of grass hay. Similar results have also been reported by [12] and [32] who found that supplementation of sheep with *M. stenopetala* fed with a basal diet of grass hay improved intake of DM, OM, and CP. The increased DM, OM, and CP intake with increasing levels of *M. stenopetala* supplementation may possibly be due to the lower fiber and higher CP contents of the leaves thereby enhancing the palatability of the feed. This could lead to an increase in microbial population and efficiency, thereby facilitating the rate of breakdown of the digesta which eventually leads to increment in feed intake. The low total nutrient intakes observed in non-supplemented goats might be due to the low level of CP in the hay.

4.3. Body Weight and Gain

There was improvement in final body weight, total weight gain, average daily weight gain, and feed conversion efficiency in the treatment groups with high levels of MSL supplementations (T3 and T4). The findings are consistent to those of [5] and [12] who reported increased body weight values in sheep receiving supplement containing *M. stenopetala* leaf meals. Studies by [33] also indicated linear increase in body weight of sheep supplemented with *Gliricdia sepium* leaf meal when fed a basal diet of elephant grass and Rhodes grass.

The average daily weight gain as were observed among the goats reared on T3 and T4 diets were in the range of values reported by [34] in sheep fed on a basal diet of Rhodes grass hay and those of [35] where the *M. oleifera* leaf was used as a supplementary feed for goats. The average daily weight gains are also in accordance with the reports of [36], who supplemented M. oleifera leaf meal to goats. Increment in body weight was also reported in goats fed with cassava leaf meal as compared to those reared on grass hay alone [37].

The difference in feed conversion ratios among the treatment groups are comparable with the observations of [38], who reported significantly higher values of efficiency in rabbits fed with *M. oleifera* leaf meal than those of the control diet. In this study, goats supplemented with high levels of MSL (170 and 220 g/day and goat) consumed less feed but gain more weight as compared to those reared in the control and less supplemented (120 g/head/day) groups. This implies that due to the year round availability and easy access of *M. stenopetala* tree by smallholder farmers, the leaf can be used as a very useful protein supplement to small ruminants, particularly during the dry season, where most of local feed resources become scarce.

5. CONCLUSIONS

The supplementation of *Moringa stenopetala leaf* to natural grass hay has considerably improved the nutrient intake and body weight gain of Aris-Bale goats as compared to those of the non-supplemented goats. Therefore, supplementation of low quality feed resources such as natural grass hay with Moringa leaves is a viable option for improving the productivity of local goats under smallholder farmer conditions where conventional protein supplements are beyond the reach of local farmers.

ACKNOWLEDGMENTS

This research project was fully supported by the research fund granted by the Hawssa University, Research and Development Directorate Office (now Vice President for Research and Technology Transfer) for which the authors are highly grateful. The support received from Tadesse Bekore in analyzing the feed nutrient composition is highly acknowledged.

AUTHOR CONTRIBUTIONS

The first author prepared the research proposal and solicited fund. He together with the third author closely monitored and supervised the second author who has been mainly involved in data collection and organization as a part of his Master thesis. Moreover, the first author analyzed the data using appropriate statistical packages and prepared the manuscript. The forth and third authors assisted the first author during the preparation of the research proposal.

CONFLICTS OF INTEREST

The authors declare no conflict of interest.

REFERENCES

1. Gala, S. Biodiversity in goats. Small Rumin. Res. 2005, 60, 75–81.
2. Central Statistical Agency; Federal Democratic Republic of Ethiopia. Agricultural Sample Survey, Report on Livestock and Livestock Characteristics; CSA: Addis Ababa, Ethiopia, 2014; volume 2, pp. 12–25.
3. Rumosa-Gwaze, F.G.; Chimonyo, M.; Dzama, K. Communal goat production in Southern Africa: A review. Trop. Anim. Health Prod. 2009, 41, 1157–1168.
4. Simela, L.; Merkel, R. The contribution of chevon from Africa to global meat production. Meat Sci. 2008, 80, 101–109.
5. Manaye, T.; Tolera, A.; Zewdu, T. Feed intake, digestibility and body weight gain of sheep fed Napier grass mixed with different levels of Sesbania sesban. Livest. Sci. 2009, 122, 24–29.
6. Thurber, M.D.; Fahey, J.W. Adoption of Moringa oleifera to combat under-nutrition viewed through the lens of the "Diffusion of Innovations" theory. Ecol. Food Nutr. 2009, 48, 212–225.
7. Melesse, A.; Bulang, M.; Kluth, H. Evaluating the nutritive values and in vitro degradability characteristics of leaves, seeds and seedpods from Moringa stenopetala. J. Sci. Food Agric. 2009, 89, 281–287.

8. Melesse, A.; Tiruneh, W.; Negesse, T. Effects of feeding Moringa stenopetala leaf meal on nutrient intake and growth performance of Rhode Island Red chicks under tropical climate. Trop. Subtrop. Agroecosyst. 2011, 14, 485–492.
9. Negesse, T.; Makkar, H.P.S.; Becker, K. Nutritive value of some non-conventional feed resources of Ethiopia determined by chemical analyses and an in vitro gas method. Anim. Feed Sci. Technol. 2009, 154, 204–217.
10. Melesse, A.; Steingass, H.; Boguhn, J.; Schollenberger, M.; Rodehutscord, M. Effects of elevation and season on nutrient composition of leaves and green pods of Moringa stenopetala and Moringa oleifera. Agroforest. Syst. 2012, 86, 505–518.
11. Sánchez, R.; Spröndly, E.; Ledin, I. Effect of feeding different level of foliage of Moringa oliefera to creole dairy cows on intake, digestibility, milk production and composition. Livest. Sci. 2006, 101, 24–31.
12. Gebregiorgis, F.; Negesse, T.; Nurfeta, A. Feed intake and utilization in sheep fed graded levels of dried moringa (Moringa stenopetala) leaf as a supplement to Rhodes grass hay. Trop. Anim. Health Prod. 2011, 44, 511–517.
13. Melesse, A.; Getye, Y.; Berihun, K. Effect of feeding different levels of Moringa stenopetala leaf meal on growth performance, carcass traits and some serum biochemical parameters of Koekoek chickens. Livest. Sci. 2013, 157, 498–505.
14. Banerjee, S.; Melesse, A.; Dotamo, E.; Berihun, K.; Beyan, M. Effect of feeding different dietary protein levels with Iso-Caloric ration on nutrients intake and growth performances of dual-purpose koekoeck chicken breeds. Int. J. Appl Poult. Res. 2013, 2, 27–32. Available online: http://ijscience.com/Journal_of_Applied_Poultry_Research/pdf-files/Issue-2-2013/27-32.pdf (accessed on 27 November 2015).
15. McDonald, P.; Edwards, R.A.; Greenhalgh, J.F.D.; Morgan, C.A.; Sinclair, L.A.; Wilkinson, R.G. Animal Nutrition, 7th ed.; Pearson Education Limited: Edinburgh, UK, 2010; p. 692.
16. AOAC (Association Official Analytical Chemists). Official Methods of Analysis, 16th ed.; AOAC Inc.: Arlington, VA, USA, 1995; p. 1298.
17. Van Soest, P.J.; Robertson, J.B.; Lewis, B.A. Methods of dietary fiber, neutral detergent fiber and non-starch polysaccharides in relation to animal nutrition. J. Dairy Sci. 1991, 74, 3583–3597.
18. Kakengi, A.M.V.; Kaijage, J.T.; Sarwatt, S.V.; Mutayoba, S.K.; Shem, M.N.; Fujihara, T. Effect of Moringa oleifera leaf meal as a substitute for sunflower seed meal on performance of laying hens in Tanzania. Livest. Res. Rural Dev. 2007, 19, 120. Available online: http://www.lrrd.org/lrrd19/8/kake19120.htm (accessed on 27 November 2015).
19. Dechasa, J.; Sonder, K.; Alemayehu, L.; Mekonen, Y.; Anjulo, A. Leaf yield and nutritive value of Moringa stenopetala and Moringa oleifera accessions: Its potential role in food security in constrained dry farming agro forestry system. In Proceedings of the Moringa and Other Highly Nutritious Plant Resources: Strategies, Standards and Market for Better Impact on Nutrition in Africa, Accra, Ghana, 16–18 November 2006; pp. 1–14.
20. Yang, R.Y.; Tsou, S.C.S.; Lee, T.C.; Chang, L.C.; Kuo, G.; Lai, P.Y. Moringa a novel plant rich in antioxidants bio-available iron and nutrients. Am. Chem. Soc. Symp. Ser. 2006, 925, 224–239.
21. Odeyinka, S.M.; Oyedele, O.J.; Adeleke, T.O.; Odedire, J.A. Reproductive performance of rabbits fed Moringa oleifera as a replacement for Centrosema Pubescens. In Proceedings of the 9th World Rabbit Congress, Verona, Italy, 10–13 June 2008; pp. 411–416.
22. Asaolu, V.O.; Jarju, A.K.; Joof, A.; Odeyinka, S.M.; Manne, J.; Darboe, D.; Jallow, M. Moringa Oleifera Horticulture-Livestock Integration Approach to Improving Incomes and Livelihoods of Women Farmers: The Gambian Experience. Available online: http://hdl.handle.net/10625/44856 (accessed on 15 July 2015).
23. Olugbemi, T.S.; Mutayoba, S.K.; Lekule, F.P. Moringa oleifera leaf meal as a hypocholesterolemic agent in laying hen diets. Livest. Res. Rural Dev. 2010, 22, 84.

Available online: http://www.lrrd.org/lrrd22/4/olug22084.htm (accessed on 27 November 2015).
24. Olugbemi, T.S.; Mutayoba, S.K.; Lekule, F.P. Effect of Moringa (Moringa oleifera) inclusion in Cassava based diets fed to broiler chickens. Int. J. Poult. Sci. 2010, 9, 363–367.
25. Nouala, F.S.; Akinbamijo, O.O.; Adewumi, A.; Hoffmann, E.; Muetzel, S.; Becker, K. The influence of Moringa oleifera leaves as substitute to conventional concentrate on the in vitro gas production and digestibility of groundnut hay. Livest. Res. Rural Dev. 2006, 18, 121. Available online: http://www.lrrd.org/lrrd18/9/noua18121.htm (accessed on 27 November 2015).
26. Babiker, M.S. Chemical composition of some non-conventional and local feed resources for poultry in Sudan. Int. J. Poult. Sci. 2012, 11, 283–287.
27. Van Soest, P.J. Nutritional Ecology of the Ruminant, 2nd ed.; Cornell University Press: Ithaca, Greece, 1994.
28. Banerjee, S.; Demo, K.; Abebe, A. Some serum biochemical and carcass traits of Arsi bale rams reared on graded levels of Millettia ferruginea leaf meal. World Appl. Sci. J. 2013, 28, 532–539.
29. Seyoum, B.; Zinash, S. The Composition of Ethiopian Feedstuffs; IAR: Addis Ababa, Ethiopian, 1989.
30. Gizachew, L. Crude Protein and Mineral Status of Forages Grown on Pelvic Vertisol of Ginchi, Central Highlands of Ethiopia. Ph.D. Thesis, University of the Free State, Bloemfontein, South Africa, November 2002.
31. Eshetie, T. Compatiblity of Quality Protein Maize and Sesame Seed Meal as a Substitute for Synthetic Amino Acids in Broiler Rations. Master's Thesis, Alemaya University of Agriculture, East Harerge, Ethiopia, March 2005.
32. Foster, J.L.; Adesogan, A.T.; Carter, J.N.; Blount, A.R.; Myer, R.O.; Phata, K. Intake, digestibility and N retention by sheep supplemented with warm-season legume hays or soybean meal. J. Anim. Sci. 2009, 87, 2891–2898.
33. Mpairwe, D.R.; Sabiiti, E.N.; Mugerwa, J.S. Effect of dried Gliricidia sepium leaf supplement on feed intake, digestibility and nitrogen retention in sheep fed dried KW4 elephant grass (Pennisetum purpureum) ad libitum. Agroforest. Syst. 1998, 41, 139–150.
34. Nurfeta, A. Feed intake, digestibility, nitrogen utilization and body weight change of sheep consuming wheat straw supplemented with local agricultural and agro-industrial by-products. Trop. Anim. Health Prod. 2010, 42, 815–824.
35. Aregheore, E.M. Intake and digestibility of Moringa oleifera-batiki grass mixtures by growing goats. Small Rumin. Res.2002, 46, 23–28.
36. Moyo, B.; Masika, P.J.; Muchenje, V. Effect of supplementing crossbred Xhosa lop-eared goat castrates with Moringa oleifera leaves on growth performance, carcass and non-carcass characteristics. Trop. Anim. Health Prod. 2012, 44, 801–809.
37. Tilahun, S.; Animut, G.; Urge, M. Effects of supplementing cassava leaf meal, brewers' dried grain and their mixture on body weight change and carcass traits of local goats fed urea treated tef straw. J. Livest. Sci. 2013, 4, 31–43.
38. Nuhu, F. Effect of Moringa Leaf Meal (MOLM) on Nutrient Digestibility, Growth, Carcass and Blood Indices of Weaner Rabbits. Ph.D. Thesis, Kwame Nkrumah University of Science and Technology, Kumasi, Ghana, February 2010.

CHAPTER 14

Histopathological Study of Cyclosporine Pulmonary Toxicity in Rats

Said Said Elshama,[1,2] Ayman El-Meghawry EL-Kenawy,[2,3] and Hosam-Eldin Hussein Osman[2,4]

[1]*Department of Forensic Medicine and Clinical Toxicology, College of Medicine, Suez Canal University, P.O. Box 3457, Ismailia, Egypt*

[2]*Taif University, Taif, Saudi Arabia* [3]*Department of Molecular Biology, GEBRI, University of Sadat City, P.O. Box 79, Sadat City, Egypt*

[3]*Department of Anatomy, College of Medicine, Al-Azhar University, P.O. Box 345, Cairo, Egypt*

ABSTRACT

Cyclosporine is considered one of the common worldwide immunosuppressive drugs that are used for allograft rejection prevention. However, articles that address adverse effects of cyclosporine use on the vital organs such as lung are still few. This study aims to investigate pulmonary toxic effect of cyclosporine in rats by assessment of pulmonary histopathological changes using light and electron microscope examination. Sixty male adult albino rats were divided into three groups; each group consists of twenty rats. The first received physiological saline while the second and third groups received 25 and 40 mg/kg/day of cyclosporine, respectively, by gastric gavage for forty-five days. Cyclosporine reduced the lung and body weight with shrinkage or pyknotic nucleus of pneumocyte type II, degeneration of alveoli and interalveolear septum beside microvilli on the alveolar surface, emphysema, inflammatory cellular infiltration, pulmonary blood vessels congestion, and increase of fibrous tissues in the interstitial tissues and around alveoli with negative Periodic Acid-Schiff staining. Prolonged use of cyclosporine induced pulmonary ultrastructural and histopathological changes with the lung and body weight reduction depending on its dose.

1. INTRODUCTION

Cyclosporine A (CsA) is considered one of the common immunosuppressive agents which are used during organ transplantation preventing allograft rejection and increasing the rates of patients' survival and as a treatment for autoimmune diseases. It is a cyclic endecapeptide whereas it is extracted from the fungus calledTolypocladium inflatum [1].

Recent studies reported that long-term use of cyclosporine A may lead to many systemic toxic effects such as nephrotoxicity, hepatotoxicity, and cardiovascular affection based on its oxidative stress mechanism that generates free radicals inducing lipid peroxidation in the different organs [2]. Other studies referred that oxidative damage may develop pulmonary disorders during systemic or local administration of cyclosporine; therefore the safety of its use is still considered a controversial issue [3].

Organ transplantation such as a lung transplantation is a lifesaving procedure whereas the rate of its use is increasing in the different countries such as the United States and Japan and then the rate of cyclosporine use is also rising [4] although the incidence risk of bronchogenic carcinoma and pulmonary fibrosis development with the use of cyclosporine A is ranging from 2 to 4% in some cases [5].

So, the present study aims to investigate the toxic effect of cyclosporine on the lung tissues in rats via assessment of pulmonary histopathological changes by using light and electron microscope examination.

2. MATERIAL AND METHODS

Sixty healthy male adult albino rats weighing 200 ± 20 g were obtained from the animal house of King Abdulaziz University, Jeddah, exposed to 12 hr day and night cycles with a free access to water and the standard rat pellet during the experimental period. They were divided into three groups; each group consists of twenty rats. The first group (control) received physiological saline only while the second received 25 mg/kg/day of cyclosporine dissolved in physiological saline [6], whereas the third group received 40 mg/kg/day of cyclosporine dissolved in physiological saline [7]. Daily administration of physiological saline and cyclosporine was done for forty-five days by using gastric gavage whereas cyclosporine was available in a soft gelatin capsule (50 mg) that was manufactured by R. P. Scherer GmbH & Co. KG, Eberbach/Baden, Germany, for Novartis Pharma AG, Basel, Switzerland.

2.1. Histopathological Studies

Rats were weighed and then they were sacrificed under diethyl ether anesthesia after 24 hours from the last administration of cyclosporine. An incision was carried out in the chest for a lung excision that was weighed and washed several times with a normal saline and then fixed by an intratracheal instillation of 10% neutral buffered formalin. The preserved tissues were dehydrated, embedded in paraffin, sectioned at 3-4 μm thickness, and stained by haematoxylin and eosin, Periodic Acid-Schiff (PAS), and Mallory stain [8]. The pulmonary histological slides were examined and scored under a light microscope by a blinded pathologist to the experimental groups for quantifying the extent of lung histopathological changes using a scoring scale of 0 to 4 for each lung damage parameter (oedema, hemorrhage, cell infiltration, and alveolar septal thickening) with total score of 0–16 [9, 10].

Ultrastructural studies on the lung cells were carried out by using an electron microscope whereas the lung tissue specimens were prepared via fixation in 2.7% glutaraldehyde solution in 0.1 M phosphate buffer for 1.5 hours at 4°C and then they were washed in 0.15 M phosphate buffer (pH 7.2) and postfixed in 2% osmic acid solution in 0.15 M phosphate buffer for one hour at 4°C. Dehydration was carried out in acetone while the inclusion was in the epoxy embedding resin Epon 812. The blocks were cut with an ultramicrotome type LKB at 70 nm thickness. The sections were doubly contrasted with the solutions of uranyl acetate and lead citrate for analysis by using an electron microscope [11].

2.2. Ethical Considerations

The most appropriate animal species was chosen for this research. Promotion of the high standard of the care and animal well-being at all times was done. Appropriate sample size was calculated by using the fewest number of animals to obtain the valid results statistically. Painful procedures were performed under anesthesia to avoid distress and pain. Our standards of the animal care and administration met those required by applicable international laws and regulations [12].

2.3. Statistical Analysis

Statistical analysis was performed by using SPSS version 17. Variability of the results was expressed as mean ± SD. Results were analyzed by using one-way ANOVA and post hoc multiple comparisons test (TUKEY) to investigate the difference among groups. A value of 0.05 was considered statistically significant.

3. RESULTS

3.1. Effect of Cyclosporine Different Doses on the Rats' Lung Weight, Body Weight, and the Lung/Body Weight Ratio

There is a significant decrease in the rats' lung and body weight of second group which received 25 mg/kg/day of cyclosporine in comparison with the control group while there is a significant decrease in the rats' lung and body weight of third group which received 40 mg/kg/day of cyclosporine in comparison with the second and control groups.

Figure 1 shows that mean + SD values of the rats' lung weight in control, second, and third groups are ; ; and , respectively, whereas value is 86.473 indicating a difference between groups which is statistically significant ().

Table 1 shows that mean + SD values of the rats' body weight in control, second, and third groups are ; ; and , respectively. The value of is 179.85 indicating the difference between the groups while the value of is <0.001 indicating that the difference between the groups is statistically significant. The lung/body weight ratio in control, second, and third groups is 1.14%, 1.22%, and 1.31%, respectively.

Table 1: Comparison between effects of cyclosporine different doses on the rats' lung weight, body weight, and the lung/body weight ratio.

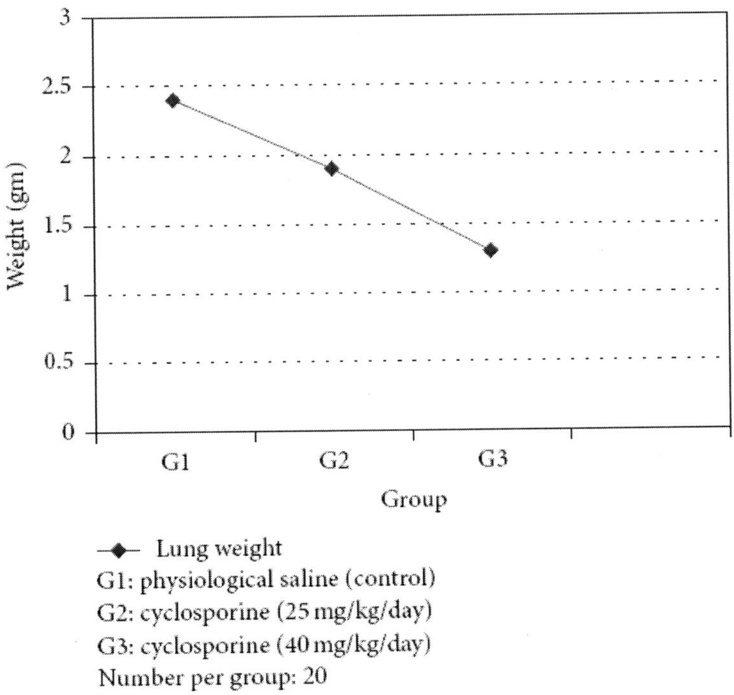

Figure 1: The effect of cyclosporine different doses on the rats' lung weight.

3.2. Histopathological Findings

3.2.1. Pulmonary Histopathological Findings by Using a Light Microscope

Examination of the lung tissues in the first group rats (control) which received physiological saline showed a normal lung architecture, thin interalveolar septa, folded columnar epithelial cells of bronchiole, clearly seen alveolar sacs, normal pulmonary vessels with positive stain of Periodic Acid-Schiff (PAS), and normal fibrous tissues distribution (Figures 2(a), 3(a), and 4(a)). Rats of the second group which received 25 mg/kg/day of cyclosporine showed thickened wall of bronchiole with moderately congested blood capillaries, macrophages infiltration in the alveolar spaces, degenerated and thickened interalveolar septa with moderate positive Periodic Acid-Schiff (PAS) stain, and mild fibrous tissues distribution (Figures 2(b), 3(b), and 4(b)). The lung tissues in the third group rats which received 40 mg/kg/day of cyclosporine showed degeneration of alveoli and interalveolar septum, emphysema, marked thinning of interstitial tissues, aggregations areas of inflammatory cellular infiltration beside thickened and congested pulmonary blood vessels, negative Periodic Acid-Schiff (PAS) stain, and the increase of fibrous tissues in the interstitial tissues and around alveoli (Figures 2(c), 3(c), and 4(c)).

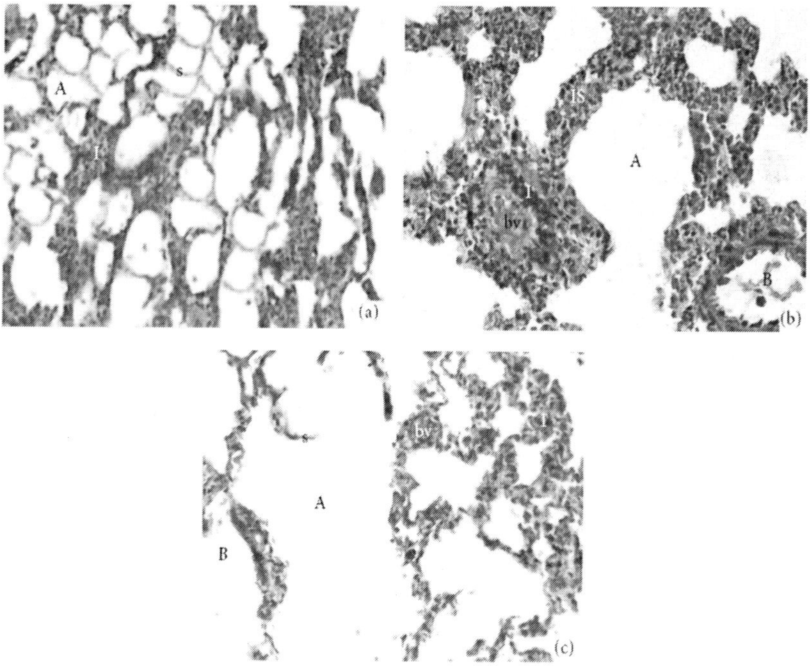

Figure 2: (a) A photomicrograph of transverse section in the control rat lung shows normal architecture of alveoli (A) with thin interalveolar septa (s) and normal interstitial tissues (I) (H&E ×400). (b) A photomicrograph of transverse section in the second group rat lung shows marked thickening of interstitial tissues (IS) with numerous areas of cellular infiltration (I), congested blood capillaries (bv), and bronchiole (B) with widening alveoli (A) (H&E ×400). (c) A photomicrograph of transverse section in the third group rat lung shows loss of normal architecture, cellular infiltration (I), fragmentation and degeneration of alveoli and interalveolar septum (s) with a subsequent compensatory dilatation of alveoli (A), thickened and congested pulmonary blood vessels (bv), and degenerated bronchiolar wall (B) (H&E ×400).

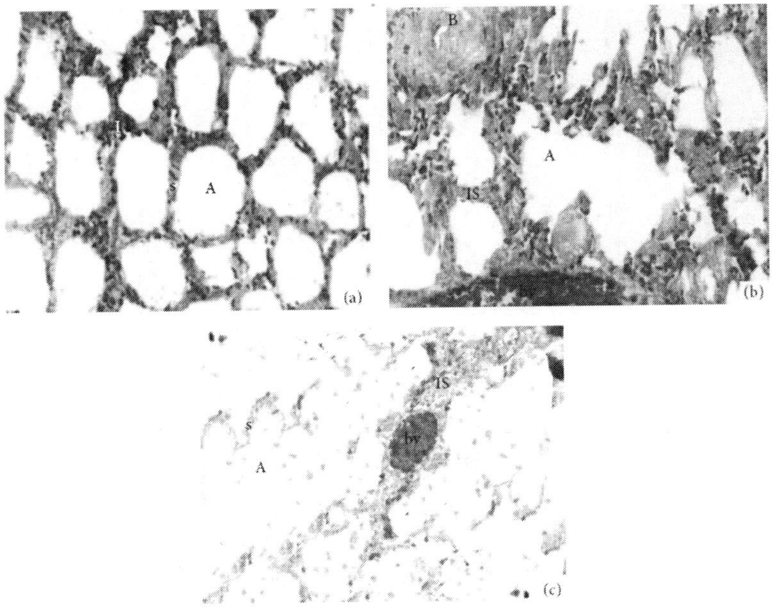

Figure 3: (a) A photomicrograph of transverse section in the control rat lung shows normal distribution of fibrous tissues and architecture of alveoli (A) with thin interalveolar septa (s) and normal interstitial tissues (I) (Mallory ×400). (b) A photomicrograph of transverse section in the second group rat lung shows mild distribution of fibrous tissues around alveoli (A) with thickening of the interalveolar septa (IS) and bronchiole (B) (Mallory ×400). (c) A photomicrograph of transverse section in the third group rat lung shows loss of normal architecture, increase of fibrous tissues around alveoli (A), subsequent compensatory dilatation of other alveoli, marked thinning of the interalveolar septum (s) with an increase of fibrous tissues in the areas of interstitial tissues (IS), and thickened and congested pulmonary blood vessels which contain hemorrhagic blood cells (bv) (Mallory ×400).

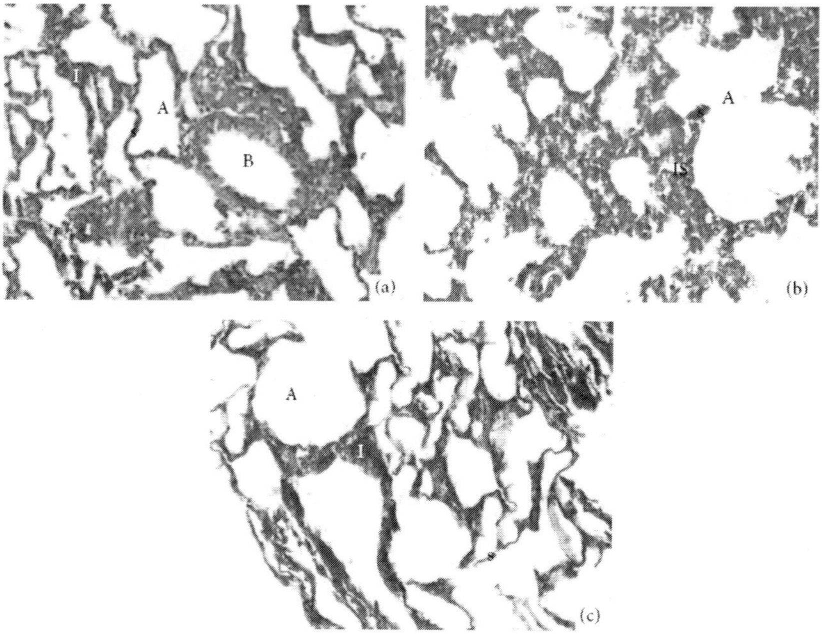

Figure 4: (a) A photomicrograph of transverse section in the control rat lung shows positive PAS staining with normal architecture of alveoli (A) with thin interalveolar septa (s) and normal interstitial tissues (I) (PAS ×400). (b) A photomicrograph of transverse section in the second group rat lung shows moderate positive PAS staining with nearly normal architecture of alveoli (A), thick interalveolar septa (s), and interstitial tissues (I) (PAS ×400). (c) A photomicrograph of transverse section of third group rat lung shows negative PAS staining, accumulation of inflammatory cells near the blood vessels with abnormal architecture of alveoli (A), and destruction of the interalveolar septa (s) and interstitial tissues (I) (PAS ×400).

Table 2 showed the severity of pulmonary histopathological changes according to scoring scale, whereas a statistical significant difference in the overall scores of the lung histopathological parameters (oedema, hemorrhage, cell infiltration, and alveolar septal thickening) was observed in the second group in comparison with the control group and in the third group in comparison with the second and control groups.

Table 2: Comparison between effects of cyclosporine different doses on the rats' lung histopathological parameters.

Parameter	Group		
	First M ± SD	Second M ± SD	Third M ± SD
Body weight (gm)	210.9 ± 15.11	155.35 ± 12.15*	99.55 ± 10.26**
Lung weight (gm)	2.4 ± 0.18	1.9 ± 0.19*	1.3 ± 0.24**
Lung/body weight ratio %	1.14	1.22*	1.31**

First group: physiological saline (control).
Second group: cyclosporine (25 mg/kg/day).
Third group: cyclosporine (40 mg/kg/day).
M ± SD: mean ± standard deviation. Number per group: 20
* $P < 0.001$: significant difference in comparison with the control group
** $p < 0.001$: significant difference in comparison with the second group.

3.2.2. Pulmonary Histopathological Findings by Using an Electron Microscope

The ultrastructure of lung cells in the first group rats (control) which received physiological saline showed a rounded nucleus of pneumocyte type II that is surrounded by lamellar bodies with normal mitochondria and normal microvilli on the alveolar surface (Figure 5(a)). But the ultrastructure of lung cells in the second group rats showed a shrinkage nucleus of pneumocyte type II which is surrounded by vacuoles and congested blood capillaries filled with the blood in cytoplasm with degenerated microvilli on the surface of alveoli (Figure 5(b)). The ultrastructure of lung cells in the third group rats showed a pyknotic nucleus of pneumocyte type II that is surrounded by vacuoles which are variable in the amount and size with more congested blood capillaries filled with the blood in cytoplasm with degenerated microvilli in the alveolar surface (Figure 5(c)).

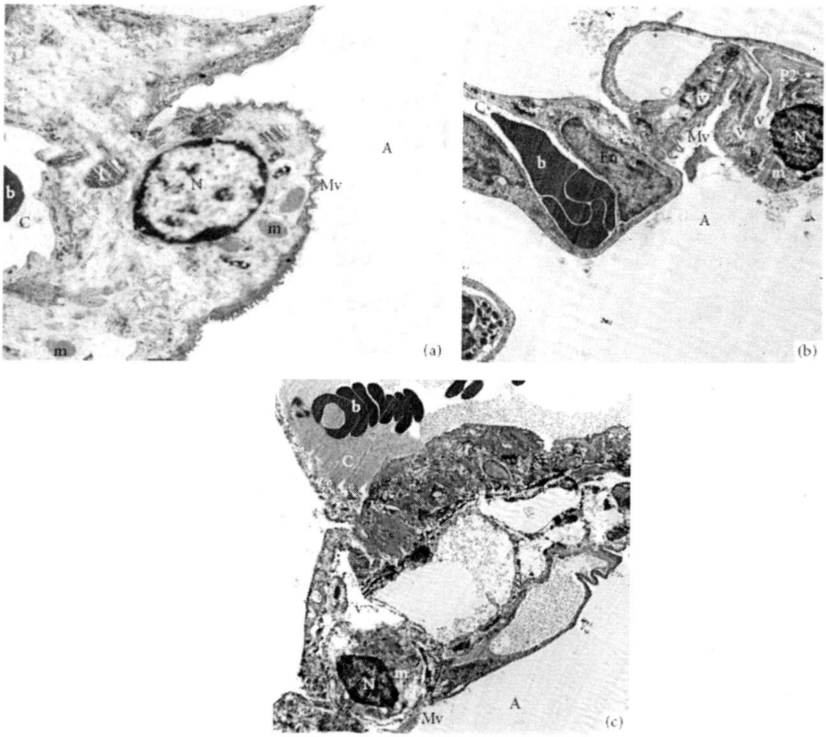

Figure 5: (a) Electronic microscopic picture of the control rat lung shows rounded nucleus (N) of pneumocyte type II, lamellar bodies (L) around the nucleus, and mitochondria (m) in cytoplasm with microvilli (Mv) on the surface of alveoli (A) (×10000). (b) Electronicmicroscopic picture of the second group rat lung shows a shrinkage nucleus (N) of pneumocyte type II, vacuoles and congested blood capillaries (C) filled with the blood (b) in the cytoplasm and around the nucleus, degenerated mitochondria (m) in cytoplasm with degenerated microvilli (Mv) on the surface of alveoli (A) (×10000). (c) Electronic microscopic picture of the third group rat lung shows pyknotic nucleus (N) of pneumocyte type II, a variable number of vacuoles (v) and more congested blood capillaries (C) filled with the blood (b) around the nucleus, and degenerated mitochondria (m) in cytoplasm with degenerated microvilli (Mv) on the surface of alveoli (A) (×10000).

4. DISCUSSION

Many previous studies indicated that there is a correlation between the prolonged use of cyclosporine among transplant organ patients and vital organs toxicity development. However, its adverse effects on vital organ such as lung are still obscure because of the paucity of researches that focus on it. So, our study tries to investigate pulmonary toxicity of cyclosporine by assessment of the probable histopathological changes via using light and electron microscope examination.

The current study showed that lung and body weights are significantly decreased in the second group which received 25 mg/kg/day of cyclosporine in comparison with the control group while the lung and body weights of third group which received 40 mg/kg/day of cyclosporine showed a significant decrease in comparison with the second group; thus the lung/body weight ratio is also affected. This is in agreement with Wongmekiat et al. [13], who indicated that cyclosporine toxicity affects the appetite beside its catabolic effect inducing the body weight loss, and in consistency also with Chakravarthi et al. [14], who referred to the fact that cyclosporine induces cell apoptosis which is correlated with the affected organ weight reduction and this is in contrast with Sato et al. [15], who confirmed that normal pulmonary cells are protected from cyclosporine induced apoptosis.

Our results revealed that the third group which received 40 mg/kg/day of cyclosporine induced marked pulmonary histopathological abnormalities such as a pyknotic nucleus of pneumocyte type II, degeneration of alveoli with its microvilli, and emphysema with inflammatory cellular infiltration and pulmonary vessels congestion in consistency with Yousef and ALRajhi [16] while the second group which received 25 mg/kg/day of cyclosporine showed mild pulmonary ultrastructures and histopathological changes such as a shrinkage nucleus of pneumocyte type II, degenerated microvilli on the alveolar surface, thickened wall of bronchiole with moderately congested blood capillaries and macrophages infiltration in the alveolar spaces, and nearly normal architecture of alveoli with thickened interalveolar septa. Furthermore, the overall lung lesion severity scores were differed significantly between the second and third groups indicating that the severity of cyclosporine induced pulmonary histological changes depending on its dose.

Most of the published articles focused on a fact which shows that generation of free radicals inducing oxidative stress leads to molecular and cellular damage which are considered the cause of cyclosporine toxic effects on the different body organs [17]. The effects of released reactive oxygen species (ROS) by normal respiratory system are counteracted by glutathione and antioxidants enzymes such as catalase and peroxidase; therefore more generation of ROS via cyclosporine leads to the balance disturbance with antioxidants defense mechanism inducing toxic cellular substances which lead to histopathological changes [18].

The present study showed that cyclosporine increases fibrous tissues formation in the lung interstitial tissues and around alveoli depending on its dose in consistency with Katrin et al. [19], who referred to the fact that cyclosporine can be trigger to stimulate fibroblast proliferation via mediators which are induced by the epithelial cells of airway passages. According to Esposito et al. [20], there is a correlation between cyclosporine cytotoxicity and mitochondrial enzyme activity disturbance and its ability to react with the nucleus receptors to prevent genetic transcription of proteins that are secreted by fibroblasts, macrophages, monocytes, and endothelial cells. Therefore, cyclosporine affects the cell of lung tissues as a result of multiple effects such as carbohydrates depletion in the cytoplasm of lung cell that leads to the lung structure disturbance which is supported in the current study by a negative stain of Periodic Acid-Schiff.

5. CONCLUSION

The prolonged use of cyclosporine may induce pulmonary histopathological and ultrastructural changes such as shrinkage or pyknotic nucleus of pneumocyte type II, degeneration of alveoli, emphysema with an inflammatory cellular infiltration, pulmonary vessels congestion, and lung and body weight reduction depending on its dose and based on reactive oxygen species generation. Therefore, further researches on human should be done in the future to verify our results and confirm pulmonary toxicity of cyclosporine prolonged use.

CONFLICT OF INTERESTS

The authors declare that there is no conflict of interests regarding the publication of this paper.

REFERENCES

1. J. R. Chapman and B. J. Nankivell, "Nephrotoxicity of ciclosporin A: short-term gain, long-term pain?" Nephrology Dialysis Transplantation, vol. 21, no. 8, pp. 2060–2063, 2006.
2. R. Rezzani, "Exploring cyclosporine A-side effects and the protective role-played by antioxidants: the morphological and immunohistochemical studies," Histology and Histopathology, vol. 21, no. 1–3, pp. 301–316, 2006.
3. D. M. Lyu and M. R. Zamora, "Medical complications of lung transplantation," Proceedings of the American Thoracic Society, vol. 6, no. 1, pp. 101–107, 2009.

4. T. Shiraishi, Y. Okada, Y. Sekine et al., "Registry of the Japanese society of lung and heart-lung transplantation: the official Japanese lung transplantation report 2008," General Thoracic and Cardiovascular Surgery, vol. 57, no. 8, pp. 395–401, 2009.
5. J. Mathew and R. A. Kratzke, "Lung cancer and lung transplantation: a review," Journal of Thoracic Oncology, vol. 4, no. 6, pp. 753–760, 2009.
6. A. J. Cologna, L. V. D. S. Lima, S. Tucci Jr. et al., "Cyclosporine action on kidneys of rats submitted to normothermic ischaemia and reperfusion," Acta Cirúrgica Brasileira, vol. 23, no. 1, pp. 36–41, 2008.
7. Uz. Ebru, Uz. Burak, A. Kaya et al., "Protective effect of erdosteine on cyclosporine induced chronic nephrotoxicity in rats," Nephro-Urology, vol. 3, no. 4, pp. 280–284, 2011.
8. J. D. Bancroft and M. Gamble, Theory and Practice Histological Techniques, Churchill Livingstone, New York, NY, USA, 5th edition, 2002.
9. R. Akcılar, A. Akcılar, H. Şimşek et al., "Hyperbaric oxygen treatment ameliorates lung injury in paraquat intoxicated rats," International Journal of Clinical and Experimental Pathology, vol. 8, no. 10, pp. 13034–13042, 2015.
10. L. Yamanel, U. Kaldirim, Y. Oztas et al., "Ozone therapy and hyperbaric oxygen treatment in lung injury in septic rats," International Journal of Medical Sciences, vol. 8, no. 1, pp. 48–55, 2011.
11. L. Graham and J. M. Orenstein, "Processing tissue and cells for transmission electron microscopy in diagnostic pathology and research," Nature Protocols, vol. 2, no. 10, pp. 2439–2450, 2007.
12. M. M. Naderi, A. Sarvari, A. Milanifar, S. B. Boroujeni, and M. M. Akhondi, "Regulations and ethical considerations in animal experiments: international laws and islamic perspectives," Avicenna Journal of Medical Biotechnology, vol. 4, no. 3, pp. 114–120, 2012.
13. O. Wongmekiat, N. Leelarugrayub, and K. Thamprasert, "Beneficial effect of shallot (Allium ascalonicumL.) extract on cyclosporine nephrotoxicity in rats," Food and Chemical Toxicology, vol. 46, no. 5, pp. 1844–1850, 2008.
14. S. Chakravarthi, C. Fu Wen, and N. Haleagrahara, "Apoptosis and expression of bcl-2 in cyclosporine induced renal damage and its reversal by beneficial effects of 4′, 5′, 7′- trihydroxyflavone," Journal of Analytical Bio-Science, vol. 32, no. 4, pp. 320–327, 2009.
15. M. Sato, I. Tsujino, M. Fukunaga et al., "Cyclosporine A induces apoptosis of human lung adenocarcinoma cells via caspase-dependent pathway," Anticancer Research, vol. 31, no. 6, pp. 2129–2134, 2011.
16. O. M. Yousef and W. I. ALRajhi, "The probable protective role of vitamin C against cyclosporine an induced pulmonary changes in mice," Journal of Life Sciences and Technologies, vol. 1, no. 1, pp. 1–6, 2013.
17. H. Argani, A. Ghorbanihaghjo, N. Rashtchizadeh, S. Seifirad, and Y. Rahbarfar, "Effect of cyclosporine-a on paraoxonase activity in wistar rats," International Journal of Organ Transplantation Medicine, vol. 2, no. 1, pp. 25–31, 2011.
18. J. Lee, "Use of antioxidants to prevent cyclosporine a toxicity," Toxicological Research, vol. 26, no. 3, pp. 163–170, 2010.
19. E. Katrin, R. Michael, K. Janette, R. Peter, and A. R. Glanville, "Cyclosporine A mediates fibroproliferation through epithelial cell," Transplantation, vol. 77, no. 12, pp. 1886–1893, 2004.
20. C. Esposito, A. Fornoni, F. Cornacchia et al., "Cyclosporine induces different responses in human epithelial, endothelial and fibroblast cell cultures," Kidney International, vol. 58, no. 1, pp. 123–130, 2000.

CHAPTER 15

Protective and Anti-oxidant Activity of the Euryops Arabicus against Paracetamol Induced Hepatorenal Toxicity in Rats

Essam Mahmoud Hafez[1*], Milad Gad Paulis[1], Mostafa Abotaleb Ahmed[1], Marian NagyFathy[2], Ahmed Abdel-Lateff [3,4] and Mardi M. Algandaby[5]

[1]*Department of Forensic Medicine and Clinical Toxicology, Faculty of Medicine, Minia University, Minia 61519, Egypt*

[2]*Department of Pathology, Faculty of Medicine, Minia University, Minia 61519, Egypt*

[3]*Department of Natural Products and Alternative Medicine, Faculty of Pharmacy, King Abdulaziz University, Jeddah, Saudi Arabia*

[4]*Department of Pharmacognosy, Faculty of Pharmacy, Minia University, Minia 61519, Egypt*

[5]*Department of Biological Sciences, Faculty of Science, King Abdulaziz University, Jeddah, Saudi Arabia*

ABSTRACT

Paracetamol overdose is a predominant cause of hepatorenal toxicity in both humans and experimental animals. The extract of Euryops arabicus was evaluated for its protective and anti-oxidant effect against paracetamol-induced hepatic and renal injuries in rats. The extract of

Euryops arabicus was administered at100 mg/kg and 200 mg/kg followed by paracetamol at 1g/kg for seven days. The animals were sacrificed after 24 hrs of paracetamol challenge .Indices of liver such as aspartate aminotransferase (AST), alanine aminotransferase (ALT), alkaline phosphatase (ALP), total bilirubin (TB), serum albumin (SA) and renal indices blood urea (BU) and serum creatinine (SC) were measured. Liver and kidney homogenates were analyzed for oxidative stress biomarkers namely Thiobarbituric acid reactive substances TBARS. Finally, histopathological examinations of both previous organs were examined. The significantly disturbed liver and kidney functions by paracetamol toxicity were restored to almost normal values by administration of Euryops arabicus. Also, the elevated TBARS in groups received paracetamol alone decreased in groups given paracetamol and Euryops arabicus. Histological effect of paracetamol on liver and kidney was also markedly abolished by co-administration of Euryops arabicus. These results suggest that Euryops arabicus may protect from paracetamol-induced liver and kidney toxicity.

Keywords: *Euryops arabicus*; paracetamol; Thiobarbituric acid reactive substances; hepatorenal histopathology; protective and anti-oxidant

INTRODUCTION

Asteraceae is a large family of herbaceous plants contains about 1100 genera and 25,000 species, widely distributed in the tropical and subtropical regions. It is economically important as a source of food, such as lettuce and artichokes, cooking oils, sweetening agents, and tea infusions [1]. Among 100 species of the genus *Euryops*, only *Euryops arabicus* (*E. arabicus*) is known in Saudi Arabia [2].

E. arabicus (Jabur) is dome shaped shrub ranged from three to five feet tall with interesting two inches long narrow lobed leathery leaves which are clustered at the tips of the branches. The fall appears in summer and the composite heads of spaced out yellow ray flowers and orange-yellow disk flowers are appeared. It is attractive plant that it has been described as a dwarf pine tree because of its elongated and narrow leaves. It is native to the Arabian Peninsula south to Somalia where the heated leaves and stems were once used in the treatment of wounds. The first description had been done by Ernst Gottlieb von Steudel, a German physician in 1852. The name for the genus comes from the Greek words 'eury' (or 'eurys') meaning "large" or "broad" and 'ops' (or 'opos') meaning "resemblance", "sight" or "the eye" probably in reference to the large eye-like flowers [1-3]. Metabolites from *E. arabicus* were studied for their anti-oxidant, antimicrobial and antiproliferative effects. For instances; flavonoids, secofuroermophilanes, furoermophilanes and eremophilanolides were identified from the genus Euryops [3-6].

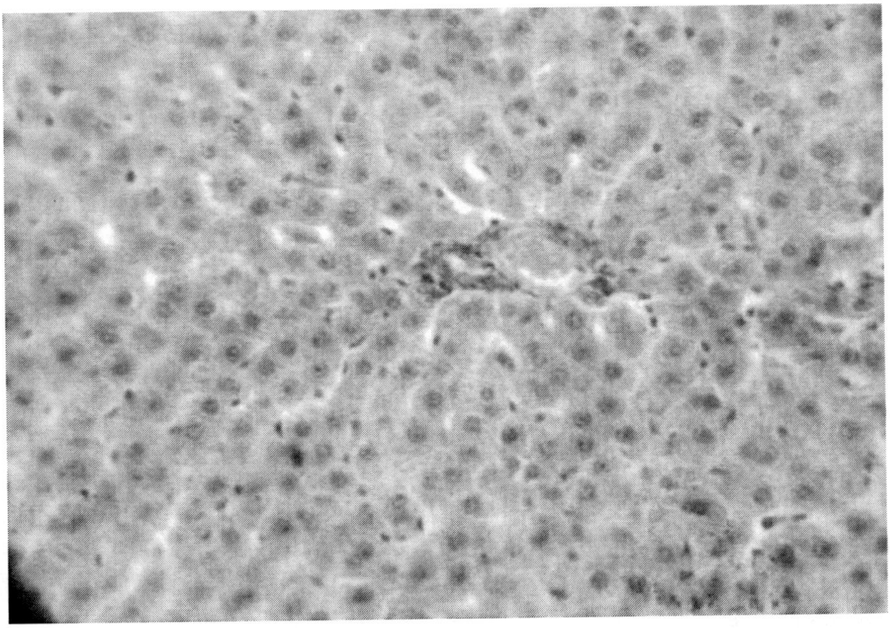

Figure 1. Group I rat liver served as normal negative control and they received normal saline 1 mL/kg bw, orally, normal architecture (H and E x 200).

Drug-induced liver injury is a potential complication of virtually every prescribed medication, because the liver occupies the main role in the metabolic disposition of all drugs and foreign substances. Most of the hepatotoxic chemicals damage liver cells mainly by lipid peroxidation and other oxidative damages and this applies also to paracetamol which is a widely used analgesic and antipyretic which is safe when used at therapeutic levels [4]. Paracetamaol is representing the drug of choice in children. However, paracetamol hepatotoxicity is the leading cause of drug induced liver failure and an acute or cumulative overdose can cause severe liver injury with the potential to progress to liver failure [5].

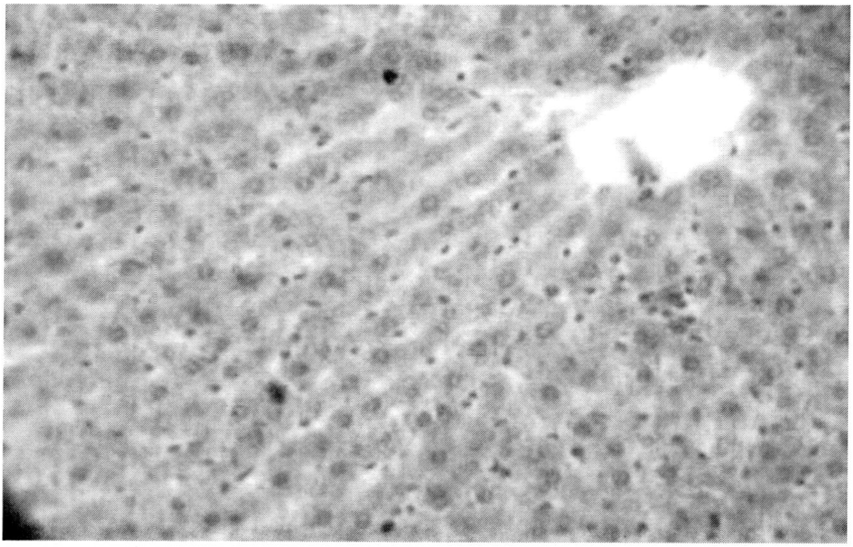

Figure 2: Group II rat liver served as normal positive control and they received 1%.carboxy methyl cellulose 1 mL/kg bw orally, normal architecture (H and E x 200).

Paracetamol induced toxicity in rats is one of the widely used experimental model to evaluate the hepatoprotective nature of herbal extracts [6]. A computer survey including different databases, especially, Scifinder, indicated no publication had been reported about in-vivo anti-oxidant activity of *E. arabicus*.

The main goal of the current study is to evaluate whether the extract of *E. arabicus* has in-vivo antioxidant and protective effect against hepatorenal toxicity induced by paracetamol. The in-vitro antioxidant effect was studied employing the volatile oil which obtained from the areal part of *E. rabicus* and showed moderate activity [7]. Interesting is the chemical composition of *E. arabicus*, is mainly polyphenolic compounds. Thus the current study is interesting in the in-vivo study its antioxidant effect. A combination of biochemical and histological parameters were used to investigate the protective potential of *E. arabicus* extract on paracetamol induced hepatorenal toxicity.

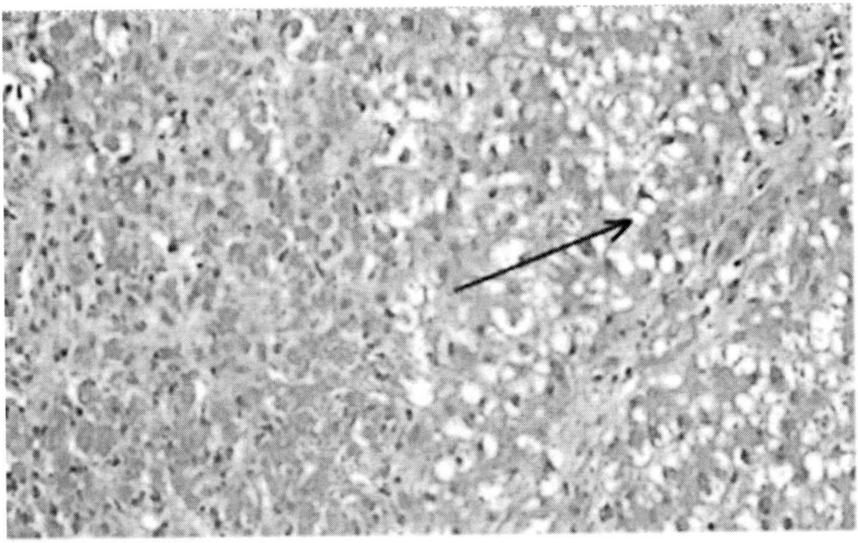

Figure 3. Group III rat liver received paracetamol at dose of 1 g/kg bw p.o. (paracetamol) showing extensive areas of hepatocellular necrosis and inflammatory cell infiltration. Most of the centrilobular hepatocytes were swollen with marked cytoplasmic vacuolation and pyknotic nuclei with obliterated intervening hepatic sinusoids (H and E x 100).

MATERIALS AND METHODS

CHEMICALS

All the materials used for this experiment were of analytical grade. Paracetamol (EPICO Company), thiobarbituric acid and potassium phosphate buffer used for determination of Thiobarbituric acid reactive substances (TBARS) level, carboxy methyl cellulose (CMC) were purchased from Sigma and Merck Chemical Companies. Diagnostic kits for the estimation of aspartate aminotransferase (AST) and alanine aminotransferase (ALT), alkaline phosphatase (ALP), serum albumin (SA), total bilirubin (TB), blood urea (BU), and serum creatinine (SC) were manufactured by Ranbaxy Diagnostics Ltd. (UK).

PLANT MATERIAL

The aerial parts *E. arabicus* (Family Asteraceae) were collected from Al-Taeif, 200 Km south Jeddah, Saudi Arabia. The powdered air-dried *E. arabicus* was extracted and prepared according to Walied et al. [7]. The extract was dissolved in 1% CMC just before administration.

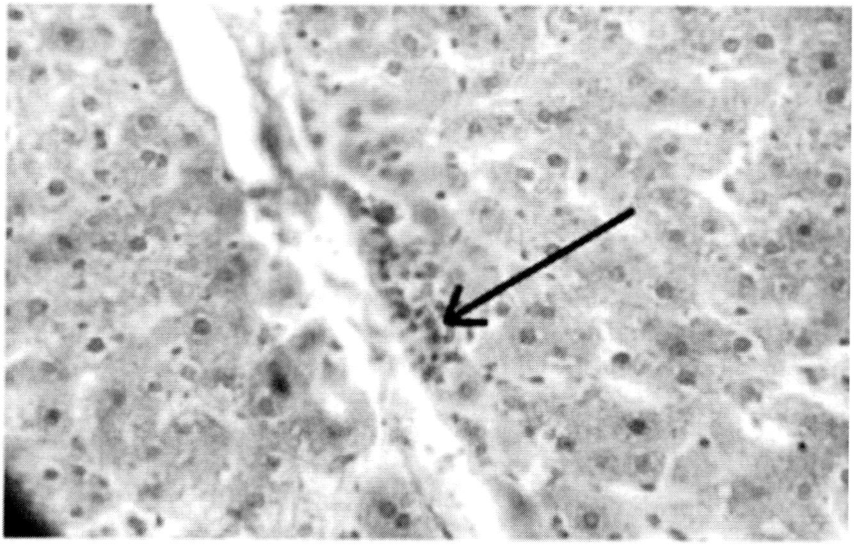

Figure 4: Group IV rat liver received paracetamol and 100 mg/kg bw, p.o. of E. arabicus leaves extract. H and E staining of liver howing scattered inflammatory cells (H and E x 200).

ANIMALS

Healthy adult male albino rats weighting about 150-170 grams were obtained from the animal house in Faculty of Science Minia University. All animals were allowed free access to distilled water and laboratory chow ad libitum. To avoid stress of isolation or overcrowdings, 6 rats were housed per cage. They were left freely wandering in their cage for two weeks with 12 hour dark: light cycle for acclimatization before starting the experiment. They were fasted 12h before administration of paracetamol. All protocols used in this study were approved by the Committee of Minia University.

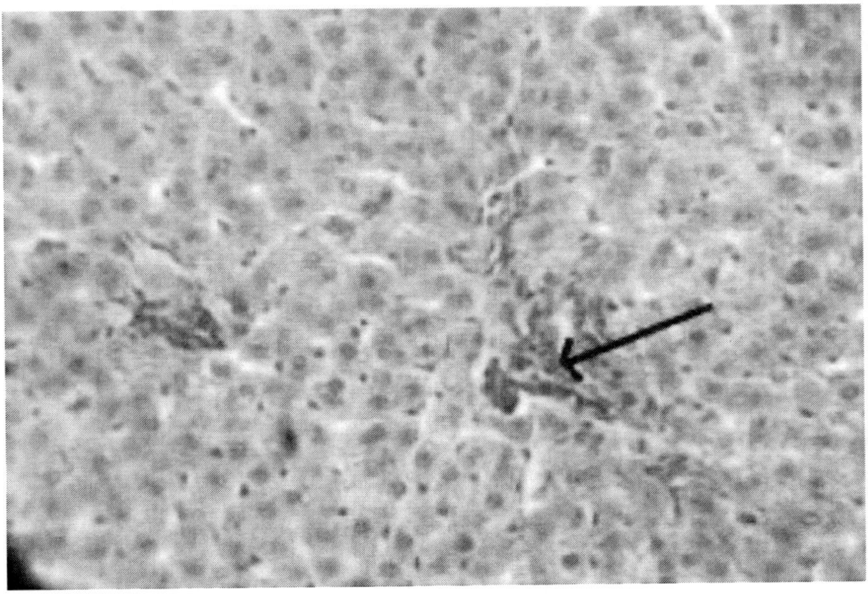

Figure 5: Group V rat liver received paracetamol and 200 mg/kg bw, p.o. of E. arabicus leaves extract. H and E staining of liver showing few scattered inflammatory cells (H and E x 200).

ACUTE TOXICITY OF THE EXTRACT

The acute toxicity of the extract was performed as per OECD-425 guidelines. Five rats of similar weight were chosen. One animal was fasted overnight with access to drinking water. They were given 2000 mg/kg of the extract and observed for 24h for mortality. The animal survived and then four additional animals were tested sequentially so that a total of five animals were tested. All the animals were observed closely for 24 h and daily for 14 days, no mortality was observed. Hence, we selected 200 mg/kg (1/10th of 2000 mg/kg) as maximum safety dose with descending dose levels with 2 fold interval i.e., 100 mg/kg and 200 mg/kg body weight of the test animal.

Rats were divided into five groups of 6 animals each and were given orally the following treatment for seven days.

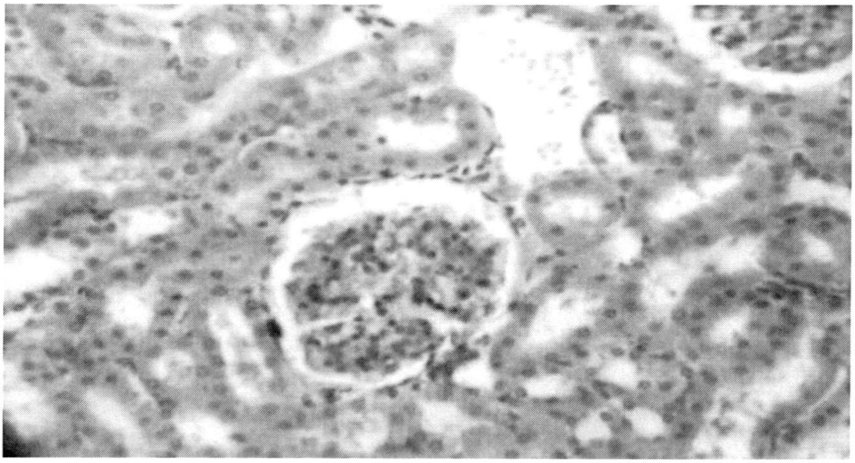

Figure 6: Group I rat renal tissue served as normal negative control and they received normal saline 1 mL/kg bw, orally normal architecture (H and E x 200).

GROUP I: served as normal negative control and they received normal saline 1 ml/kg body weight, orally (bw, p.o).

GROUP II: served as normal positive control and they received 1%.carboxy methyl cellulose 1 mL/kg bw, p.o.

GROUP III: received paracetamol at dose of 1 g/kg bw, p.o.

GROUP IV: received 100 mg/kg *E. arabicus* leaves extract followed by 1g /kg paracetamol bw, p.o

GROUP V: received 200 mg/kg *E. arabicus* leaves extract followed by 1g /kg paracetamol bw, p.o

After 24 h of the last treatment, blood was collected from retro-orbital plexus, allowed to clot for 1 h at room temperature and serum was separated by centrifugation at 2500 rpm for 15 min. The serum was then collected and analyzed for various biochemical parameters.

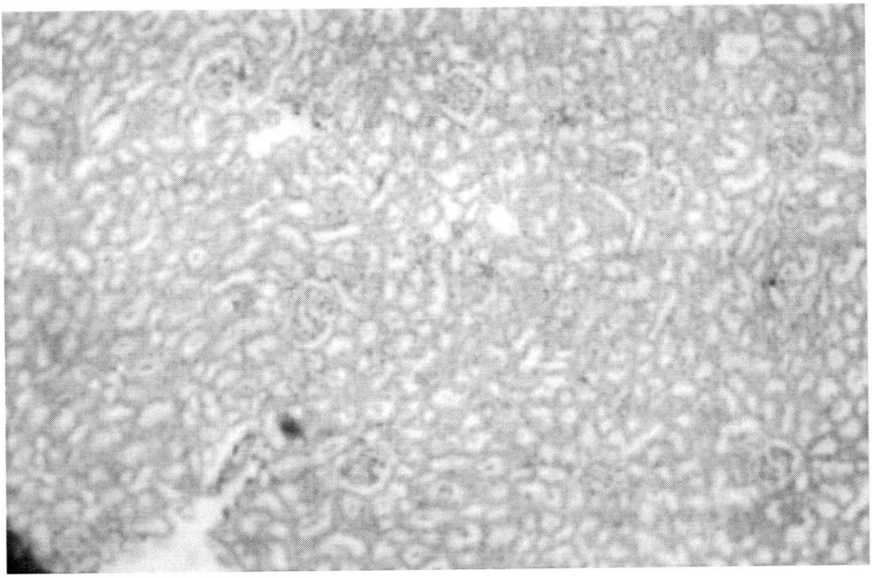

Figure 7: Group II rat renal tissue served as normal positive control and they received 1%.carboxy methyl cellulose 1 mL/kg bw orally (H and E x 200).

SERUM BIOCHEMICAL ASSAYS

The following tests were done to assess the hepatoxicity of paracetamol AST, ALT, TB, and SA. Kidney functions were looked for by measuring BU and SC. Tests were used according to the standard procedures using commercially available diagnostic kits.

OXIDATIVE STRESS

Twenty four hour after the last treatment; rat were decapitated. The liver and kidney tissues were dissected from the surrounding fat and connective tissue. Left lobe of liver and left kidney specimens were longitudinally sectioned, kept at -20°C and subsequently homogenized in cold potassium phosphate buffer (0.05 M, pH 7.4). The ratio of tissue weight to homogenization buffer was 1:10. The tissue specimens' homogenates were centrifuged at 5000 r.p.m for 10 minutes at 4°C. The resulting supernatant was used for determination of TBARS level according to the method of Ohkawa et al. [8].

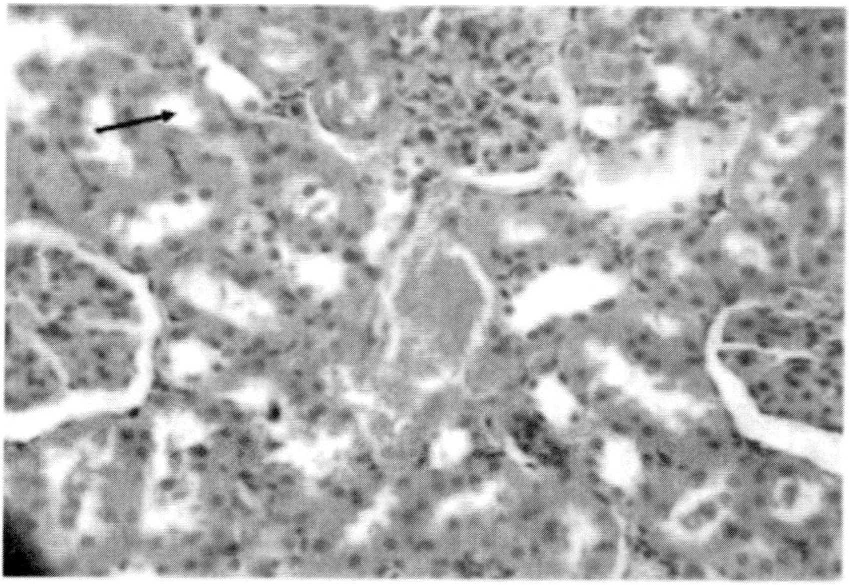

HISTOLOGICAL STUDIES

The right lobe of the liver and right kidney was fixed in 10% buffered formalin and was processed for paraffin sectioning. Sections of about 5 ml thickness were stained with haematoxylin and eosin to be examined and photographed [9].

STATISTICAL ANALYSIS

All data were represented as mean ± standard deviation (M ± SD). For comparison between groups one-way analysis of variance was used. P value was considered significant if it was less than 0.05.

RESULTS

As shown in table 1 there is significant difference in hepatic and renal biochemistry parameters between the two control groups and paracetamol group indicating toxicity induced by paracetamol. There was significant elevation of AST, ALT, ALP, TB, BU, and SC in group that received paracetamol. Also, there was significant decrease in SA in paracetamol group compared with control ones.

In pretreatment groups with *E. Arabicus* there is significant lowering of the hepatic and renal injury indicators as there is a decrease of the increased level of AST, ALT, ALP, bilirubin, and increase albumin; also, decrease blood urea and serum creatinine. There is also, much improvement in hepatic and renal function in group pretreated with 200 mg/kg *E. Arabicus* than in that treated with 100 mg/kg.

Table 2 demonstrates thiobarbituric acid reactive substances levels in liver and kidney tissues, there are statistically significant difference between TBARS level between different groups. It much increase in paracetamol group indicating oxidative damage produced by paracetamol but it decrease in rats pretreated with *E. Arabicus*, although it did not reach normal level. There are also little changes between 100 mg/kg pretreated group and 200 mg/kg one.

Histopathological examination also provided supportive evidence for the results obtained from the enzyme analysis. Microscopically, liver slices from control animals stained with haematoxylin and eosin showed normal parenchymal architecture with cords of hepatocytes, portal tracts and terminal veins without noticeable alterations (Figures 1 and 2).

Liver section of paracetamol treated rats showing extensive areas of hepatocellular necrosis and inflammatory cell infiltration. Most of the centrilobular hepatocytes were swollen with marked cytoplasmic vacuolation and pyknotic nuclei with obliterated intervening hepatic sinusoids (Figure 3).

Liver section of rats treated with paracetamol and 100 mg/kg bw of *E. Arabicus* showed less inflammatory cells around central vein and absence of necrosis (Figure 4). Liver section of rats treated paracetamol and 200 mg/kg bw of *E. Arabicus* showed minimal inflammatory cellular infiltration, regeneration of hepatocytes around central veinwas also observed and almost near normal liver architecture (Figure 5) , indicating the hepatoprotective activity of *E. Arabicus.*

Regarding to the renal histology, group I and II show normal kidney architecture (Figures 6 and 7). In group III; rats received toxic dose of paracetamol, there are many tubules with cloudy swelling manifested as enlarged tubules with conical shaped cells, abundant eosinophilic cytoplasm, satellite shaped lumen and luminal cast (Figure 8). In group IV and V; there was much decrease in the inflammation and cast than paracetamol group with more improvement in group V comparing with group IV (Figures 9 and 10).

DISCUSSION

Paracetamol induced hepatic injury is considered as one of the most commonly used model for hepatoprotectivity drug screening through estimation of serum cytoplasmic enzymes; aspartate aminotransferase (AST), alanine aminotransferase (ALT), alkaline phosphatase (ALP), serum albumin (SA), total bilirubin (TB), blood urea (BU) and serum creatinine (SC) activities is a useful as quantitative markers of the extent, type of hepatocellular and renal affection by paracetamol [10,11]. In the current study, the significant elevation of the enzymes levels, particularly, serum aspartate aminotransferase, alanine aminotransferase, alkaline phosphatase, total bilirubin blood urea, and serum creatinine in the rats which treated with paracetamol, indicate the deterioration of the hepatic functions due to the toxic effects of the drug and consequently, they have been attributed to the damage of structural integrity of the liver, because these enzymes released into the circulation after autolytic breakdown or cellular necrosis [12].

The reduction in the level of hepatic markers by the *E. arabicus* is an indicator of stabilization of the plasma membrane as well as repair of hepatic tissue damage caused by paracetamol. This effect is in agreement with the commonly accepted view that serum levels of transaminases return to normal with the healing of hepatic parenchyma and the regeneration of hepatocytes [13].

There is statistically significant TBARS level difference as it increases in paracetamol group indicating oxidative damage produced by paracetamol but it decreases in rats pretreated with *E. Arabicus* although it did not reach normal level. There are also little changes between 100 mg/kg pretreated group and 200 mg/kg one. All these results are in agreement with published data [14-16], that indicated the elevation in TBARS level in paracetamol group is an indicator of lipid peroxidation which is due to paracetamol induced tissue damage.

Histological findings of tissue damage induced by paracetamol in the liver confirmed previous laboratory results that showing extensive areas of hepatocellular necrosis and inflammatory cell infiltration and this coincides with Roomi et al. [17] who found mild and significant lobular focal hepatitis in paracetamol studied group.

Rregarding renal histological finding there are many tubules with cloudy swelling manifested as enlarged tubules with conical shaped cells, abundant eosinophilic cytoplasm, satellite shaped lumen and luminal cast. These results are in agreement with Madhukiran et al. [18] who reported that kidneys of paracetamol treated rats, the inflammatory cell infiltration in the interstitium and significant tubular damage were detected. In contrast,

Ahmed et al. [19] who observed that the ingestion of paracetamol (500 mg/kg/day) did not produce papillary necrosis nor interstitial nephritis.

Flavonoids isolated from *E. arabicus* had been proved to increase the expression of the rate limiting enzyme in the synthesis of c-glutamylcysteine synthetase with a concomitant increase in the intracellular glutathione concentrations [20-21].

The reactive species mediated hepatotoxicity can be effectively managed upon administration of agents possessing anti-oxidant [22], free radical scavenger [23] and anti-lipid per oxidant activities [24]

A potential of hepatoprotective property underlying *E. arabicus* may be attributed to the polyphenolic principles, particularly, flavonoids. This finding could be explained by the comparing results with those plants that mainly used for treatment of liver disorders. For instance, *Curcuma longa* (turmeric), *Glycyrrhiza glabra* (licorice), and *Camellia sinensis* (Green tea), are reported to be hepatoprotective due to the powerful anti-oxidative properties [25-28]. In addition, the antioxidant properties of *Trichosanthes cucumerina* are attributed to flavonoids, carotenoids, lycopene, phenolics, and β-carotene [29].

CONCLUSION

E. arabicus has the ability to protect the liver and kidneys from the damaging effects of paracetamol in acute toxic doses and stimulation of endogenous anti-oxidant defense system.

REFERENCES

1. Underberg AA, Baldwin BG, Bayer RG, Compositae (2007) The Families and Genera of Vascular Plants, In: Kadereit JW, Jeffrey C (eds.), Flowering Plants, Dicots, Asterales, VIII, Springer, Berlin, Heidelberg, New York 61.
2. Ozçelik B, Orhan I, Toker G (2006) Antiviral and antimicrobial assessment of some selected flavonoids. Z Naturforsch C 61: 632-638.
3. Mothana RA, Alsaid MS, Al-Musayeib NM (2011) Phytochemical analysis and in vitro antimicrobial and free-radical-scavenging activities of the essential oils from Euryopsarabicus and Laggeradecurrens. Molecules 16: 5149-5158.
4. Rumack BH (2004) Acetaminophen misconceptions. Hepatology 40: 10-15.
5. Lee WM (2004) Acetaminophen and the U.S. Acute Liver Failure Study Group: lowering the risks of hepatic failure. Hepatology 40: 6-9.
6. Nirmala M, Girija K, Lakshman K, Divya T (2012) Hepatoprotective activity of Musa paradisiaca on experimental animal models. Asian Pac J Trop Biomed 2: 11-15.

7. Walied M Alarif, Ahmed Abdel-Lateff, Ahmed M Al-Abd (2013) Selective cytotoxic effects on human breast carcinoma of new methoxylated flavonoids from Euryopsarabicus grown in Saudi Arabia. European Journal of Medicinal Chemistry 66: 204-210.
8. Ohkawa H, Ohishi N, Yagi K (1979) Assay for lipid peroxides in animal tissues by thiobarbituric acid reaction. Anal Biochem 95: 351:8.
9. Romero-Sarmiento Y, Soto-Rodrguez I, Arzaba-Villalba A, Garca HS, Alexander-Aguilera A (2012) Effects of conjugated linoleic acid on oxidative stress in rats with sucrose-induced non-alcoholic fatty liver disease. J. Funct. Food 4: 450–458.
10. Dash Deepak K, YeligarVeerendra C, Nayak Siva S, TirthaGhosh, Rajalingam D (2007) Evaluation of hepatoprotective and antioxidant activity of Ichnocarpusfrutescens (Linn.) R.Br. on paracetamol - induced hepatotoxicity in rats. Trop J Pharm Res 6: 755–765.
11. EhsanKheradpezhouh, Mohammad-Reza Panjehshahin, RaminMiri , KatayounJavidnia et al. (2010) Curcumin protects rats against acetaminophen-induced hepatorenal damages and shows synergistic activity with N-acetyl cysteine . European Journal of Pharmacology 628: 274–281.
12. Zhang R, Hu Y, Yuan J, Wu D (2009) Effects of Puerariae radix extract on the increasing intestinal permeability in rat with alcohol-induced liver injury. J Ethnopharmacol 126: 207-214.
13. Thabrew MI, Joice PD, Rajatissa W (1987) A comparative study of the efficacy of Pavettaindica and Osbeckiaoctandra in the treatment of liver dysfunction. Planta Med 53: 239-241.
14. Knight TR, Kurtz A, Bajt ML, Hinson JA, Jaeschke H (2001) Vascular and hepatocellular peroxynitrite formation during paracetamol-induced liver injury: Role of mitochondrial oxidant stress. Toxicol. Sci 62: 212–220.
15. SenerG, Sehirli AO, AyanDG (2003) Protective effects of melatonin, vitamin E and N-acetylcysteine against acetaminophen toxicity in mice: a comparative study. J Pineal Res 35: 61-68.
16. Khan RA, Khan MR, Sahreen S, Shah NA (2000) Hepatoprotective activity of Sonchusasper against carbon tetrachloride-induced injuries in male rats: a randomized controlled trial. BMC Complem. Altern. Med 201: 1-90.
17. Roomi MW, Kalinovsky T, Ivanov V, Rath M, Niedzwiecki A (2008) A nutrient mixture prevents acetaminophen hepatic and renal toxicity in ICR mice. Hum ExpToxicol 27: 223-230.
18. Madhukiran P, Ganga Rao B (2011) Invitro evaluation for free radical scavenging activity of methanolic leaf extract of Cyathea gigantean (wall Ex.hook). Int J Pharm Res Dev 3: 2.
19. Ahmed MH, Ashton N, Balment RJ (2003) Renal function in a rat model of analgesic nephropathy: effect of chloroquine. J PharmacolExpTher 305: 123-130.
20. Myhrstad MC, Carlsen H, Nordstrom O, Blomhoff R, Moskaug JJ (2002) Flavonoids increase the intracellular glutathione level by transactivation of the gamma-glutamylcysteinesynthetasecatalytical subunit promoter. Free Radic. Biol. Med 32: 386–393.
21. Moskaug JØ, Carlsen H, Myhrstad MC, Blomhoff R (2005) Polyphenols and glutathione synthesis regulation. Am J ClinNutr 81: 277S-283S.
22. Attri S, Rana SV, Vaiphei K, Sodhi CP, Katyal R, et al. (2000) Isoniazid- and rifampicin-induced oxidative hepatic injury--protection by N-acetylcysteine. Hum ExpToxicol 19: 517-522.
23. Sadanobu S, Watanabe M, Nakamura C, Tezuka M (1999) In vitro tests of 3-dithia-2-thioxo-cyclopent-4-ene to evaluate the mechanisms of its hepatoprotective action. J ToxicolSci 24: 375-381.
24. Lim HK, Kim HS, Choi HS, Oh S, Jang CG, et al. (2000) Effects of acetylbergenin against D -galactosamine-induced hepatotoxicity in rats. Pharmacol Res 42: 471-474.

25. Donatus IA, Sardjoko, Vermeulen NP (1990) Cytotoxic and cytoprotective activities of curcumin. Effects on paracetamol induced cytotoxicity, lipid peroxidation and glutathione depletion in rat hepatocytes.BiochemPharmacol 39: 1869–1875.
26. Soni KB, Lahiri M, Chackradeo P, Bhide SV, Kuttan R (1997) Protective effect of food additives on aflatoxin-induced mutagenicity and hepatocarcinogenicity. Cancer Lett 115: 129-133.
27. Wang GS, Han ZW (1993) The protective action of glycyrrhiza flavonoids against carbon tetrachloride hepatotoxicity in mice. Yao XueXueBao 28: 572-576.
28. Miyagawa C, Wu C, Kennedy DO, Nakatani T, Ohtani K, et al. (1997) Protective effect of green tea extract and tea polyphenols against the cytotoxicity of ,4-naphthoquinone in isolated rat hepatocytes. BiosciBiotechnolBiochem 61: 1901-1905.
29. Lavelli V, Peri C, Rizzolo A (2000) Antioxidant activity of tomato products as studied by model reactions using xanthine oxidase, myeloperoxidase, and copper induced lipid peroxidation. J Agric Food Chem 48: 1442–1448.

CHAPTER 16

Adverse Effects of Larkspur (*Delphinium* spp.) on Cattle

Kevin D. Welch [1,*], Daniel Cook [1], Benedict T. Green [1], Dale R. Gardner [1], James A. Pfister [1], Tara G. McDaneld [2] and Kip E. Panter [1]

[1] *USDA-ARS Poisonous Plant Research Laboratory, Logan, UT 84341, USA*

[2] *USDA-ARS Meat Animal Research Center, Clay Center, NE 69833; USA*

ABSTRACT

There are numerous species of larkspur (*Delphinium* spp.) in North America. Larkspurs are a major cause of cattle losses on western ranges in the USA, especially on foothill and mountain rangelands. The toxicity of larkspur species is due to various norditerpenoid alkaloids. In this article, we review the current knowledge regarding larkspur ecology and distribution, analytical technologies to study and quantify the toxins in larkspur, the toxicology of the larkspur plants and their individual toxins, known genetic variations in larkspur susceptibility, and current management recommendations to mitigate losses from larkspur poisoning.

Keywords: larkspur; delphinium; methyllycaconitine; cattle

1. INTRODUCTION

There are over 60 wild species of larkspurs (*Delphinium* spp.) in North America [1] and a number of domestic horticultural varieties. Wild larkspurs are divided into three general categories based primarily on mature plant height and distribution: low, tall and plains larkspurs [2]. The larkspurs are a major cause of cattle losses on western ranges [3,4], especially on foothill and mountain rangelands [5]. Larkspurs have been shown to be toxic to horses, although horses will not voluntarily consume sufficient quantities of larkspurs to become poisoned [6,7]. Sheep have been shown to be quite resistant to larkspurs [6,7,8]. Consequently, cattle are the primary livestock species associated with larkspur toxicosis and thus the effect of larkspurs on cattle is the focus of this review. The toxicity of larkspurs is due to norditerpenoid alkaloids including two predominant types, the *N*-(methylsuccinimido) anthranoyllycoctonine (MSAL)-type and the non MSAL-type including the 7,8-methylenedioxylycoctonine (MDL)-type (Figure 1) [7,9]. In this article, we review the current knowledge regarding larkspurs, their toxicity, and current management strategies to minimize losses from livestock poisoning.

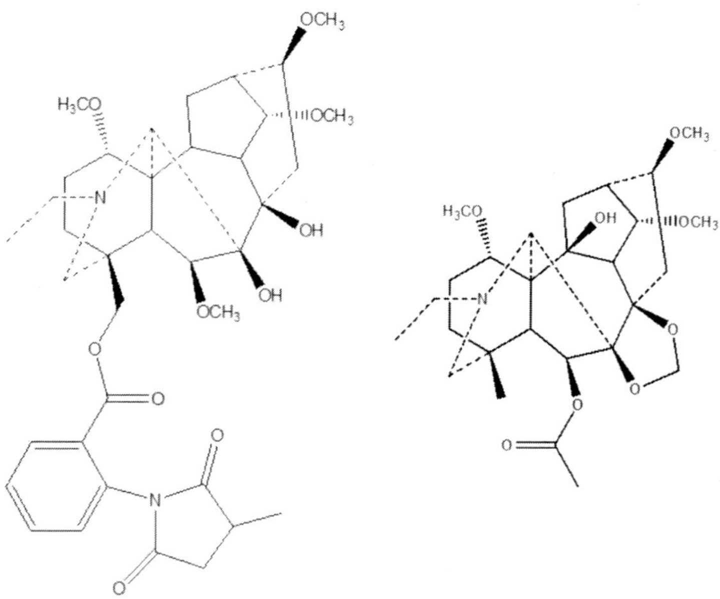

(a) (b)

Figure 1. (**a**) Methyllycaconitine (MLA) (MSAL-type alkaloid); (**b**) deltaline (non MSAL-type alkaloid). Structure of a MSAL-type alkaloid methyllycaconitine (MLA) and a non MSAL-type alkaloid deltaline.

2. PLANT ECOLOGY AND DISTRIBUTION

Taxonomic treatments of the genus *Delphinium* recognize greater than 60 species in North America [10,11,12]. Larkspurs have been divided into three categories based upon their height at maturity and their distribution: tall, low, and plains larkspur. The larkspur species in each category responsible for most of the reported livestock losses in the western United States are listed in Table 1. *Delphinium barbeyi, D. glaucescens, D. glaucum* and *D. occidentale* represent the dominant tall larkspur species [7]. Tall larkspurs are found typically at high elevation, snow-covered sites in varied plant communities (Figure 2). *Delphinium andersonii, D. bicolor, D. nuttallianum, D.parishii* and *D. scaposum* represent the dominant low larkspurs [7]. Low larkspurs are found in a variety of plant communities, for example *D. nuttallianum* may be found in mountain meadows while *D. andersonii* may be found in a desert shrub plant community. Lastly, the plains larkspur is represented by a single species, *D. geyeri*, and is intermediate in height compared to the low and tall larkspurs [7].

Delphinium species contain numerous norditerpenoid alkaloids which represent two structural types the *N*-(methylsuccinimido) anthranoyllycoctonine (MSAL)-type and the non MSAL-type, which includes the 7,8-methylenedioxylycoctonine (MDL)-type alkaloids. Purified alkaloids of these two structural types differ significantly in toxicity based upon an LD_{50} in a mouse model [9,13,14,15]. In general, the non MSAL-type alkaloids are 20–30 times less toxic compared to the MSAL-type in the mouse model. The structural type of alkaloids and the individual alkaloids differ among species and populations of the same species. For example, *D. glaucescens* is composed predominantly of methyllycaconitine, an MSAL-type alkaloid with a small amount of MDL-type alkaloids, while many populations of *D. barbeyi* are often found to contain a three-to-one ratio of non MSAL to MSAL-type alkaloids [16,17]. Additionally, another tall larkspur species, *D. stachydeum* only contains non MSAL-type alkaloids [18]. Likewise in *D. occidentale* there are two chemotypes, one that contains both MSAL and non MSAL-type alkaloids and one that does not contain MSAL-type alkaloids [19]. Each chemotype has a distinct geographic distribution shown in Figure 3. The low larkspurs in general contain a more diverse suite of MSAL-type alkaloids in addition to the non MSAL-type alkaloids [20]. These differences in the qualitative and quantitative composition of the MSAL and non MSAL-type alkaloids are implicated in differences in toxicity in cattle [16,18,21,22].

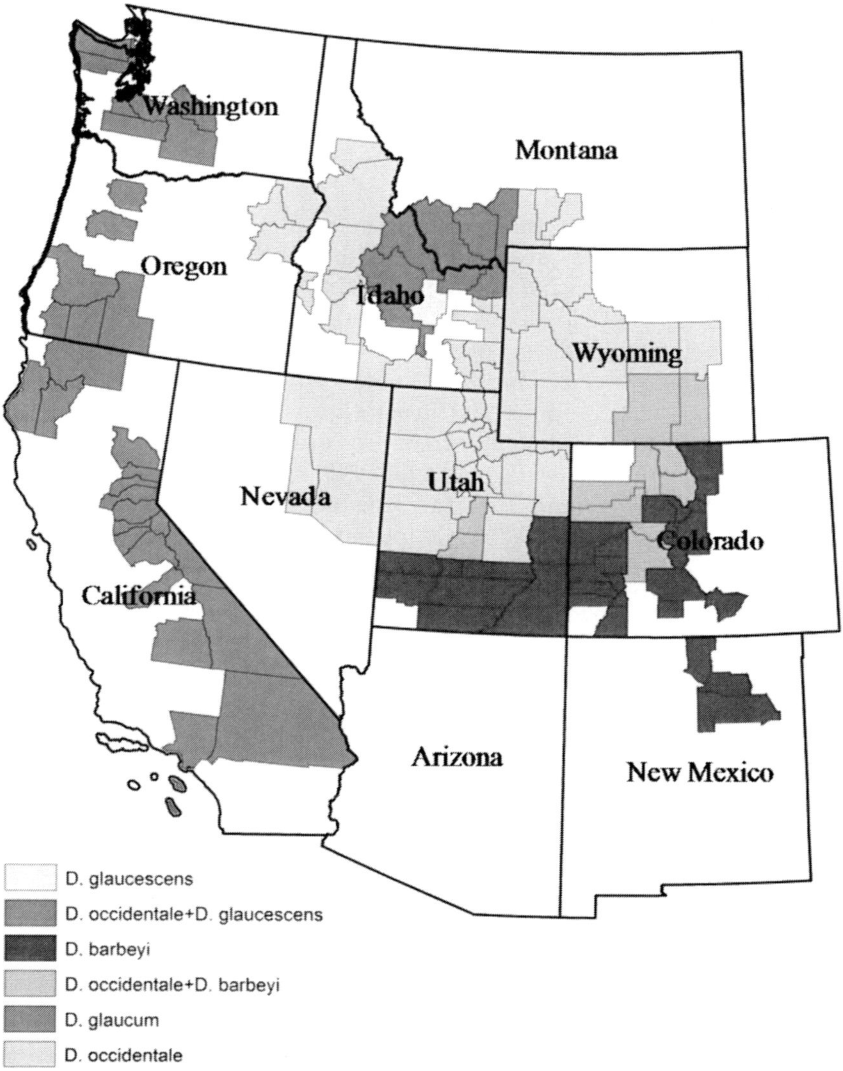

Figure 2. Distribution map of the major tall larkspur species in the western United States.

Table 1. Larkspur classes and major species causing livestock losses associated with each class, with associated plant communities and possible dominant alkaloids.

Larkspur Class	Species	Major Alkaloids	Plant Communities
Tall Larkspurs	*D. barbeyi*	Deltaline	Alpine, Aspen, Conifer, Mountain Brush, Mountain Meadows, Sagebrush
	D. glaucescens	14-*O*-Acetyldictyocarpine	
	D. glaucum		
	D. occidentale	Methyllycaconitine	
	D. ramosum		
Low Larkspurs	*D. andersonii*	Methyllycaconitine	Aspen, Conifer, Desert Shrub, Mountain Brush, Mountain Meadows, Pinon-Juniper, Sagebrush
	D. bicolor	Geyerline	
	D. nuttallianum	Nudicauline	
	D. parishii	14-Deacetylnudicauline	
	D. scaposum		
Plains Larkspur	*D. geyeri*	Methyllycaconitine	Desert Shrub, Mountain Brush, Sagebrush, Shortgrass Prairie

The accumulation of the norditerpenoid alkaloids varies seasonally and between plant parts among the different species of larkspur [23,24,25,26]. In individual parts, alkaloid concentrations are greatest in immature tissues and decrease as those tissues mature in tall and low larkspurs [27]. On a whole-plant basis, in tall larkspurs, alkaloids are the greatest early in the season and decrease as the plants matures while in low larkspurs alkaloid amounts are generally similar over the season [24,25,26,27]. As plants senesce, alkaloid concentrations continue to decrease.

The density of a given larkspur species on the range will influence consumption and thus its relative toxicity. Tall larkspur species are generally long-lived species where a similar number of plants are present on any given year [7]. In tall larkspur species weather factors such as temperature and precipitation will influence the abundance of flowers and pods. Anecdotal information suggests that on years where there are more flowers and pods, there is more relative consumption with subsequently increased cattle losses. In contrast, low larkspur species are also long-lived, but the number of plants can vary considerably from year to year based upon precipitation and temperature. Data suggests that increased precipitation in the preceding winter and spring months and cooler temperatures will increase the density of low larkspur species [28,29]. As low larkspur species are small in stature, sufficient plant density must be present in order to have sufficient material available for consumption to impart toxicity.

Figure 3. Distribution map of two chemotypes of *D. occidentale.* Chemotype A plants contain the MSAL-type alkaloids whereas chemotype B plants do not contain the MSAL-type alkaloids (modified from [19]).

3. ANALYTICAL TECHNIQUES

There have been a number of chemical methods used for the analysis of the toxic norditerpenoid alkaloids found in larkspurs. These methods have included classical wet chemical methods and various chromatographic and spectroscopic methods or a combination of chromatographic and spectroscopic methods. Earlier methods typically measured what was termed Total Alkaloid Concentration (TAC). This was measured initially by chemical titration methods [30,31], then gravimetrically [32,33,34], and then through calibrated Near Infrared Spectroscopy (NIRS) [35] and Infrared (IR) spectroscopy [36]. Methods for TAC are attractive in that they are typically less costly and can provide a relative rapid preliminary indicator of toxicity, especially if the individual alkaloid components of the larkspur being tested are previously known. Titration and gravimetric measurements require simple laboratory equipment but can be lengthy procedures. NIRS and IR methods can be rapid, but require initial investment in instrumentation and up front detailed calibrations. However, TAC methods do not distinguish between individual types of alkaloids (MSAL *vs.* non MSAL-type alkaloids). In order to partially address the need to selectively analyze for the MSAL-type alkaloids, a Fourier transform infrared (FT-IR) spectroscopy method was developed to measure the concentration of all MSAL-type alkaloids as well as the TAC in larkspurs [36]. This method has been used to analyze large number of samples obtained from rangeland investigations [37] and is still the method of choice for rapid evaluation of larkspur toxicity. The largest drawback of the FT-IR method is the extensive initial calibration of the instrument with a complete set of known samples for which the total MSAL and TAC has been measured using other standard methods.

For the analysis of specific individual larkspur alkaloids, a number of more detailed and sensitive methods have been developed based on modern chromatographic and spectroscopic methods. A relative simple method for the analysis of non MSAL-type alkaloids is based on the use of capillary gas chromatography [38]. Capillary gas chromatography can be an excellent method for the analysis of a number of individual alkaloids of the lycoctanine, or the MDL-type, alkaloids that are thermally stable and still volatile enough to be chromatographed in the gas phase. However, the non MSAL-type alkaloids tend to be the less toxic compounds and thus analysis of larkspur using solely this technique may not produce a complete evaluation of the potential toxicity of the larkspurs being analyzed.

The MSAL-type alkaloid methyllycaconitine (MLA) was initially indentified as one of the major and most toxic alkaloids found in larkspurs. MLA is not particularly amenable for analysis by capillary gas chromatography, but a number of different methods using liquid chromatography (HPLC) with UV detection have been developed that work well for these larger, thermally labile compounds. HPLC methods developed have used both normal phase [13] and reverse phase [39,40] type separations. When combined with atmospheric ionization and mass spectrometry (*i.e.*, HPLC-esi(+)MS) the methods can become extremely sensitive and selective [20,41,42,43,44]. Even in the absence of chromatography, a simple flow injection mass spectrometry method can be a valuable tool in assessing the alkaloid composition of larkspurs [42]. Such flow injection methods have been valuable in chemo-typing the various larkspurs based on alkaloid profiles, geographical location and species identifications [17,18,19]. With development of fully hyphenated methods, such as LC-MS/MS, the sensitivity and selectivity have become sufficient for detection of very low concentration of alkaloids (*i.e.*, ng/mL) in such samples as sera or tissue (liver) for the purposes of defining toxicokinetics [22,45,46] or for diagnostic purposes in the analysis of contaminated feed supplies, tissues, rumen content or fluid samples of poisoned animals (Figure 4). The use of MS/MS techniques is also beneficial in the structure elucidation of new norditerpenoid alkaloids [23,42,44].

4. TOXICOLOGY

The primary result of larkspur toxicosis is neuromuscular paralysis from inhibition of normal function of the post-synaptic neuromuscular junction [47,48]. The toxicity of larkspurs is primarily due to the norditerpenoid alkaloids [7,22]. Larkspur alkaloids are competitive inhibitors of acetylcholine, particularly at nicotinic acetylcholine receptors (nAChRs) [49]. The binding of larkspur alkaloids to nAChRs in various tissues is correlated to toxicity [50], and may explain the tolerance of sheep to larkspur, in that larkspur toxins appear to bind less avidly to sheep nAChRs [51,52].

Clinical signs of larkspur intoxication include muscular weakness and trembling, straddled stance, periodic collapse into sternal recumbency, respiratory difficulty, and finally death while in lateral recumbency [53]. A non-lethal dose of larkspur causes labored breathing, rapid and irregular heartbeat, and muscle weakness to the point of collapse [54]. A lethal dose tall larkspur in Hereford steers (one of the more susceptible breeds of steers to tall larkspur toxicity) has been shown to be approximately 470 g of plant material (dry weight of a population of *D. barbeyi* containing 4 mg/g MSAL-type alkaloids and 16 mg/g total alkaloids) for a 500 kg animal, or 3.1 kg wet weight [22]. However, it should be kept in mind that due to the fact that the larkspur alkaloids are neuromuscular toxins, anything that causes additional stress and/or exertion can reduce the toxic/lethal dose.

The majority of dose-dependent research studies that have been performed followed a dosing regimen that consisted of a single bolus dose of finely ground plant material. These studies were extremely informative, but they lacked the multiple exposure scenario that likely occurs to cattle in a grazing environment [55]. Therefore a recent study was conducted to evaluate the acute toxicity of tall larkspur (*D. barbeyi*) administered in multiple doses, with the intent of identifying a No Observable Adverse Effect Level (NOAEL). The adverse effect was muscle weakness to the point the cattle could no longer walk, as would be required in a grazing environment. The research demonstrated that the overall daily intake is critical, and should be maintained under 2 mg MSAL-type alkaloids/kg BW/day [56]. Consequently, based upon a population of tall larkspur that contains 4 mg MSAL-type alkaloids/g plant material (16 mg total alkaloid/g plant material), a dose of 2 mg/kg/day would correspond to a daily intake of 250 g of tall larkspur on a dry weight basis for a 500 kg animal, which, assuming that tall larkspur is approximately 85% water, would correspond to 1.67 kg of fresh plant material each day. Therefore, based upon the results of that study, a 500 kg cow should be able to consume up to 1.67 kg of fresh tall larkspur per day without becoming severely poisoned.

Due to the fact that the MSAL-type alkaloids are much more toxic than the MDL-type alkaloids [13,14], management recommendations for grazing cattle on larkspur-containing ranges are based primarily on the concentration of MSAL-type alkaloids in larkspur [5,37]. However, in many species of tall larkspur the non MSAL-type alkaloids can be more abundant [7,17]. Research work using a mouse model suggested that MDL-type alkaloids enhance the overall acute toxicity of MLA in an additive manner [15]. In subsequent studies with cattle, the MSAL-type alkaloids such as MLA were the primary factors responsible for the toxicity of larkspurs [16,21,22]. However, populations of larkspurs that contained large amounts of MDL-type alkaloids, in addition to high MSAL-type alkaloid content, were found to pose a greater risk to cattle than plants with only high MSAL-type alkaloids [16]. Consequently, in general for larkspurs to be toxic to livestock, a sufficient quantity of MSAL-type alkaloids is required. However, MDL-type alkaloids appear to potentiate the overall toxicity of the MSAL-type alkaloids and should be considered when predicting potential toxicity of larkspur populations [22]. However, recent investigations into the toxicity of *D. stachydeum* populations demonstrated that some populations could potentially have adverse effects in cattle even though they do not contain MSAL-type alkaloids [18]. Based upon the results of the last few years of research, it has become evident that the toxicity of tall larkspur populations is greatly dependent upon the alkaloid profile of each species and even specific populations within a species [16,18,19,22]. Consequently, it is impossible to provide a general statement as to how much larkspur a cow can eat without dying, as a lethal dose of larkspur is completely dependent upon the alkaloid profile of the plant and the genetic susceptibility of the animal.

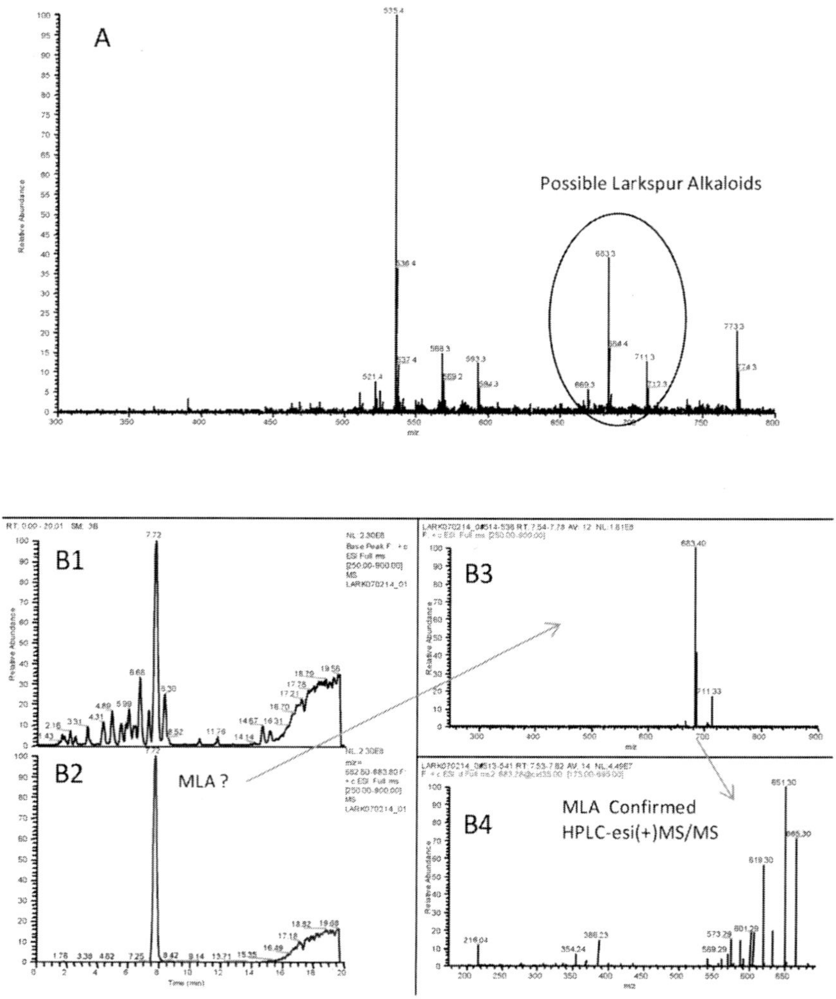

Figure 4. (**A**) MS spectrum of crude rumen extract obtained from a dead cow involved in possible rangeland plant poisoning case and analyzed by flow-injection esi(+)MS. Possible larkspur alkaloids in the rumen were indicated by ions at *m/z* 669, 683 and 711 that are consistent with those found in the low larkspurs (**B1**); HPLC-esi(+)MS/MS ion chromatogram obtained after isolation of the crude alkaloid fraction by acid/base extraction (**B2**); extracted ion chromatogram for *m/z* 683 and possible detection of the alkaloid methyllycaconitine (MLA) (**B3**); and confirmed by its MS/MS spectrum (**B4**).

5. GENETIC SUSCEPTIBILITY

Recently, there has been a great deal of research effort put into genomic-based selection methods for the improvement of economically important traits such as milk yield and meat tenderness in cattle [57]. However, the genetic basis of livestock responses to plant toxins, such as larkspur alkaloids, has received little attention. The identification of larkspur-resistant animals would be beneficial for animal welfare, thus potentially improving the economics of livestock production of cattle on larkspur-containing ranges. Initial research on the selection of animals for resistance to larkspur toxins at this laboratory used inbred mouse strains as a model to determine if there was a potential to select for differences in larkspur toxicity. Results from this research showed that there was a difference in susceptibility of nine mouse strains to MLA toxicosis with an approximate two-fold difference between the A/J and 129 strains (3.3 mg/kg *vs.* 5.8 mg/kg respectively) [58]. The difference in the LD_{50} values between the two strains was attributed to the amount of nAChR subunits expressed by the two strains. The 129 strain of mice had roughly twice the amount of alpha 7 nAChR subunit expression as the A/J strain, as well as higher expression of the alpha 3 and alpha 5 nAChR subunits. The presence of an alpha 7 nAChR reserve in the 129 strain could explain the resistance to MLA toxicity, as MLA is a competitive antagonist of the nAChRs. With more nAChRs, the 129 strain would be able to tolerate a higher dose of MLA before the percentage of blocked receptors would reach toxic concentrations compared to the A/J mice. However, more recent studies have demonstrated that there is no difference in the susceptibility of mice lacking the alpha 7 nAChR to MLA toxicosis compared to wild type mice [59]. The results of that study suggest that either the alpha 7 subunit does not play an integral role in the acute toxicosis of MLA or that alpha 7 nAChR knockout mice developed a compensatory response to the lack of the alpha 7 subunit. Consequently, other nAChR subunits found in the neuromuscular junction are likely the primary target for MLA, resulting in acute toxicosis. The success of these experiments provided the basis for ongoing experiments to identify genetic factors that correlate with susceptibility to larkspur toxicity in cattle.

Recently, we reported the phenotyping of cattle with a standardized dose of tall larkspur, and the associated muscle weakness (Figure 5) [60]. The measurement of phenotypic differences between cattle is the first step in the identification of genetic markers associated with resistance to larkspur poisoning. Based on the results of that experiment, cattle were classified as susceptible, intermediate or resistant in each breed of cattle. We have identified at least one individual resistant to larkspur poisoning in all breeds tested to date. The number of larkspur resistant animals may vary in each individual herd/breed due to the genetic background of that herd/breed. These differences are likely due to each individual animal's genetic predisposition for larkspur poisoning, which may be predictable from its DNA sequence and genomic analysis.

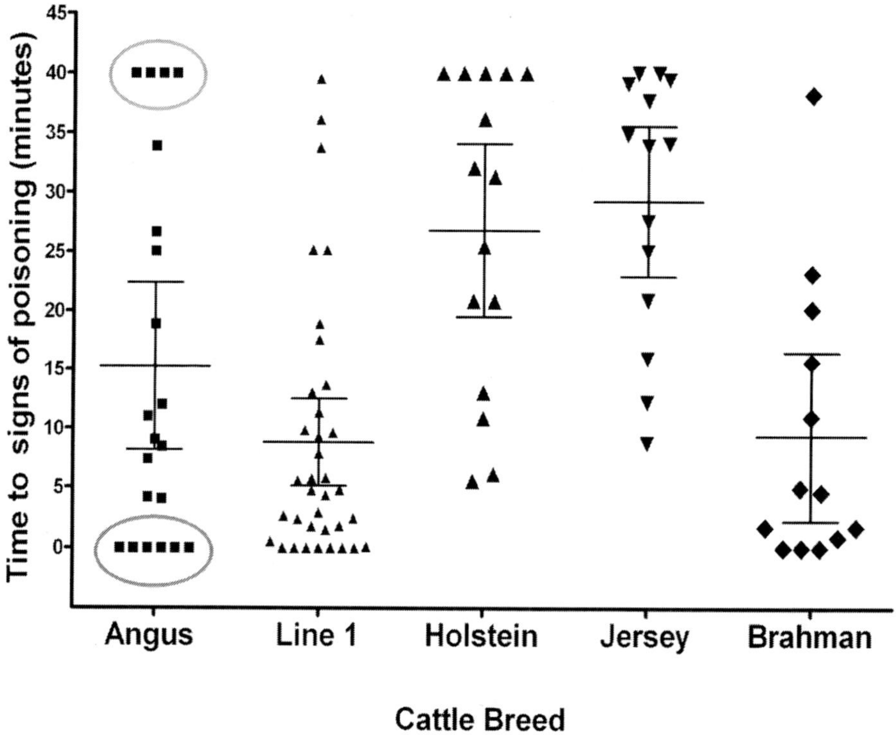

Figure 5. A comparison of different breeds of cattle to larkspur toxicosis. Data represent a scatter plot with the mean and 95% confidence interval for each breed. The data depict the cattle responses to a standardized dose of tall larkspur (8 mg/kg total toxic alkaloid) at 24 h after oral dosing. Line 1 represents Line 1 Herefords from Fort Keogh Livestock and Range Research Laboratory, Miles City, Montana. To determine the time to signs of poisoning, the cattle were exercised behind a tractor at 5–6 kph on a dirt track for 40 min or until they exhibited larkspur-induced muscle weakness. Examples of resistant (**green circle**) and susceptible (**red circle**) animals are depicted for Angus (From [60]).

Genomic analyses allow for the capture of information from larkspur-challenged animals that can be subsequently applied to select for larkspur-resistant cattle. The calves from these selected animals would also be expected to inherit tolerance to larkspur. In order to capture this genetic information, the first aim was to identify regions of the genome harboring genetic variation in larkspur resistance. To accomplish this, we have initially applied a genome wide association study approach based on a large DNA marker assay (BovineHD beadchip assay, over 770,000 markers, Illumina Inc., San Diego, CA, USA). Sixty-eight cattle (21 Angus, 16 Line 1 Hereford, 3 Brahman, 26 Holstein, and 2 Jersey) with divergent phenotypes for larkspur resistance were genotyped with the BovineHD beadchip assay. From these genotyping results, regions across the genome have been identified as significantly associated with larkspur resistance. To further define these regions of interest across the genome, an exome sequence (gene coding regions of the genome; ~50 million bases) was generated for 32 Angus cattle (21 with BovineHD genotypes). The combination of the genotyping and sequence data allows one to evaluate every base of DNA in the genomic regions of interest and identify possible candidate genes and DNA markers that can be used by the industry to select for larkspur tolerant cattle. These experiments will allow us to work towards development of commercially available DNA markers for larkspur tolerance in cattle.

6. FOOD SAFETY

Studies have been performed to characterize the kinetic profile of larkspur alkaloids in cattle [46,61,62,63]. In these studies, the elimination half-life of MLA from cattle dosed once with dried and finely ground larkspur via oral gavage was approximately 20 h. Additionally, the T_{max}, or time to maximal toxin concentration, was approximately 10 h. These data indicate that cattle that have consumed larkspur will be most susceptible to poisoning 10 h after consumption and that after the cattle have stopped consuming larkspur >99% of the toxins will be eliminated after 6 days. Consequently no food products should be obtained from cattle for at least 7 days after they have been exposed to larkspurs. It is important to note here that these data are for serum only, and not samples such as meat, milk, liver or other biological samples. More complete kinetic studies have been performed in mice wherein the elimination rate of MLA from numerous tissues was determined [15,45]. The results from those studies indicate that serum MLA concentrations are a good representation of body burden for MLA, and that once serum larkspur alkaloid concentrations are zero the alkaloid concentrations in the rest of the tissues are close to zero as well.

7. MANAGEMENT RECOMMENDATIONS

Two critical elements in risk management to reduce cattle losses from larkspur are the amount of alkaloid consumed (*i.e.*, alkaloid concentration and amount of plant eaten) and the rate of ingestion. Both of these factors are encompassed in two major conclusions from numerous field studies: (1) cattle typically eat most larkspur after the plants have begun to flower [55]; and (2) weather conditions (e.g., short-lived storms) may promote rapid consumption of relatively large amounts of larkspur [7,64]. Cattle generally begin consuming tall larkspur after flowering racemes are elongated, and consumption peaks during the pod stage of growth in late summer, when cattle may eat large quantities (25% to 30% of diet as herd average; >60% on some days by individual animals). Toxicity of tall larkspurs declines with maturation, but acceptability to cattle, and the potential rate of ingestion, typically increases as the grazing season advances [32,65]. These conclusions led to the development of the "toxic window" as a management scheme to reduce losses (Figure 6). Grazing during spring before larkspur flowers engenders relatively low risk because consumption is low. Further, later in summer once pods begin to shatter, larkspur-infested rangelands can usually be grazed with impunity because pod toxicity declines rapidly, and leaf toxicity is low.

There is one exception to the observation that cattle begin consuming tall larkspurs after flowering racemes elongate—*Delphinium glaucescens*, which grows in more xeric habitats compared to other tall larkspur species—and this may explain why cattle consume this species before flowering [66]. In addition, low larkspurs do not show a large decline in toxic alkaloid concentrations over time [24], and acceptability after flowering remains mostly static until the plants senesce [28,67].

Previous studies showed no relationship between concentration of alkaloids and larkspur palatability to cattle [68]. Thus, higher alkaloid concentrations alone appear to be unimportant in altering flavor and reducing short-term consumption by cattle. Although short-term larkspur consumption by cattle is not negatively influenced by high alkaloid concentrations, sub-lethal intoxication and multi-day consumption patterns are partially dependent on toxic alkaloid concentrations. Larkspur ingestion by cattle above a toxic threshold is regulated by post-ingestive consequences in a dose-dependent manner [55,69]. Cattle limit ingestion of larkspur (*i.e.*, dose of toxic alkaloids) so that periods of high consumption (1–2 days) are followed by periods of reduced consumption (3–4 days). This cyclic consumption pattern allows for detoxification after episodes of intoxication [55,69]. Cattle apparently increase consumption because larkspur is very nutritious (e.g., 15%–

20% crude protein; [32]), but then decrease consumption due to toxic feedback in a dose-dependent fashion. Cyclic consumption generally enables cattle to regulate larkspur consumption below a subclinical toxic threshold (~8 mg alkaloid/kg; [54,60]), and allows cattle to periodically graze this toxic but nutritious plant. On larkspur-infested rangelands during periods when cattle are eating larkspur, all grazing cattle will consume some quantity of larkspur, yet over 85%–95% of the typical herd will not be fatally poisoned. Our observations also indicate that many cattle deaths occur after brief (~90 min) periods of over-ingestion [32,64], and death likely occurs before negative post-ingestive feedback can act to temporarily reduce larkspur consumption. Further, consumption of smaller amounts of larkspur for several consecutive days may keep animals from exceeding a toxic threshold such that overt symptoms are not evident, but only brief exercise (e.g., running to consume water) is sufficient to push animals over a toxic threshold such that they become recumbent [70]. Thus, cattle may quickly be fatally intoxicated when intensely exercised even in the absence of overt clinical signs [28,67,70]. The observations regarding exercise have been consistent for cattle grazing either tall or low larkspurs.

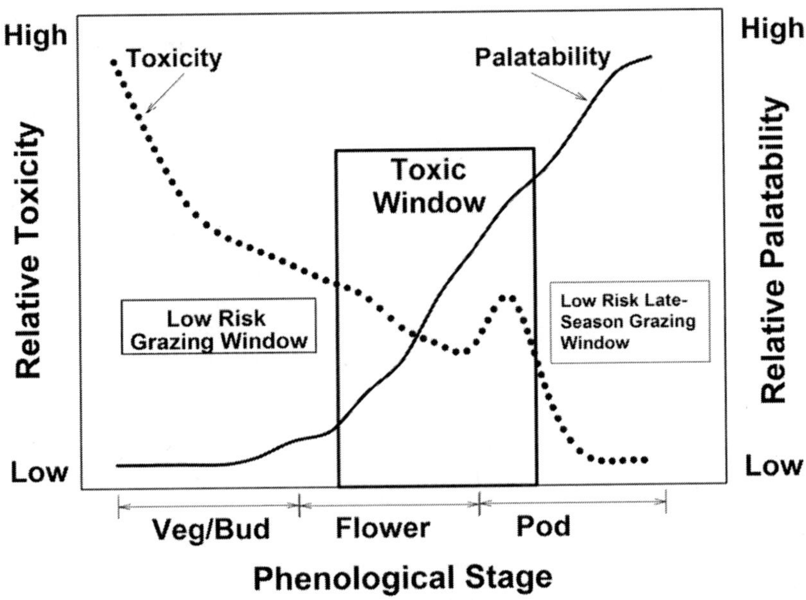

Figure 6. Relationship between toxicity and palatability in tall larkspurs. Most cattle deaths are predicted to occur during the toxic window when the concentration of norditerpenoid alkaloids is substantial and consumption by cattle increases after flowering racemes elongate.

Diagnosis of larkspur poisoning may be difficult particularly under extensive rangeland circumstances; grazing animals are often found dead and the only link to larkspur as the cause may be the presence of the plants in the pasture or analysis of tissues or rumen contents from dead animals (Figure 4). Recently, polymerase chain reaction (PCR) has been used to detect larkspur remnants within the partially-digested rumen matrix [71]; the PCR-based technique appears promising to diagnose larkspur poisoning.

8. GRAZING AND OTHER MANAGEMENT OPTIONS

Delphinium glaucum is the most toxic of the tall larkspurs in western North America, containing the highest levels of the toxic alkaloids MLA and 14-deacetylnudicauline (DAN; [17]. The mean concentration of MLA in *D. glaucum* from California was 19.5 mg/g, compared to 9 mg/g in *D. barbeyi*, and 3.5 mg/g in *D. occidentale* over the summer growing season. Even though the MSAL alkaloid concentration declined throughout the growing season, toxic alkaloid concentrations in *D. glaucum* remained relatively high (*i.e.*, above 3 mg/g) until after seed shatter [72], and ripe pods contained 4.9 mg/g [27]. Therefore, *D. glaucum* appears to remain highly toxic throughout most of the grazing season. However, no grazing studies have been conducted with *D. glaucum* to determine when it is preferred by cattle. Nonetheless, historical losses of cattle have been large where it grows in thick patches [6]. Furthermore, heavy winter snow in the Sierra Mountains may retard summer flowering in *D. glaucum* until late summer, thus extending the toxic window to late August and September [73]. After pods ripen and shatter, toxic alkaloid concentrations decrease rapidly and subsequent risk to cattle is low [23]. Our recommendation is to restrict cattle access to this species throughout the growing season until seed shatter, which typically occurs in late August or September.

Delphinium barbeyi makes up 2%–5% of total plant cover throughout much of its range, but may constitute up to 50% of the total plant cover in areas with large patches. *D. barbeyi* (along with *D. occidentale*) is also one of the most conspicuous species of the tall forb community in snow drift sites, at the edge of wet meadows, and along streams in various mountain habitats (Figure 2). *Delphinium barbeyi* is also a dominant component of the tall forb plant community growing in aspen groves throughout its distribution. Grazing studies with *D. barbeyi* indicate that cattle will not graze it until flowering begins [55]. Thus, we recommend grazing these mountain pastures during the early/late season to avoid the high risk of poisoning during the toxic window from bud elongation to the mature pod stage.

Delphinium occidentale occurs in the northern Rocky Mountains from northern Utah and Colorado into western Wyoming, Idaho, Montana, and throughout the basin and range mountains of Nevada (Figure 2). It grows mostly under aspen and in more mesic areas in sagebrush communities. *Delphinium occidentale* grows in areas where summer drought is common, thus it typically matures and senesces by mid to late summer. However, there are extensive populations of *D. occidentale* in the western U.S. that contain little or no MSAL alkaloids (Figure 3; [19]). The *D. occidentale* populations that contain little or no MSAL alkaloids appear to provide safe and nutritious forage at any stage of growth. However, MLA concentrations in populations of *D. occidentale* in central and southern Idaho generally remain above 4 mg/g until the pods mature and shatter [27,72], and thus pose a moderate risk until late in the grazing season.

Low larkspurs contain varying concentrations of toxic alkaloids, depending on location [24]. Concentrations of MLA in *D. nuttallianum* ranged from 0.8–4.5 mg/g plant, compared to a range of 3–6 mg/g plant for *D. andersonii* [24]. Management to reduce losses due to low larkspur begins with recognition of the plant during spring. Low larkspurs will typically begin growth before other forage. Low larkspur populations (*i.e.*, density) fluctuate with environmental conditions, with cool, wet late winter and spring climatic conditions favoring the growth of dense populations [28]. Risk of losing cattle is much higher during years with dense populations, as cattle typically consume low larkspurs in proportion to the availability of the plant [7]. During periods of high density, deferring grazing for 4–6 weeks until the low larkspur has senesced will virtually eliminate cattle losses.

Sheep can be herded into or bedded onto discreet tall larkspur patches to reduce availability or acceptability to cattle [74]. In other areas where tall larkspur is uniformly spaced over a pasture, sheep must eat immature larkspur and leave sufficient feed for subsequent grazing by cattle. This can be problematic, since early growth tall larkspur may not be palatable to sheep. Our observations [75] indicate that sheep eat little low larkspur (*D. nuttallianum*) unless a high stock density is used.

A variety of remedies have been applied in the field when ranchers find intoxicated animals (e.g., bleeding by cutting the tail), but most are without a solid scientific rationale [76]. Any imagined success with these treatments was probably related to the dose. If a sub-lethal dose was ingested, the animal would likely recover in spite of any treatment, unless bloat or aspiration pneumonia occurred during recumbency.

Drugs that increase the acetylcholine concentration in the neuromuscular junction have potential for reversing larkspur toxicity. The cholinergic drugs, physostigmine and neostigmine (0.08 mg/kg i.v.), have been successfully used under field and pen conditions to reverse clinical larkspur intoxication [54,77,78]. This reversal lasts about 2 h, and thus repeated injections of the drug may be required. Under field conditions, physostigmine temporarily abates clinical signs and animals quickly (~15 min) become ambulatory. Depending on the larkspur dose, the intoxication may recur. Consequently, the use of drug therapy may aggravate losses in the absence of further treatment if suddenly ambulatory animals later develop increased muscular fatigue, dyspnea, and death.

Larkspur losses can be reduced greatly if dense larkspur populations are reduced by herbicides. Picloram, metsulfuron, and glyphosate have proven to be effective in killing tall larkspurs when applied at specific growth stages [79]. Such treatments may be efficacious for >15 years. However, caution should be used as these herbicides do not reduce toxic alkaloid concentrations in treated larkspurs, and in fact metsulfuron may increase toxicity. Additionally herbicide application may increase the palatability of larkspurs. Therefore, sprayed areas should not be grazed until the following growing season. Herbicide use as a technique to control low larkspur species should be used with some caution as the number of plants in a population that may be flowering on a given year can vary significantly.

9. CONCLUSIONS

Larkspur poisoning in cattle is a serious problem for many livestock producers in the western ranges of North America. Years of research efforts have resulted in the development of numerous management strategies to reduce the number of cattle deaths from larkspurs. Future research efforts will continue to evaluate the role of individual alkaloids and the effects of various mixtures of alkaloids on the potency of the larkspurs. Studies will continue in order to better understand the distribution of toxic and non-toxic larkspur populations. Additionally, considerable efforts will be made to identify genetic markers of susceptibility and/or resistance to larkspur toxicosis.

ACKNOWLEDGMENTS

This work was supported by direct congressional appropriations via the United State Department of Agriculture to the Poisonous Plant Research Laboratory, which is part of the Agricultural Research Service.

AUTHOR CONTRIBUTIONS

All authors contributed to the writing of this review article.

CONFLICTS OF INTEREST

The authors declare no conflict of interest.

REFERENCES

1. Kingsbury, J.M. Poisonous Plants of the United States and Canada; Prentice-Hall, Inc.: Englewood Cliffs, NJ, USA, 1964.
2. Burrows, G.E.; Tyrl, R.J. Toxic Plants of North America; Iowa State University Press: Ames, IA, USA, 2001.
3. Nielsen, D.B.; Ralphs, M.R. Larkspur: Ecconomic considerations. In The Ecology and Ecconomic Impact of Poisonous Plants on Livestock Production; James, L.F., Ralphs, M.H., Nielsen, D.B., Eds.; Westview Press: Boulder and London, UK, 1988; pp. 119–129.
4. Nielsen, D.B.; Ralphs, M.H.; Evans, J.O.; Call, C.A. Economic feasibility of controlling tall larkspur on rangelands. J. Range Manag. 1994, 47, 369–372.
5. Pfister, J.A.; Ralphs, M.H.; Gardner, D.R.; Stegelmeier, B.L.; Manners, G.D.; Panter, K.E.; Lee, S.T. Management of three toxic Delphinium species based on alkaloid concentrations. Biochem. Syst. Ecol. 2002, 30, 129–138.
6. Marsh, C.D.; Clawson, A.B. Larkspur poisoning of livestock. USDA Bull. 1916, 365, 1–91.
7. Pfister, J.A.; Gardner, D.R.; Panter, K.E.; Manners, G.D.; Ralphs, M.H.; Stegelmeier, B.L.; Schoch, T.K. Larkspur (Delphinium spp.) poisoning in livestock. J. Nat. Toxins 1999, 8, 81–94.
8. Olsen, J.D. Tall larkspur poisoning in cattle and sheep. J. Am. Vet. Med. Assoc. 1978, 173, 762–765.
9. Panter, K.E.; Manners, G.D.; Stegelmeier, B.L.; Lee, S.; Gardner, D.R.; Ralphs, M.H.; Pfister, J.A.; James, L.F. Larkspur poisoning: Toxicology and alkaloid structure—Activity relationships. Biochem. Syst. Ecol. 2002, 30, 113–128.
10. Ewan, J. Synopsis of the North American Species of Delphinium; University of Colorado: Boulder, CO, USA, 1945.
11. Warnock, M. A taxonomic conspectus of North American Delphinium. Phytologia 1995, 78, 73–101.
12. Warnock, M. Delphinium. Flora N. Am. North Mex. 1997, 3, 196–240.

13. Manners, G.D.; Panter, K.E.; Ralphs, M.H.; Pfister, J.A.; Olsen, J.D.; James, L.F. Toxicity and chemical phenology of norditerpenoid alkaloids in the tall larkspurs (Delphinium species). J. Agric. Food Chem. 1993, 41, 96–100.
14. Manners, G.D.; Panter, K.E.; Pelletier, S.W. Structure-activity relationships of norditerpenoid alkaloids occurring in toxic larkspur (Delphinium) species. J. Nat. Prod. 1995, 58, 863–869.
15. Welch, K.D.; Panter, K.E.; Gardner, D.R.; Green, B.T.; Pfister, J.A.; Cook, D.; Stegelmeier, B.L. The effect of 7,8-methylenedioxylycoctonine-type diterpenoid alkaloids on the toxicity of methyllycaconitine in mice. J. Anim. Sci.2008, 86, 2761–2770.
16. Welch, K.D.; Green, B.T.; Gardner, D.R.; Cook, D.; Pfister, J.A.; Stegelmeier, B.L.; Panter, K.E.; Davis, T.Z. Influence of 7,8-methylenedioxylycoctonine-type alkaloids on the toxic effects associated with ingestion of tall larkspur (Delphinium spp.) in cattle. Am. J. Vet. Res. 2010, 71, 487–492.
17. Gardner, D.R.; Ralphs, M.H.; Turner, D.L.; Welsh, S.L. Taxonomic implications of diterpene alkaloids in three toxic tall larkspur species (Delphinium spp.). Biochem. Syst. Ecol. 2002, 30, 77–90.
18. Cook, D.; Welch, K.D.; Green, B.T.; Gardner, D.R.; Pfister, J.A.; Constantino, J.R.; Stonecipher, C.A. The relative toxicity of Delphinium stachydeum in mice and cattle. Toxicon 2015, 99, 36–43.
19. Cook, D.; Gardner, D.R.; Pfister, J.A.; Welch, K.D.; Green, B.T.; Lee, S.T. The biogeographical distribution of duncecap larkspur (Delphinium occidentale) chemotypes and their potential toxicity. J. Chem. Ecol. 2009, 35, 643–652.
20. Gardner, D.R.; Pfister, J.A. HPLC-MS analysis of toxic norditerpenoid alkaloids: Refinement of toxicity assessment of low larkspurs (Delphinium spp.). Phytochem. Anal. 2009, 20, 104–113.
21. Cook, D.; Green, B.T.; Welch, K.D.; Gardner, D.R.; Pfister, J.A.; Panter, K.E. Comparison of the toxic effects of two duncecap larkspur (Delphinium occidentale) chemotypes in mice and cattle. Am. J. Vet. Res. 2011, 72, 706–714.
22. Welch, K.D.; Green, B.T.; Gardner, D.R.; Cook, D.; Pfister, J.A.; Panter, K.E. The effect of 7,8-methylenedioxylycoctonine-type diterpenoid alkaloids on the toxicity of tall larkspur (Delphinium spp.) in cattle. J. Anim. Sci. 2012, 90, 2394–2401.
23. Gardner, D.R.; Manners, G.D.; Panter, K.E.; Lee, S.T.; Pfister, J.A. Three new toxic norditerpenoid alkaloids from the low larkspur Delphinium nuttallianum. J. Nat. Prod. 2000, 63, 1127–1130.
24. Gardner, D.R.; Pfister, J.A. Toxic alkaloid concentrations in Delphinium nuttallianum, Delphinium andersonii, and Delphinium geyeri in the intermountain region. Rangel. Ecol. Manag. 2007, 60, 441–446.
25. Ralphs, M.H.; Gardner, D.R. Alkaloid levels in duncecap (Delphinium occidentale) and tall larkspur (D. barbeyi) grown in reciprocal gardens: Separating genetic from environmental influences. Biochem. Syst. Ecol. 2001, 29, 117–124.
26. Ralphs, M.H.; Gardner, D.R. Distribution of norditerpene alkaloids in tall larkspur plant parts through the growing season. J. Chem. Ecol. 2003, 29, 2013–2021.
27. Gardner, D.R.; Pfister, J.A. Late season toxic alkaloid concentrations in tall larkspur (Delphinium spp.). J. Range Manag. 2000, 53, 329–334.
28. Pfister, J.; Cook, D. Influence of weather on low larkspur (Delphinium nuttallianum) density. J. Agric. Sci. 2011, 3, 36.
29. Saavedra, F.; Inouye, D.W.; Price, M.V.; Harte, J. Changes in flowering and abundance of Delphinium nuttallianum (ranunculaceae) in response to a subalpine climate warming experiment. Glob. Chang. Biol. 2003, 9, 885–894.
30. Cundiff, R.H.; Markunas, P.C. Determination of nicotine, nornicotine, and total alkaloids in tobacco. Anal. Chem. 1955, 27, 1650–1653.
31. Williams, M.C.; Cronin, E.H. Effect of silvex and 2,4,5-T on alkaloid content of tall larkspur. Weeds 1963, 11, 317–319.

32. Pfister, J.A.; Manners, G.D.; Ralphs, M.H.; Hong, Z.X.; Lane, M.A. Effects of phenology, site, and rumen fill on tall larkspur consumption by cattle. J. Range Manag. 1988, 41, 509–514.
33. Manners, G.D.; Pfister, J.A.; Ralphs, M.H.; Panter, K.E.; Olsen, J.D. Larkspur chemistry: Toxic alkaloids in tall larkspurs. J. Range Manag. 1992, 45, 63–67.
34. Pelletier, S.W.; Dailey, O.D., Jr.; Mody, N.V.; Olsen, J.D. Isolation and structure elucidation of the alkaloids of Delphinium glaucescens rybd. J. Org. Chem. 1981, 46, 3284–3293.
35. Clark, D.; Ralphs, M.; Lamb, R. Total alkaloid determinations in larkspur and lupine with near infrared reflectance spectroscopy. Agron. J. 1987, 79, 481–485.
36. Gardner, D.R.; Manners, G.D.; Ralphs, M.H.; Pfister, J.A. Quantitative analysis of norditerpenoid alkaloids in larkspur (Delphinium spp.) by fourier transform infrared spectroscopy. Phytochem. Anal. 1997, 8, 55–62.
37. Ralphs, M.H.; Gardner, D.R.; Turner, D.L.; Pfister, J.A.; Thacker, E. Predicting toxicity of tall larkspur (Delphinium barbeyi): Measurement of the variation in alkaloid concentration among plants and among years. J. Chem. Ecol. 2002, 28, 2327–2341.
38. Manners, G.D.; Ralphs, M.H. Capillary gas chromatography of Delphinium diterpenoid alkaloids. J. Chromatogr. 1989, 466, 427–432.
39. Majak, W.; McDiarmid, R.E.; Benn, M.H. Isolation and HPLC determination of methyllycaconitine in a species of low larkspur (Delphinium nuttallianum). J. Agric. Food Chem. 1987, 35, 800–802.
40. Coates, P.A.; Blagbrough, I.S.; Lewis, T.; Potter, B.V.; Rowan, M.G. An HPLC assay for the norditerpenoid alkaloid methyllycaconitine, a potent nicotinic acetylcholine receptor antagonist. J. Pharm. Biomed. Anal. 1995, 13, 1541–1544.
41. Turek, J.W.; Kang, C.H.; Campbell, J.E.; Arneric, S.P.; Sullivan, J.P. A sensitive technique for the detection of the alpha 7 neuronal nicotinic acetylcholine receptor antagonist, methyllycaconitine, in rat plasma and brain. J. Neurosci. Methods 1995, 61, 113–118.
42. Gardner, D.R.; Panter, K.E.; Pfister, J.A.; Knight, A.P. Analysis of toxic norditerpenoid alkaloids in Delphinium species by electrospray, atmospheric pressure chemical ionization, and sequential tandem mass spectrometry. J. Agric. Food Chem. 1999, 47, 5049–5058.
43. Wada, K.; Bando, H.; Kawahara, N.; Mori, T.; Murayama, M. Determination and quantitative analysis of alkaloids in Aconitum japonicum by liquid chromatography atmospheric pressure chemical ionization mass spectrometry. Biol. Mass Spectrom. 1994, 23, 97–102.
44. Wada, K.; Mori, T.; Kawahara, N. Stereochemistry of norditerpenoid alkaloids by liquid chromatography/atmospheric pressure chemical ionization mass spectrometry. J. Mass Spectrom. 2000, 35, 432–439.
45. Stegelmeier, B.L.; Hall, J.O.; Gardner, D.R.; Panter, K.E. The toxicity and kinetics of larkspur alkaloid, methyllycaconitine, in mice. J. Anim. Sci. 2003, 81, 1237–1241.
46. Green, B.T.; Welch, K.D.; Gardner, D.R.; Stegelmeier, B.L.; Pfister, J.A.; Cook, D.; Davis, T.Z. A toxicokinetic comparison of norditerpenoid alkaloids from Delphinium barbeyi and D. glaucescens in cattle. J. Appl. Toxicol. 2011, 31, 20–26.
47. Aiyar, V.N.; Benn, M.H.; Hanna, T.; Jacyno, J.; Roth, S.H.; Wilkens, J.L. The principal toxin of Delphinium brownii Rydb., and its mode of action. Experientia 1979, 35, 1367–1368.
48. Benn, M.H.; Jacyno, J.M. The toxicology and pharmacology of diternpenoid alkaloids. In Alkaloids: Chemical and Biological Perspective; Pelletier, S.W., Ed.; John Wiley and Sons: New York, NY, USA, 1983; pp. 153–210.
49. Dobelis, P.; Madl, J.E.; Pfister, J.A.; Manners, G.D.; Walrond, J.P. Effects of Delphinium alkaloids on neuromuscular transmission. J. Pharmacol. Exp. Ther. 1999, 291, 538–546.
50. Kukel, C.F.; Jennings, K.R. Delphinium alkaloids as inhibitors of alpha-bungarotoxin binding to rat and insect neural membranes. Can. J. Physiol. Pharmacol. 1994, 72, 104–107.

51. Ward, J.M.; Cockcroft, V.B.; Lunt, G.G.; Smillie, F.S.; Wonnacott, S. Methyllycaconitine: A selective probe for neuronal α-bungarotoxin binding sites. FEBS Lett. 1990, 270, 45–48.
52. Stegelmeier, B.L.; Panter, K.E.; Pfister, J.A.; James, L.F.; Manners, G.D.; Ralphs, M.R.; Olsen, J.D. Experimental modification of larkspur (Delphinium spp.) poisoning in livestock. In Toxic Plants and Other Natural Toxicants; Garland, T.R., Barr, A.C., Eds.; CABI: Wallingford, Oxforshire, UK, 1998; pp. 205–210.
53. Panter, K.E.; Welch, K.D.; Gardner, D.R.; Lee, S.T.; Green, B.T.; Pfister, J.A.; Cook, D.; Davis, T.Z.; Stegelmeier, B.L. Poisonous plants of the United States. In Veterinary Toxicology, Basic and Clinical Principles, 2nd ed.; Gupta, R.C., Ed.; Elsevier: New York, NY, USA, 2012; pp. 1031–1079.
54. Green, B.T.; Pfister, J.A.; Cook, D.; Welch, K.D.; Stegelmeier, B.L.; Lee, S.T.; Gardner, D.R.; Knoppel, E.L.; Panter, K.E. Effects of larkspur (Delphinium barbeyi) on heart rate and electrically evoked electromyographic response of the external anal sphincter in cattle. Am. J. Vet. Res. 2009, 70, 539–546.
55. Pfister, J.; Provenza, F.; Manners, G.; Gardner, D.; Ralphs, M. Tall larkspur ingestion: Can cattle regulate intake below toxic levels? J. Chem. Ecol. 1997, 23, 759–777.
56. Welch, K.D.; Green, B.T.; Gardner, D.R.; Cook, D.; Pfister, J.A. The effect of administering multiple doses of tall larkspur (Delphinium barbeyi) to cattle. J. Anim. Sci. 2015. in press.
57. Goddard, M.E.; Hayes, B.J. Mapping genes for complex traits in domestic animals and their use in breeding programmes. Nat. Rev. Genet. 2009, 10, 381–391.
58. Welch, K.D.; Green, B.T.; Panter, K.E.; Gardner, D.R.; Pfister, J.A.; Cook, D.; Stegelmeier, B.L. Investigation of the susceptibility of various strains of mice to methyllycaconitine toxicosis. J. Anim. Sci. 2009, 87, 1558–1564.
59. Welch, K.D.; Green, B.T.; Panter, K.E.; Pfister, J.A.; Gardner, D.R. The role of the α7 subunit of the nicotinic acetylcholine receptor in the acute toxicosis of methyllycaconitine in mice. J. Appl. Toxicol. 2013, 33, 1011–1016.
60. Green, B.T.; Welch, K.D.; Pfister, J.A.; Chitko-McKown, C.G.; Gardner, D.R.; Panter, K.E. Mitigation of larkspur poisoning on rangelands through the selection of cattle. Rangelands 2014, 36, 10–15.
61. Green, B.T.; Welch, K.D.; Gardner, D.R.; Stegelmeier, B.L.; Davis, T.Z.; Cook, D.; Lee, S.T.; Pfister, J.A.; Panter, K.E. Serum elimination profiles of methyllycaconitine and deltaline in cattle following oral administration of larkspur (Delphinium barbeyi). Am. J. Vet. Res. 2009, 70, 926–931.
62. Green, B.T.; Welch, K.D.; Gardner, D.R.; Stegelmeier, B.L.; Pfister, J.A.; Cook, D.; Panter, K.E. Toxicokinetics of norditerpenoid alkaloids from low larkspur (Delphinium andersonii) orally administered to cattle. Am. J. Vet. Res. 2012, 73, 1318–1324.
63. Green, B.T.; Welch, K.D.; Gardner, D.R.; Stegelmeier, B.L.; Lee, S.T. A toxicokinetic comparison of two species of low larkspur (Delphinium spp.) in cattle. Res. Vet. Sci. 2013, 95, 612–615.
64. Ralphs, M.H.; Jensen, D.T.; Pfister, J.A.; Nielsen, D.B.; James, L.F. Storms influence cattle to graze larkspur: An observation. J. Range Manag. 1994, 47, 275–278.
65. Pfister, J.A.; Ralphs, M.H.; Manners, G.D. Cattle grazing tall larkspur on Utah mountain rangeland. J. Range Manag.1988, 41, 118–122.
66. Pfister, J.A.; Cook, D.; Gardner, D.R.; Baker, S.D. Early season grazing by cattle of waxy larkspur (Delphinium glaucescens) in central Idaho. Rangelands 2013, 35, 2–5.
67. Pfister, J.A.; Gardner, D.R. Consumption of low larkspur (Delphinium nuttallianum) by cattle. J. Range Manag. 1999, 63, 378–383.
68. Pfister, J.A.; Manners, G.D.; Gardner, D.R.; Price, K.W.; Ralphs, M.H. Influence of alkaloid concentration on acceptability of tall larkspur (Delphinium spp.) to cattle and sheep. J. Chem. Ecol. 1996, 22, 1147–1168.
69. Pfister, J.A.; Provenza, F.D.; Manners, G.D. Ingestion of tall larkspur by cattle: Separating effects of flavor from postingestive consequences. J. Chem. Ecol. 1990, 16, 1697–1705.

70. Pfister, J.A.; Gardner, D.R.; Stegelmeier, B.L.; Hackett, K.; Secrist, G. Catastrophic cattle loss to low larkspur (Delphinium nuttallianum) in Idaho. Vet. Hum. Toxicol. 2003, 45, 137–139.
71. Cook, D.; Pfister, J.A.; Constantino, J.R.; Roper, J.M.; Gardner, D.R.; Welch, K.D.; Hammond, Z.J.; Green, B.T. Development of a PCR-based method for detection of Delphinium species in poisoned cattle. J. Agric. Food Chem. 2015.
72. Ralphs, M.H. Toxic alkaloid concentration in tall larkspur species in the western U.S. J. Range Manag. 1997, 50, 497–502.
73. Chesnut, V.; Wilcox, E. The stock poisoning plants of Montana. USDA Div. Bot. Bull. 1901, 26, 1–150.
74. Ralphs, M.H.; Olsen, J.D. Comparison of larkspur alkaloid extract and lithium chloride in maintaining cattle aversion to larkspur in the field. J. Anim. Sci. 1992, 70, 1116–1120.
75. Pfister, J.A.; Gardner, D.R.; Cheney, C.C.; Panter, K.E.; Hall, J.O. The capability of several toxic plants to condition taste aversions in sheep. Small Rumin. Res. 2010, 90, 114–119.
76. Knight, A.P.; Pfister, J.A. Larkspur poisoning in livestock: Myths and misconceptions. Rangelands 1997, 19, 10–13.
77. Nation, P.N.; Benn, M.H.; Roth, S.H.; Wilkens, J.L. Clinical signs and studies of the site of action of purified larkspur alkaloid, methyllycaconitine, administered parenterally to calves. Can. Vet. J. 1982, 23, 264–266.
78. Pfister, J.A.; Panter, K.E.; Manners, G.D.; Cheney, C.D. Reversal of tall larkspur (Delphinium barbeyi) poisoning in cattle with physostigmine. Vet. Hum. Toxicol. 1994, 36, 511–514.
79. Ralphs, M.H.; Evans, J.O.; Dewey, S.A. Timing of herbicide applications for control of larkspurs (Delphinium spp.). Weed Sci. 1992, 40, 264–269.

CHAPTER 17

Safety Evaluation of Parastar® Plus in Dogs and Assessment of Transferable Residue of Fipronil and Cyphenothrin from Dogs to Humans

Katharine M. Case[1,2], Natalia M. Vega[1,2], Ramesh C. Gupta[1*], Michelle A. Lasher[1] and Terry D. Canerdy[2]

[1]*Toxicology Department, Breathitt Veterinary Center, Murray State University, Hopkinsville, KY, USA*

[2]*Murray State University, Murray, KY, USA*

ABSTRACT

Dogs are easily infested with fleas, ticks, and other ectoparasites serving as vectors for transmitting bacterial, viral, and parasitic diseases. Therefore, the use of ectoparasiticides is inevitable and important. The present investigation was undertaken with two specific objectives: one, to evaluate the safety of fipronil and cyphenothrin in dogs after topical application of Parastar® Plus, and two, to determine the transferable residue of these insecticides from dogs to humans. Six healthy, adult dogs (medium length hair, weighing between 20.5 and 27.3 kg) received topical application of Parastar® Plus (2.68 mL; fipronil, 9.8%, and cyphenothrin, 5.2%) on the back between the shoulder blades. At predetermined intervals, dogs were given a full physical exam, and residues of fipronil and cyphenothrin were determined in dog blood and cotton glove extracts using GC/MS. Fipronil and cyphenothrin peaks eluted at 7.453 and 9.913 min, correspondingly, and the compounds were confirmed based on characteristic ions. At no time was fipronil or cyphenothrin residue detected in blood samples. In glove extracts, residues of fipronil and cyphenothrin were maximally present at 24-h posttreatment (43.84 ± 5.69 and 59.26 ± 8.97 ppm, respectively). By 48 h, the residue levels sharply declined (16.89 ± 2.82 and 17.98 ± 2.07 ppm, respectively). The insecticides' residues were detected in insignificant amounts after 1 week (5.69 ± 2.16 and 10.00 ± 1.51 ppm, respectively), and only in trace amounts after 2 weeks. At no time did any dog show side effects, except itching at the site of Parastar® Plus application. The findings suggest that Parastar® Plus was safe for dogs, and transferable residues of fipronil and cyphenothrin were minimal, posing very little or no health concern to pet owners or veterinary personnel. Of course, veterinary personnel, who handle many dogs daily, may require proper protection to avoid cumulative exposure.

Keywords: parastar® plus, fipronil, cyphenothrin, pyrethroids, ectoparasiticide safety, dogs

INTRODUCTION

Currently, there are approximately 80 million dogs residing alongside humans in North America. Dogs serve as hosts for several ectoparasites, such as fleas, ticks, lice, and mites. These ectoparasites are capable of spreading many diseases that are significant for both animal and human health. Zoonotic diseases spread by ectoparasites include Lyme disease, plague, ehrlichiosis, Rocky Mountain spotted fever, scabies, and several others. Internal parasites can also be transmitted, such as the flea tapeworm (*Dipylidium caninum*) that frequently infects children (1). Currently, in more than fifteen countries, microcephaly in thousands of babies born to mothers exposed to Zika virus spread by mosquitoes is the most serious human health concern. Because of the close contact between dogs and humans, and the variety of disease vectors for which dogs serve as the host, treatment of ectoparasites is important and inevitable.

Parastar® Plus for Dogs (subsequently referred to as Parastar® Plus) is a monthly topical ectoparasiticide developed by Novartis Animal Health. It kills fleas, several species of ticks, and it controls chewing lice and sarcoptic mange mites (2). The product contains two active ingredients: (1) fipronil:

(±)-5-amino-1-[2,6-dichloro-4-(trifluoromethyl)phenyl]-4-[(trifluoromethyl)sulfinyl]-1H-pyrazole-3-carbonitrile, 9.8% and (2) cyphenothrin: [cyano-(3-phenoxyphenyl) methyl] 2,2-dimethyl-3-(2-methylprop-1-enyl)cyclopropane-1-carboxylate, 5.2%. The chemical structures of fipronil and cyphenothrin are shown in Figure 1.

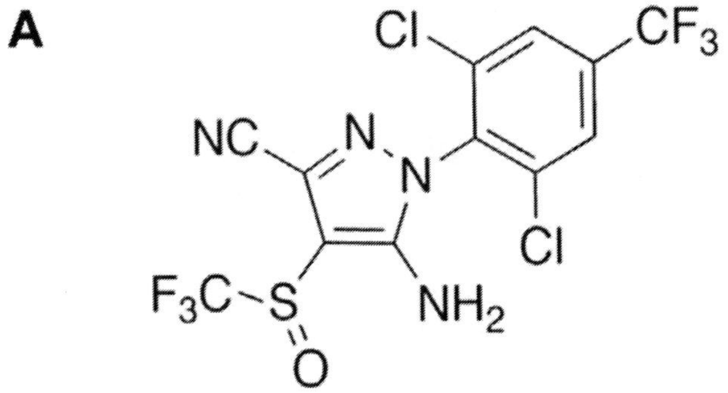

Fipronil.

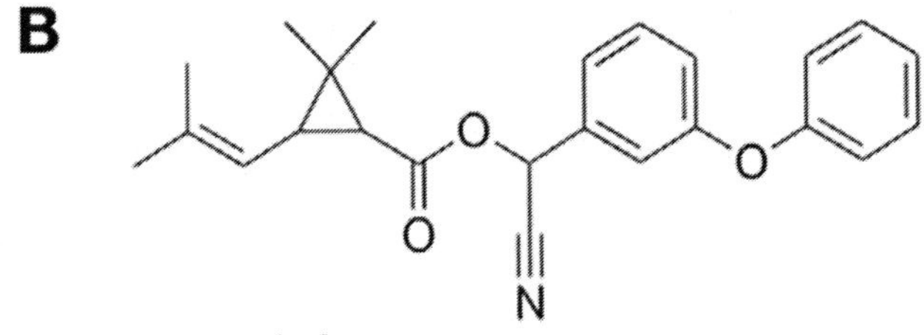

Cyphenothrin.

Figure 1. The chemical structures of (a) fipronil and (b) cyphenothrin.

According to the manufacturer, Parastar® Plus begins working within 5 min of application and maintains effectiveness for 30 days (3). It kills fleas (*Ctenocephalides* spp.), four species of ticks (*Dermacentor variabilis, Ixodes* spp., *Amblyomma americanum*, and *Rhipicephalus sanguineus*), mites (*Sarcoptes scabiei* var. *canis*), and lice (order Mallophaga), many of which are vectors for zoonotic diseases of humans.

Fipronil is a phenylpyrazole insecticide that was first developed as an agricultural pesticide in 1987. Due to its lipophilic nature, fipronil is absorbed by the sebaceous glands of the skin, leading its distribution throughout the skin and hair follicles (4, 5), thereby providing ectoparasiticide effect for a month.

Pyrethroids are semisynthetic derivatives of naturally occurring pyrethrins extracted from flowers of the chrysanthemum plant (*Chrysanthemum cinerariaefolium*). Pyrethroids are of two types (type I and type II). Type II pyrethroids are distinguished from type I by having an α-cyano group at the molecule's alcohol moiety that increases its insecticidal activity. Cyphenothrin is a type II pyrethroid that was first registered at the US Environmental Protection Agency (EPA) in 1991. Like fipronil, cyphenothrin also exhibits lipophilic tendencies that allow it to be distributed among fatty tissue (6) and in oily sebaceous gland secretions.

Safety evaluation of Parastar® Plus (active ingredients, fipronil, and cyphenothrin) in dogs and transferable residues of these insecticides from dogs to humans have not yet been reported. The current investigation was therefore undertaken to evaluate the safety of Parastar® Plus in dogs. Further, since the risk of continuous exposure due to repeated application of Parastar® Plus for children and pet owners exists, the present investigation determined the insecticides (fipronil and cyphenothrin) residues in gloves.

MATERIALS AND METHODS

Animals

Six, privately owned adult dogs (various breeds, with medium length hair coats) weighing between 20.5 and 27.3 kg (45–60 lbs) served as the participants in this investigation, representing both the control (pretreatment) and the experimental (treatment) groups. Each of the dog's owners signed a consent form detailing the experimental protocol and any risks involved with ectoparasiticide use. Two weeks prior to the beginning of the study, the dogs were taken off any current ectoparasiticides, as well as heartworm prevention. Blood and glove samples were collected on day 0 and were tested for residues of fipronil, cyphenothrin, and any other chemicals. All of the dogs were negative for chemical residues.

Chemicals

Parastar® Plus was purchased from 1 to 800 PetMeds® (Novartis Animal Health US, Inc., Greensboro, NC, USA). A single application of the product contained 2.68 mL (fipronil 9.8%; cyphenothrin 5.2%) for dogs weighing 45–88 lbs (20.5–40 kg). Analytical grade standards of fipronil and cyphenothrin were purchased from Chem Service, Inc. (West Chester, PA, USA).

Biohazards

Murray State University's policy and procedures were followed in this investigation for the disposal of dog blood and gloves.

Experimental Design

The experimental design of this investigation was based upon previous studies conducted at Murray State University and elsewhere (4, 7–10). In this study, each dog served as its own control. The study protocol was approved by Murray State University's Institutional Animal Care and Use Committee (IACUC). The IACUC committee approved the use of six dogs based on power analysis.

Parastar® Plus Application

Parastar® Plus was topically applied in the amount of 2.68 mL directly to the skin between the shoulder blades, according to the manufacturer's instructions (2).

Physical Examination

At predetermined intervals, dogs were given a physical examination for the measurement of body weight, heart rate, respiration rate, and body temperature. The dogs were examined for any change in their behavior and the skin was examined for signs of irritation (such as erythema) at the site of Parastar® Plus application. Owners were questioned about their dog's behavior and whether they observed the dogs itching at the application site.

Sample Collection

Sampling included blood and topical glove collection. Whole blood (3.5–5.0 mL) was collected from the jugular vein in EDTA tubes on day 0 and at 24, 48, and 72 h, and 1-week post-application of Parastar®Plus. Blood samples were stored in the refrigerator until analysis within 72 h. Topical samples were collected using the wipe sampling technique during which the investigator wore a 100% cotton glove on one hand and pet the dog along its sides and back continuously for 5 min, avoiding the application site. Topical samples were collected on day 0, at 24, 48, and 72 h, and at 1, 2, 3, and 4 weeks post-application of Parastar® Plus. The gloves were then placed into individual (473 mL) glass jars and stored at room temperature until analysis within 72 h.

Sample Extraction

Blood samples were transferred from EDTA tubes to individual (500 mL) separatory funnels and weighed. Twenty-five milliliter of methylene chloride: petroleum ether (1:1) was added, and the funnels were gently shaken three times and allowed to vent between shakes. The samples sat for 30 min, and then a disposable pipette was used to draw off the solvent and pass it through a sodium sulfate-filled filter into a 50 mL beaker. The solvent was swirled and passed through the filters. A syringe and needle were used to aspirate the remaining 2–3 mL and pass it through a Sep-Pak® cartridge (Waters Corp., Milford, MA, USA) into a clean tube. After it evaporated to dryness, it was reconstituted in an appropriate volume of methylene chloride: petroleum ether for fipronil and cyphenothrin residue analysis by GC/MS.

Gloves were removed from the glass jars, weighed, and then placed into individual 250 mL beakers. Hundred milliliter methylene chloride: petroleum ether (1:1) was added, and the gloves swirled and fully submerged in the solvent, after which they were allowed to sit for 30 min. The solvent was then poured through a sodium sulfate-filled filter into a new 100 mL beaker. During evaporation, the remaining 2–3 mL of solvent were aspirated with a needle and syringe and passed through a Sep-Pak® cartridge into a clean tube. After it evaporated to dryness, it was reconstituted in an appropriate volume of methylene chloride: petroleum ether for fipronil and cyphenothrin residue analysis by GC/MS.

GC/MS Analysis

The active ingredients of Parastar® Plus (fipronil and cyphenothrin) were confirmed and quantified using an agilent gas chromatograph (GC model 7890A)/mass spectrometer (MS model 5975C) coupled with a computer, and their concentrations were expressed in terms of microgram per gram (parts per million). One microliter of the reconstituted extract was injected into the GC. The capillary column used was Ultra II cross-linked with 5% phenyl methyl siloxane coating and was of the following dimensions (25 m × 0.52 µm), which was directly connected to the Mass Selective Detector *via* an interface and heated transfer line. The carrier gas was ultrapure (99.9999%) helium at a flow rate of 2.3 mL/min, and the injector temperature was 200°C. The injector was operated in the splitless mode. A temperature program for the GC-oven was used starting at a temperature of 100°C, and then increased to a final temperature of 300°C in 20°C/min increments. The final temperature was maintained for 5 min. The transfer line temperature was 280°C, and the source temperature was 230°C. The instrument was operated in electron ionization mode, and the ion energy was 70 eV. The total duration of each injection run was 16 min, with a solvent delay of 7 min. Peaks of fipronil and cyphenothrin eluted at 7.453 and 9.913 min, respectively. Sensitivity of the GC/MS for these compounds was in the range of nanogram, and the limit of detection was in the low microgram per gram (parts per million) range. Total ion chromatograms and mass spectra with characteristic ions for fipronil (77.1, 179, 213, 255, 367, 417, and 436) and cyphenothrin (81.1, 123.1, 152.1, 167.1, 181.1, 208.1, and 375.2) are presented in Figures 2 and 3, respectively. The identification and confirmation of each insecticide was based on characteristic ion-based spectrum. Percent recovery for fipronil and cyphenothrin was greater than 95%.

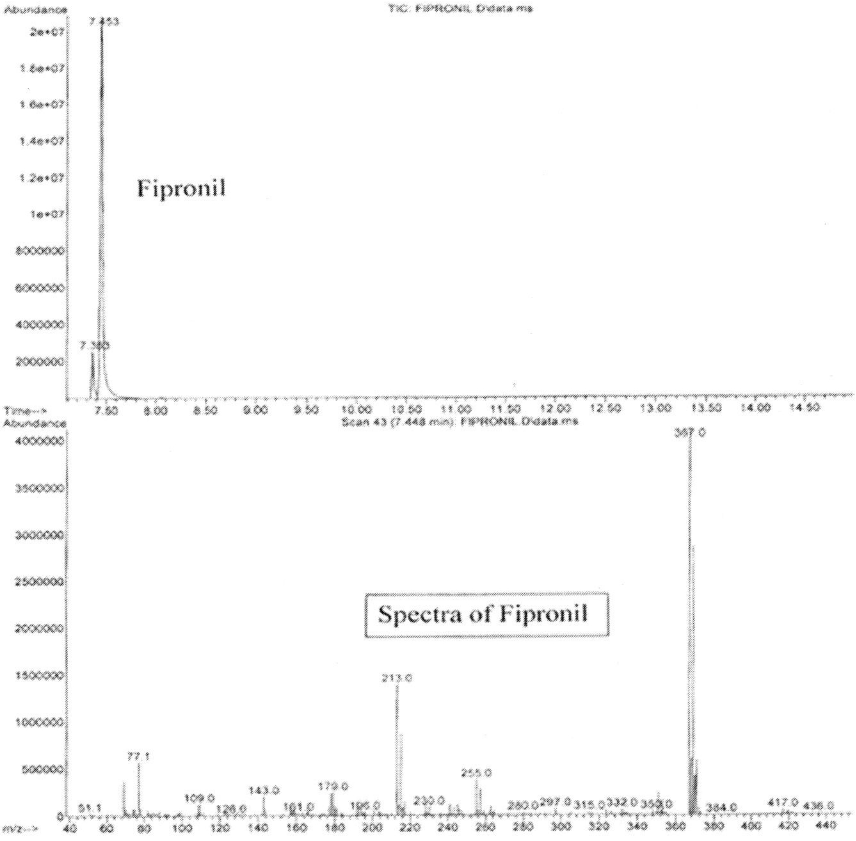

Figure 2. Total ion chromatogram (upper panel) and ion spectrum (lower panel) of fipronil.

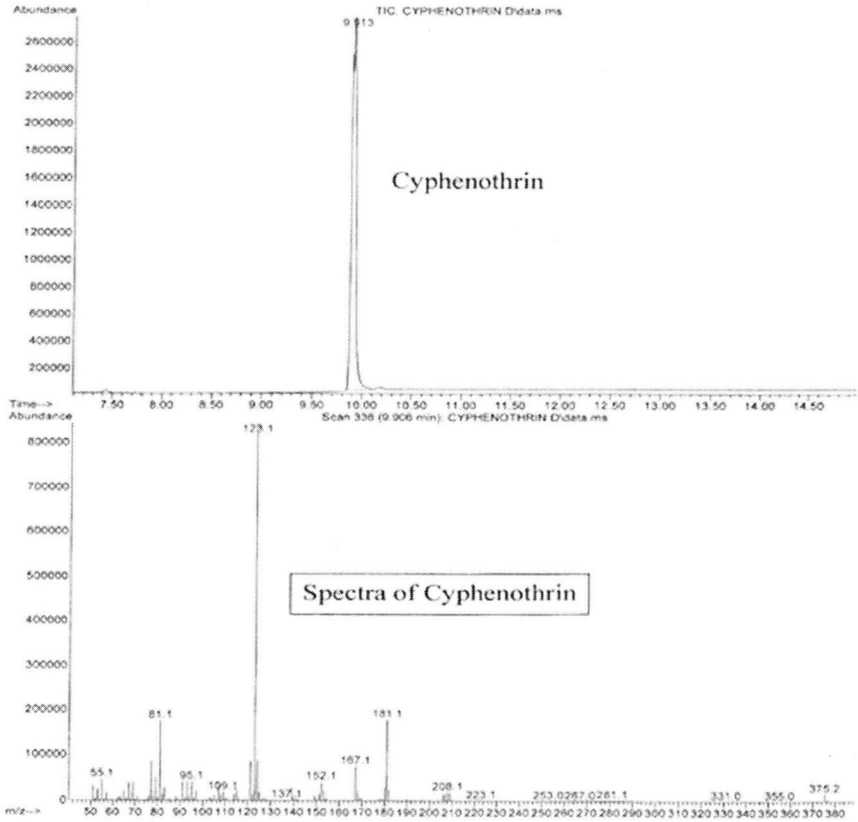

Figure 3. Total ion chromatogram (upper panel) and ion spectrum (lower panel) of cyphenothrin.

RESULTS

Adverse Reactions

At no time during this investigation were adverse reactions to Parastar® Plus observed in any of the dogs. Some participants experienced itching following application, as observed by the owners. It cannot be stated definitively that application of Parastar® Plus caused the clinical signs observed, as patient histories before the study were not entirely known. It was possible that one participant displayed scratching behavior due to an ongoing flea infestation. Flea dirt was observed in this dog's hair coat throughout the study.

Fipronil and Cyphenothrin Concentrations in Blood and Glove Samples

The skin and hair coat residue levels of Parastar® Plus's active ingredients, fipronil (9.8%) and cyphenothrin (5.2%), were expected to reach maximum concentrations at 24-h post-application. In blood, due to slow dermal absorption (11), the maximum concentrations of fipronil and cyphenothrin were expected at 24- to 48-h post-application. Because Parastar® Plus is designed to provide protection for 30 days, insecticide residues on hair coat were expected to be detectable for up to 4 weeks post-application. Residue persistence on hair coat corresponded with these expectations.

At no time throughout the study duration were fipronil and cyphenothrin detected in blood samples. Data of fipronil and cyphenothrin concentrations in glove extracts are presented in Figures 4 and 5, respectively. Fipronil showed the highest concentration levels in gloves at 24-h post-application with a range of 22.23–59.66 ppm (43.83 ± 5.69 ppm). At the same time point, cyphenothrin concentration was in the range of 26.24–90.40 ppm (59.26 ± 8.97 ppm). At 48-h post-application, residues for both fipronil and cyphenothrin exhibited a steep decline (16.89 ± 2.82 and 17.98 ± 2.07 ppm, respectively) from the 24-h values. At 72 h, the value for fipronil was 8.31 ± 1.69 ppm, and for cyphenothrin it was 16.60 ± 3.19 ppm.

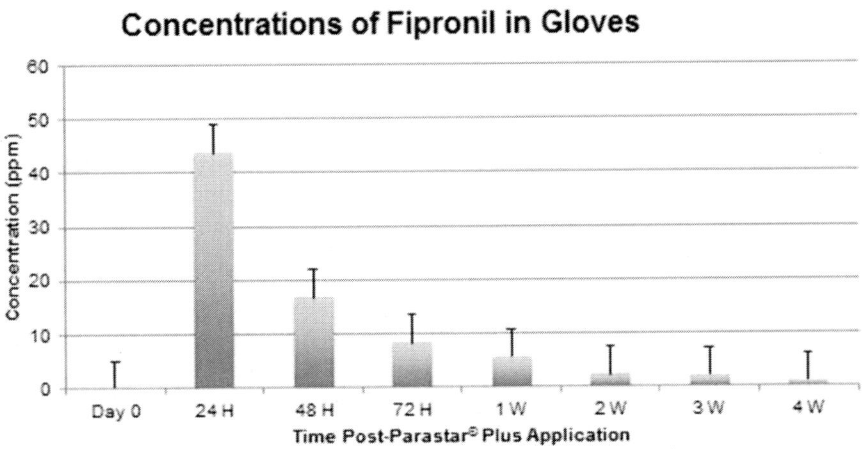

Figure 4. Concentrations of fipronil in gloves (mean ± sem).

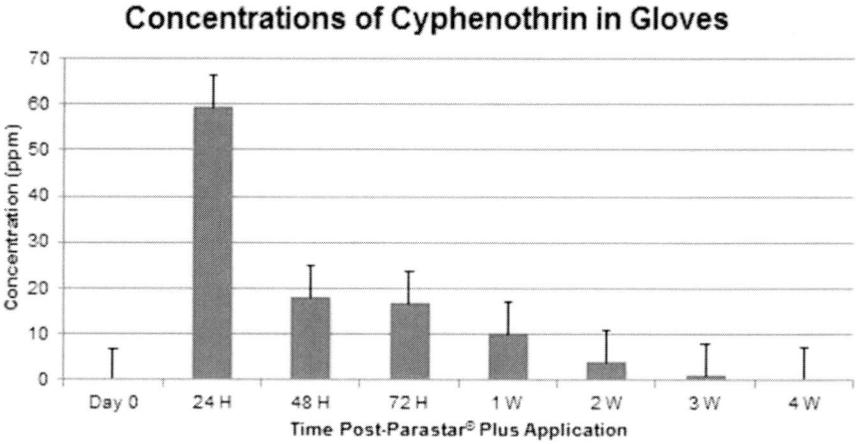

Figure 5. Concentrations of cyphenothrin in gloves (mean ± sem).

One week after Parastar® Plus application, the value of fipronil was 5.69 ± 2.16 ppm, and of cyphenothrin was 10.00 ± 1.51 ppm. With a decreasing trend at 4 weeks post-application, the value for fipronil was 1.09 ± 0.13 ppm, and for cyphenothrin was 0.43 ± 0.28 ppm.

DISCUSSION

The present investigation was undertaken with two specific objectives: (1) to evaluate the safety of Parastar® Plus in dogs; and (2) to measure the transferable residue of the active ingredients (fipronil and cyphenothrin) of Parastar® Plus from dog coats to humans. Determining the level of transferable residues of insecticides can aid in assessing the level of exposure and risks to pet owners, veterinary professionals, dog trainers, police, and military K9 units from dogs topically treated with Parastar® Plus.

The results of this study suggest a high level of safety in the use of Parastar® Plus for routine ectoparasite control in dogs, if the manufacturer's instructions are followed. At no time during this investigation, were adverse reactions to Parastar® Plus observed in any of the dogs. Also, at no time were residues of fipronil and cyphenothrin detected in the dogs' blood. This indicates that the dermal absorption of fipronil and cyphenothrin is minimal or insignificant, and systemic exposure to these insecticides for dogs is extremely limited. Further, no risk of exposure to insecticides *via* contact with blood exists for veterinary personnel. Residues of fipronil and cyphenothrin on the dogs' hair coats were found to be maximal at 24 h (43.83 ± 5.69 and 59.26 ± 8.97 ppm, respectively) post-Parastar® Plus application. The manufacturer advises that pet owners limit contact with their dogs during the first 24 h to allow the product to fully dry at the site of application. If this instruction is heeded, the risk of accidental exposure to Parastar® Plus through coat residues is greatly diminished. Of course, local irritation and local reactions could still be observed but have been investigated in this study. This aspect of investigation will be pursued in future studies. Insecticide residues drastically reduced by 48 h from 24 h (fipronil, 61.5% and cyphenothrin, 69.7%). If this 24-h time frame is observed by pet owners, accidental, oral, or dermal exposure to the product would be unlikely, even in children or pet cats.

In the setting of a veterinary clinic, a number of factors can influence the level of insecticide transfer from dogs to veterinary personnel, such as the type of patients seen, the volume of the practice, the number of exposure events occurring per day, and whether members of the staff are observing safety measures like wearing gloves while contacting patients.

In 2009 and 2010, the EPA investigated an increase observed in reports of adverse reactions to spot-on ectoparasiticides from 2007 to 2008. Cyphenothrin, in addition to other pyrethrins/pyrethroids, was among one of the active ingredients highly represented across reports. Small breed dogs and cats were more commonly reported to experience adverse effects to spot-on products (12). The EPA suggested three explanations for why small animals were more susceptible than large animals to the use of these products. One, the weight range for spot-on treatments might vary too greatly. For example, the dose (2.68 mL) of Parastar® Plus used in the current study was for the weight range of 20.5–40 kg. Dogs on the lower end of the range might be receiving twice the dose compared to those at the higher end. Two, owner misuse may be to blame for the increase of adverse events in smaller animals. In an attempt to save money, owners might purchase a dose meant for large dogs and then divide that dose up among multiple small dogs, resulting in a dose that is still too high. In the case of cats, the EPA found a high number of adverse events associated with owners using pyrethrin/pyrethroid-containing products meant for dogs only, despite warnings on the product labels that the treatment was not to be used on cats. Three, small dogs might be more naturally susceptible to the use of spot-on treatments because of biological characteristics specific to the individual animal, even when the correct dose is used. Therefore, it is important for the veterinarian–client–patient relationship to be maintained, as the EPA has suggested that pet owners who purchase spot-on treatments from their veterinarians receive more guidance for their correct usage.

Veterinarians may also be able to aid in choosing an ectoparasiticide that is in fact effective. Some insects are becoming resistant to certain insecticides. This characteristic is referred to as knockdown resistance (*kdr*) and seems to be prevalent in the case of pyrethroids (13). Some species of lice and ticks have already shown to be resistant to pyrethroids, and research suggests that mutations in the genes that form the voltage-gated Na^+ ion channels are responsible for *kdr*. However, owners might believe that resistance exists when in fact a spot-on treatment does work, especially in the case of fipronil. According to Michael W. Dryden, "It can take time for a flea infestation to be gone, and that has nothing to do with resistance" (14), when speaking of owners claiming that their spot-on treatments that contain fipronil were no longer working. Currently, the practice of using a product having multiple insecticides is common because they provide broad spectrum activity. Ectoparasiticides, such as Parastar® Plus, need to be used as directed by the manufacturer and veterinarians and reapplied at the correct (as indicated earlier) and consistent times (once a month) to obtain the highest efficacy and safety.

Although findings of the current study have shown that the use of Parastar® Plus is safe for both dogs and humans, further investigation is warranted on a larger sample size, dogs of different weight categories, and long-term exposure to Parastar® Plus to observe for any adverse effects. Long-term exposure to Parastar®Plus may result in detectable residue of fipronil and cyphenothrin in dog blood and adverse effects in dogs and greater amounts of residues transferred from dogs to humans. This may require detailed biological monitoring of fipronil and cyphenothrin in dogs as well as humans in our future studies.

Poisonings in dogs and cats due to fipronil or cyphenothrin commonly occur as a result of accidental or intentional ingestion. In humans, poisoning is mainly due to accident or suicide attempt (15–17). In this context, authors are compelled to provide a brief discussion of mechanism of action and toxicity of fipronil and cyphenothrin in this paper. Common clinical signs of fipronil toxicosis are of CNS hyper excitability, including tremors, convulsions, seizures, and death (5). Fipronil is a selective non-competitive blocker of the γ-aminobutyric acid (GABA)-gated chloride channel, manifesting high target specificity between insects versus mammals (18, 19). GABA is an inhibitory neurotransmitter in the insect and mammalian CNS. Following blockage of the GABA-gated chloride channels, hyperexcitation occurs due to abated GABA-induced inhibitory effect, leading to characteristic convulsions, followed by paralysis and death (5). Fipronil-induced clinical signs in humans may include headache, dizziness, fatigue, and itching and irritation of the skin and eyes. Interestingly, a most recent study suggests that a fipronil metabolite fipronil sulfone is more toxic than fipronil itself (20).

Most type I and type II pyrethroids are classified based on the symptoms they produce, but cyphenothrin represents an intermediate pyrethroid. While it is a type II pyrethroid, the symptoms it produces are a combination of whole body tremors (T) and choreoathetosis–salivation (CS) syndrome (21–23). This combination of clinical signs is referred to as TS syndrome. Pyrethroids target the voltage-sensitive sodium (Na^+) ion channels, causing a depolarization-dependent blockage that holds the channel open for an extended period of time, thereby suppressing the action potential (24, 25). Prolonged opening of the Na^+ ion channels in the presence of a type II pyrethroid causes type II syndrome or CS syndrome, characterized by burrowing and pawing behaviors, hypersalivation, and ataxia, that progresses to writhing spasms, tonic seizures, and death (22, 24, 26). Type I pyrethroids are recognized for causing a similar but still distinguishable condition referred to as type I or T syndrome. Symptoms of T syndrome are aggressive sparring, clonic seizures, rigor, and death (22).

While both fipronil and cyphenothrin exhibit high target specificity for insects as compared to mammals, due to differences in the CNS, body temperature, and body size (18, 19, 27, 28), safety concerns dealing with accidental poisoning and exposure to humans exist, especially for people who have higher exposure rates to dogs treated with ectoparasiticides (veterinary professionals, dog trainers, military, and police professionals) compared to the general public. Fipronil is classified as a Group C: possible human carcinogen, which, according to the United States EPA (29), includes "agents with limited animal evidence and little or no human data" upon which to establish carcinogenicity. Fipronil's oral LD_{50} is 97 mg/kg, and its dermal LD_{50} is >2000 mg/kg. Studies strongly suggest that fipronil is an endocrine disruptor (30, 31). Cyphenothrin is grouped into Toxicity Class III: chemicals that represent low toxicity and are labeled with the EPA signal word CAUTION. According to the National Pesticide Information Center (32), chemicals designated under this class are "slightly toxic if eaten, absorbed through the skin, inhaled," or they cause "moderate eye or skin irritation." Cyphenothrin's oral LD_{50} is 318 mg/kg, and its dermal LD_{50} is >5000 mg/kg. These low dermal LD_{50} data suggest that mammalian toxicity *via* skin contact with Parastar® Plus is of low risk in most animals and humans. Biomarkers that could aid in detecting various facets of toxicity of fipronil and pyrethrins/pyrethroids have been described recently by Gupta and Milatovic (33).

CONCLUSION AND FUTURE DIRECTION

Taken together, findings of this study suggests that if used in accordance with manufacturer's instructions and allowed 24 h to dry, Parastar® Plus is a safe topical ectoparasiticide and does not pose a threat to the health of dogs or the people who come into contact with them. Since veterinary personnel have direct physical contact with multiple dogs every day, they should observe the proper use of PPE in order to protect themselves against cumulative exposure to Parastar® Plus, as at this time, long-term studies of transferable residues from the combination of fipronil and cyphenothrin have not been completed. In the present investigation, we did not observe eye or skin irritation; in future studies, we will perform histopathology of dermal biopsy at the site of Parastar® Plus application.

AUTHOR NOTE

Part of the Masters thesis (Katherine M. Case) presented at the 55th Annual Meeting and ToxExpo, New Orleans, LA, USA, March 13–17, 2016. Part of this investigation will also be presented at the 3rd International Veterinary Congress to be held in London, UK, August 18–20, 2016.

AUTHOR CONTRIBUTIONS

KC, major contributor; NV, assistant to KC; RG, thesis advisor; ML, data analysis/graphics; and TC, graduate advisor.

CONFLICT OF INTEREST STATEMENT

The authors declare that the research was conducted in the absence of any commercial or financial relationships that could be construed as a potential conflict of interest.

ACKNOWLEDGMENTS

Sincerest thanks are extended to the staff at Murray State University and the Breathitt Veterinary Center for their efforts in completion of this investigation. Thanks are also extended to the owners and the dogs that participated in this investigation.

FUNDING

This research project was funded by the MSU Breathitt Veterinary Center Foundation.

REFERENCES

1. Current Advice on Parasite Control. *Ectoparasites – Fleas*. Companion Animal Parasite Council (2015). Available from: https://www.capcvet.org/capc-recommendations/fleas
2. *Parastar® Plus for Dogs Brochure*. Greensboro, NC: Novartis Animal Health, U. S., Inc. (2014).
3. *Parastar® Plus for Dogs Product Info Sheet*. Greensboro, NC: Novartis Animal Health, U. S., Inc. (2014).
4. Jennings KA, Canerdy TD, Keller RJ, Atieh BH, Doss RB, Gupta RC. Human exposure to fipronil from dogs treated with Frontline. *Vet Hum Toxicol* (2002) 44(5):301–3.
5. Anadón A, Gupta RC. Fipronil. In: Gupta RC, editor. *Veterinary Toxicology: Basic and Clinical Principles*. Amsterdam: Academic Press/Elsevier (2012). p. 604–8.
6. Ensley SM. Pyrethrins and Pyrethroids. In: Gupta RC, editor. *Veterinary Toxicology: Basic and Clinical Principles*. Amsterdam: Academic Press/Elsevier (2012). p. 591–5.
7. Craig MS, Gupta RC, Canerdy TD, Britton DA. Human exposure to imidacloprid from dogs treated with advantage®. *Toxicol Mech Methods* (2005) 15:287–91.
8. Nichols H, Gupta RC, Doss RB, Bland SD, Canerdy TD, Zieren J. Residue of fipronil, s-methoprene, and amitraz in dog blood and in gloves from topical Certifect® application: toxicity and safety considerations. *J J Vet Sci Res* (2014) 1:1–9.
9. Litchfield H, Gupta RC, Doss RB, Bland SD, Canerdy TD. Safety evaluation of permethrin and indoxacarb in dogs topically exposed to Activyl® tick plus. *J Vet Sci Technol* (2015) 6:218.
10. Driver JH, Ross JH, Holden LR, Selim S, Sharp JK, Carlson D, et al. Cyphenothrin flea and tick squeeze-on for dogs: evaluation of potential health risks based on the results of observational biological monitoring. *J Toxicol Environ Health A* (2015) 78:1105–21.
11. Nerella NG, Block LH, Noonan PK. The impact of lag time on the estimation of pharmacokinetic parameters. One compartment open model. *Pharm Res* (1993) 10(7):1031–6.
12. May K, White-Shim L. *Flea and Tick Treatments: EPA's Investigation of Spot-On (FAQ)*. American Veterinary Medical Association (2009). Available from: https://www.avma.org/public/PetCare/Pages/Flea-and-Tick-Products-EPA-FAQs.aspx
13. Soderlund DM, Knipple DC. The molecular biology of knockdown resistance to pyrethroid insecticides. *Insect Biochem Mol Biol* (2003) 33:563–77.
14. Parker-Pope T. *Vets Face Rising Worry over Fleas*. The New York Times (2014). Available from: http://well.blogs.nytimes.com/2014/09/29/vets-face-rising-worry-over-fleas/?_r=0
15. Gasmi A, Chataigner D, Garnier R, Lagier G. Toxicity of fipronil-containing insecticides. Report of 81 cases from the Paris Poison Center. *Vet Hum Toxicol* (2001) 43:247.
16. Chodorowski Z, Anand JS. Accidental dermal and inhalation exposure with fipronil-a case report. *J Toxicol Clin Toxicol* (2004) 42:189–90.
17. Lee S-J, Mulay P, Diebolt-Brown B, Lackovic MJ, Mehler LN, Beckman J, et al. Acute illnesses associated with exposure to fipronil-surveillance data from 11 states in the United States, 2001-2007. *Clin Toxicol* (2010) 48:737–44.

18. Narahashi T, Zhao X, Ikeda T, Nagata K, Yeh JZ. Differential actions of insecticides on target sites: basis for selective toxicity. *Hum Exp Toxicol* (2007) 26:361–6.
19. Narahashi T, Zhao X, Ikeda T, Salgado VL, Yeh JZ. Glutamate-activated chloride channels: unique fipronil targets present in insects but not in mammals. *Pestic Biochem Physiol* (2010) 97:149–52.
20. Romero A, Ramos E, Ares I, Castellano V, Martínez M, Martínez-Larrañaga MR, et al. Fipronil sulfone induced higher cytotoxicity than fipronil in SH-SY5Y cells: protection by antioxidants. *Toxicol Lett* (2016) 252:42–9.
21. Wright CDP, Forshaw PJ, Ray DE. Classification of the actions of ten pyrethroid insecticides in the rat, using the trigeminal reflex and skeletal muscles as test systems. *Pestic Biochem Physiol* (1988) 30(1):79–86.
22. Wolansky MJ, Harrill JA. Neurobehavioral toxicology of pyrethroid insecticides in adult animals: a critical review. *Neurotoxicol Teratol* (2007) 30:55–78.
23. Anadón A, Martínez-Larraãga MR, Martínez MA. Use and abuse of pyrethrins and synthetic pyrethroids in veterinary medicine. *Vet J* (2009) 182:7–20.
24. Shafer TJ, Meyer DA, Crofton KM. Developmental neurotoxicity of pyrethroid insecticides: critical review and future research needs. *Enivron* (2005) 113(2):123–36. doi:10.1289/ehp.7254
25. Dong K. Insect sodium channels and insecticide resistance. *Invert Neurosci* (2007) 7:17–30.
26. Narahashi T. Cellular and molecular mechanisms of action in insecticides: neurophysiological approach. *Neurobehav Toxicol Teratol* (1982) 4(6):753–8.
27. Ginsburg K, Narahashi T. Time course and temperature dependence of allethrin modulation of sodium channels in rat dorsal root ganglion cells. *Brain Res* (1999) 847:38–49.
28. Narahashi T. Neuroreceptors and ion channels as the basis for drug action: past, present, and future. *J Pharmacol Exp Ther* (2000) 294(1):1–26.
29. United States Environmental Protection Agency. *Risk Assessment for Carcinogens* (2016). Available from: https://www.epa.gov/fera/risk-assessment-carcinogens
30. Leghait J, Gayrard V, Picard-Hagen N, Camp M, Perdu E, Toutain PL, et al. Fipronil-induced disruption of thyroid function in rats is mediated by increased total and free thyroxine clearances concomitantly to increased activity of hepatic enzymes. *Toxicology* (2009) 255:38–44.
31. Roques BB, Lacroix MZ, Puel S, Gayrard V, Picard-Hagen N, Jouanin I, et al. CYP 450-dependent biotransformation of the insecticide fipronil into fipronil sulfone can mediate fipronil-induced thyroid disruption in rats. *Toxicol Sci* (2012) 127:29–41.
32. National Pesticide Information Center. *Signal Words Topical Fact Sheet.* Corvallis, OR: Oregon State University (2008). Available from: http://npic.orst.edu/factsheets/signalwords.pdf
33. Gupta RC, Milatovic D. Insecticides. In: Gupta RC, editor. *Biomarkers in Toxicology.* Amsterdam: Academic Press/Elsevier (2014). p. 389–407.

CHAPTER 18

Nonclinical Safety Evaluation of a Transforming Growth Factor β Receptor I Kinase Inhibitor in Fischer 344 Rats and Beagle Dogs

Anja J Stauber[1]*, Kelly M Credille[1], Lewis L Truex[1], William J Ehlhardt[1] and Jamie K Young[2]

[1] *Lilly Research Laboratories, Toxicology and Pathology, Eli Lilly and Company, Indianapolis, IN, 46285, USA*

[2] *Covance Laboratories Inc., Greenfield, Indiana, 46140, USA*

ABSTRACT

OBJECTIVE: The transforming growth factor β (TGF-β) pathway regulates diverse cellular functions and plays a prominent role in diseases such as cancer, autoimmune disorders and cardiovascular disease. LY2157299 monohydrate (LY2157299) is a potent and selective inhibitor of TGF-β receptor I kinase that is under clinical evaluation for the treatment of advanced cancer.

METHODS: This paper characterizes the toxicity profile of LY2157299 in Fischer 344 rats and beagle dogs for up to six months of daily oral dosing. LY2157299 is well tolerated in the rat and dog for up to one month of daily dosing at doses of 150 and 20 mg/kg, respectively.

RESULTS: In the rat, LY2157299 is well tolerated after three months of 2 weeks on/2 weeks off intermittent dosing schedule at 50 mg/kg. Chronic (≥3 months) oral administration results in multiple target organ toxicities involving the cardiovascular, gastrointestinal, immune, bone/cartilage, reproductive, and renal systems.

CONCLUSION: Defining the appropriate dose and schedule led to a better understanding of how to define safety margins and thus enable the clinical investigation of LY2157299 in cancer patients.

Keywords: TGF-β Inhibitor; Kinase; Drug-associated cardiovascular toxicity; Rats; Dogs

INTRODUCTION

Transforming growth factor β (TGF-β) signaling leads to pleiotropic activation, including regulation of cell growth, migration, and differentiation [1-3]. TGF-β ligands are differentially expressed across a variety of tissues including hematopoietic, cartilage, bone, cardiac, and neuronal tissues in both endothelial and epithelial cells [4]. The ligands TGF-β1, TGF-β2, and TGF-β3 signal through binding to the TGF-β type I and type II Ser/Thr kinase receptors. Upon ligand binding-mediated activation, phosphorylation of the type II receptor activates the Type I receptor resulting in phosphorylation and activation of downstream Smad proteins, which are the intracellular effectors of TGF-β signaling [5].

Because of its pleiotropic role, genetically altered animals can provide useful insights on the relevance of TGF-β signaling during development and normal physiology. Genetic manipulation of the TGF-β signaling pathway in mice demonstrates its crucial role in embryonic development in controlling vasculogenesis, angiogenesis, and inflammation. The phenotypes of TGF-β1, β2, or β3 knockout models all result in high embryonic lethality or mortality shortly after birth. TGF-β1 and TGF-β2 knockouts have vascular, aortic arch, and cardiac septal defects [6,7]. TGF-β1 knockout mice die shortly after birth due to cardiopulmonary complications with severe tissue inflammation [6]. Ablation of TGF-β receptors I and II results in embryonic lethality with vascular and angiogenesis defects [8,9]. The Smad proteins are the downstream signal transducers of TGF-β signaling, and genetic knockdown of key regulators such as Smad 2 and 3 are similarly embryonic lethal or affected mice experience a reduced lifespan with abnormalities consisting of mucosal inflammation, skeletal defects, and colon cancer [10].

Based upon the key role that TGFβ signaling plays in embryonic development and survival as well as its central role in certain cancers and other disease states, inhibition of this important signaling pathway may be anticipated to have adverse consequences on animals.

Blocking TGF-β signaling can be achieved via different mechanisms, including neutralizing antibodies and small molecule inhibitors that interrupt the phosphorylation of the TGF-β receptor I kinase (RIK) [11]. Within the last 5 years, a small number of publications have disclosed the effects of TGF-β inhibition in nonclinical toxicology species. Inhibition of TGF-β signaling results in a physeal dysplasia that is observed in rats administered small molecule inhibitors of TGF-β RIK [12,13]. Rats administered repeated doses of small molecule TGF-β RIK inhibitors develop a valvulopathy that is characterized by inflammation, hemorrhage and stromal hyperplasia [12-14]. Valvulopathy in rats is likely a result of direct inhibition of the TGF-β signaling pathway, rather than an "off-target" effect, because multiple structurally diverse small molecule inhibitors have produced similar effects. In contrast, large molecule inhibitors of this pathway appear to have a somewhat different toxicity profile. Monkeys administered GC1008, a human IgG4 pan-neutralizing TGF-β antibody for six months developed a dose- and time-dependent increase in epithelial hyperplasia of the gingiva and bladder [15]. These nonclinical findings from pharmaceutical companies have led to termination of several small molecule drug development programs targeting the TGF-β signaling pathway based on an apparent lack of acceptable safety margins for clinical testing [16].

Here, we disclose the nonclinical toxicology profile of LY2157299, a potent and selective oral TGF-β RIK inhibitor in Fisher 244 rats and Beagle dogs. This inhibitor is being developed for the treatment of patients with advanced cancer. Administration of LY2157299 was associated with several toxicological findings that involved multiple organ systems including cardiovascular, gastrointestinal, immune, bone/cartilage, reproductive, and renal. Toxicities were dose and duration dependent, some evident within days at high doses, whereas others were only evident in studies of 6 months' duration. Despite these serious adverse nonclinical toxicities, careful evaluation of various dosing regimens provided sufficient data to ensure the safety of patients in clinical trials.

MATERIALS AND METHODS

TEST COMPOUND

LY2157299, (4-[2-(6-methyl-2-pyridyl)-5,6-dihydro-4H-pyrrolo[1,2-b]pyrazol-3-yl]quinoline-6-carboxamide hydrate), a specific and potent inhibitor of TGF-β receptor 1 kinase, was synthesized by Lilly Research Laboratories, a division of Eli Lilly and Company. Dosing suspensions were prepared and stored within established stability parameters in 1% (w/v) carboxymethylcellulose, 0.25% (v/v) Polysorbate 80, and 0.05% (v/v) Dow Corning® Antifoam 1510-US in purified water. LY2157299 monohydrate was prepared in 500mM phosphate buffer (pH 1.8 to 2.2) for administration in dogs.

ANIMALS AND HUSBANDRY

Eight oral nonclinical toxicology studies are described in this manuscript: a 2-week pilot study; 1-month, two 3-month studies, and a 6-month study in rats; two 1-month studies and a 6-month study in dogs. With the exception of the 2-week rat pilot study, all studies were conducted to conform to the United States Food and Drug Administration (CFR 21 – Part 58) and Organization for Economic Cooperation and Development (OECD) Good Laboratory Practice standards in place at the time of study initiation.

Male and female Fischer 344 rats (obtained from Taconic, Germantown, NY) were used for nonclinical toxicity studies up to 6 months in duration. All animals were between 6-11 weeks old at the start of treatment with initial body weights ranging from 100 to 300 grams. Clinically acceptable male and female rats were randomly assigned to dose groups to ensure equivalent distribution of body weights. Rats were housed individually in stainless steel cages and offered Certified Rodent Diet and were provided water ad libitum. Animals were fasted overnight prior to euthanasia.

Beagle dogs obtained from Marshall Farms (North Rose, NY) for the three dog studies were approximately 7-10 months of age at the start of treatment. Clinically acceptable male and female dogs were randomly assigned to dose groups to ensure equivalent distribution of body weights. Dogs were individually housed in stainless steel cages with suspended mesh floors and offered Certified Canine Diet and were provided water ad libitum. Animals were fasted overnight prior to euthanasia.

In all studies, animal care and use were done in accordance with federal and local laws, policies, regulations, and standards in effect at the time of their conduct, e.g., Animal Welfare Act and Regulations. Laboratories conducting these studies were accredited by the Association for Assessment and

Accreditation of Laboratory Animal Care International and all study protocols were approved by each laboratory's Institutional Animal Care and Use Committee.

Table 1: Nonclinical Studies Conducted with LY2157299.

Species	Duration (weeks)	Recovery (weeks)	Dosing Regimen	N (sex/group)	Doses (mg/kg)
Rat	2	None	Daily	5	0, 50, 300, 1200
	4	4	Daily	10	0, 15, 50, 150
	12	None	Daily	10	0, 50, 150, 250
	12	None	2 weeks on 2 weeks off	10	0, 50, 150, 250
	8	6	Daily	10 (control) 35 (treated)	0, 250
	26	None	Daily	15	0, 50, 150, 250
Dog	4	4	Daily	3	0, 50, 250, 500
	4	None	Daily	3	0, 2, 8, 20
	26	None	Daily	4	0, 8, 20, 60

EXPERIMENTAL PROCEDURES

Nonclinical toxicology studies were conducted in Fisher 344 rats and Beagle dogs. Table 1 describes the study designs for all studies conducted with LY2157299 reported in this paper. Rats were administered control vehicle or compound orally via gavage, while dogs were administered control vehicle or LY2157299-filled capsules. Plasma for measuring systemic exposure were taken on the first day of dosing and on the last day of dosing in all studies in addition to periodic intervals in studies lasting 3 months or longer. Plasma was analyzed for LY2157299 concentration using the same validated liquid chromatography/mass spectrometry method used for analysis in human plasma [17], except the method was validated over the range of 0.5-500 ng/mL in rat and dog plasma.

Animals were observed at least once daily for mortality and clinical signs; body weights and food consumption were analyzed weekly. Electrocardiograms (ECGs) were collected at pretest in order to screen for preexisting ECG abnormalities and to provide baseline ECG data for later comparisons. During the live phase, ECGs were collected before dosing and after dosing during the first week of treatment and at 1 month and near the end of the reversibility phase (for 1-month study containing reversibility). For the 6-month study, ECGs were also collected at 3 months and 6 months.

Blood samples for complete blood counts, coagulation, and serum biochemical assays were collected from rats and dogs at the end of the treatment and/or reversibility phases. In the dog, these analyses were evaluated prior to initiation of treatment, around the start of the study and at other time points during the study. Cardiac troponin I was also evaluated in some studies.

Urine samples were collected by cystocentesis at necropsy. Urine was collected from rats overnight using metabolism cages at the end of treatment in the 6-month study and from a subset of rats in the 1-month study. Samples were analyzed for appearance/color, microscopic examination of sediment, bilirubin, blood, glucose, ketones, pH, protein, specific gravity, urobilinogen, and volume.

Rats were anesthetized with isoflurane or carbon dioxide inhalation, euthanized by exsanguination and necropsied. Dogs were administered a barbiturate overdose, exsanguinated and necropsied. Necropsies were conducted in all rats and dogs and any observations were recorded. Tissues collected at necropsy for microscopic evaluation included adrenals, aorta, bone marrow smear, bone with marrow, brain, brown fat (interscapular), cecum, cerebellum with pons or medulla oblongata, cerebrum, epididymides, esophagus, eyes with optic nerves, femur with knee joint, gastrointestinal tract (cecum, colon, duodenum, esophagus, ileum, jejunum, stomach), gross lesions, heart, kidneys, liver, lungs (including bronchi), lymph nodes (mandibular and mesenteric), mammary gland, ovaries with oviducts, pancreas, sciatic nerve, pituitary, prostate, salivary glands, seminal vesicles, skeletal muscle (rectus femoris), skin, spinal cord (cervical), spleen, sternum, testes, thymus, thyroids with parathyroid, tongue, trachea, urinary bladder, uterus with cervix, and vagina. Tissues were preserved in 10% buffered neutral formalin, trimmed, processed routinely, and embedded in paraffin. Paraffin blocks were microtomed and sections stained with hematoxylin and eosin (H&E). The trimming procedure for hearts consisted of longitudinally bisecting these along a plane perpendicular to the plane of the pulmonary artery to expose the right atrioventricular, left atrioventricular, and aortic valves. Each hemisection was embedded in paraffin with the cut surface down. In addition to H&E stained sections, some sections of heart immunostained for alpha-smooth muscle actin (a-SMA) using standard immunohistochemical methods. The tissue sections were examined by light microscopy by board-certified members of the American College of Veterinary Pathologists (ACVP). All studies were peer reviewed by ACVP-certified veterinary pathologists [18].

RESULTS

LY2157299 was dosed to Fischer 344 rats and Beagle dogs for up to 6 months (Table 1). Daily oral administration was the dose schedule in the rat and the dog except for a study in the rat in which an intermittent dose schedule of 2 weeks on/2 weeks off was investigated. LY2157299 was well tolerated in the rat as a single dose up to 2000 mg/kg (data not shown). In repeat dose studies in the rat, LY2157299 caused mortality at progressively lower doses as the duration of treatment increased. Mortality was observed in rats administered 1200 mg/kg of LY2157299 in a 14-day study. In the 3- and 6-month studies, mortality occurred in animals receiving daily doses of 150 mg/kg (6-month study) and 250 mg/kg (3- and 6-month study) beginning on days 83 and 54, respectively. At a dose of 1200 mg/kg in rats, a gross change of sternal deviation consistent with acquired pectusexcavatum occurred, rarely with concurrent rib deformities, as early as nine days after initiation of dosing.In the longer term (>3 months) rat studies, most of the deaths that occurred prior to the schedule termination were attributed to compound-related inflammation and rupture of the aorta and/or distributing arteries at or near the base of the heart resulting in hemothorax and/or hemoabdomen.

LY2157299 was evaluated in beagle dogs up to 500 mg/kg. All animals tolerated LY2157299 for the duration of the study with the exception of a single animal administered 500 mg/kg which was euthanized shortly following the end of the treatment period in a 1-month study due to deteriorating condition. This animal was found to have severe necrotizing cholecystitis.

LY2157299 is well absorbed following oral dosing and exposures appear to be dose linear in the rat and dog. In rats, exposures in females are higher than in males and are consistent with the toxicity data in which female rats are more affected than males. No exposure differences are observed between sexes in the beagle dog.

CARDIOVASCULAR TOXICOLOGY

In both rats and dogs, the heart and great vessels were identified as the major target organs for toxicity. Cardiovascular findings in F344 rats treated with LY2157299 included degenerative and inflammatory valvular lesions (valvulopathy), myocardial degeneration and necrosis, aortitis with rupture, vasculitis/perivasculitis, and increased heart weights. The cardiovascular lesions were dose- and duration-dependent in rats and dogs, respectively (Figures 1 and 2). These cardiovascular lesions were not accompanied by chronic pulmonary or hepatic congestion that would have been suggestive of congestive heart failure, even after 6 months of daily dosing.

VALVULOPATHY

The earliest microscopic changes included heart valve lesions and myocardial degeneration at the insertion point of the valves which occurred in less than 2 weeks in the rat at a 1200-mg/kg dose and was also observed in the 3-month and 6-month daily dosing studies at =50 mg/kg. No cardiac lesions were observed following 1 month of daily dosing study in the rat at doses of 15 mg/kg, 50 mg/kg, and 150 mg/kg. In the dog, valve lesions were observed following one month of daily dosing at =50 mg/kg or at doses as low as 8 mg/kg following 6 months of daily dosing.

In both species, the valvulopathy was characterized by valve thickening, hemorrhage, hemosiderin-laden macrophages, inflammation, stromal and endothelial hyperplasia and increased myxomatous matrix (Figure 3). The inflammation was typically a mixture of lymphocytes and macrophages, but occasionally included neutrophils. The stromal hyperplasia and valve thickening were similar to spontaneous mitral valvulopathy [19], but occurred at a high incidence in the rat and were usually associated with other lesions such as hemorrhage or inflammation. Both atrioventricular and semilunar valves had lesions, and often multiple valves were affected in a given animal. Increased smooth muscle actin (SMA) immunolabeling of the valve consistently occurred when light microscopic lesions were present and indicated valvular interstitial cell activation with transdifferentiation into myofibroblast-like cells. However, because increases in SMA expression were always concurrent with morphologic changes, this finding was not useful as a premonitory marker for the morphologic changes (data not shown).

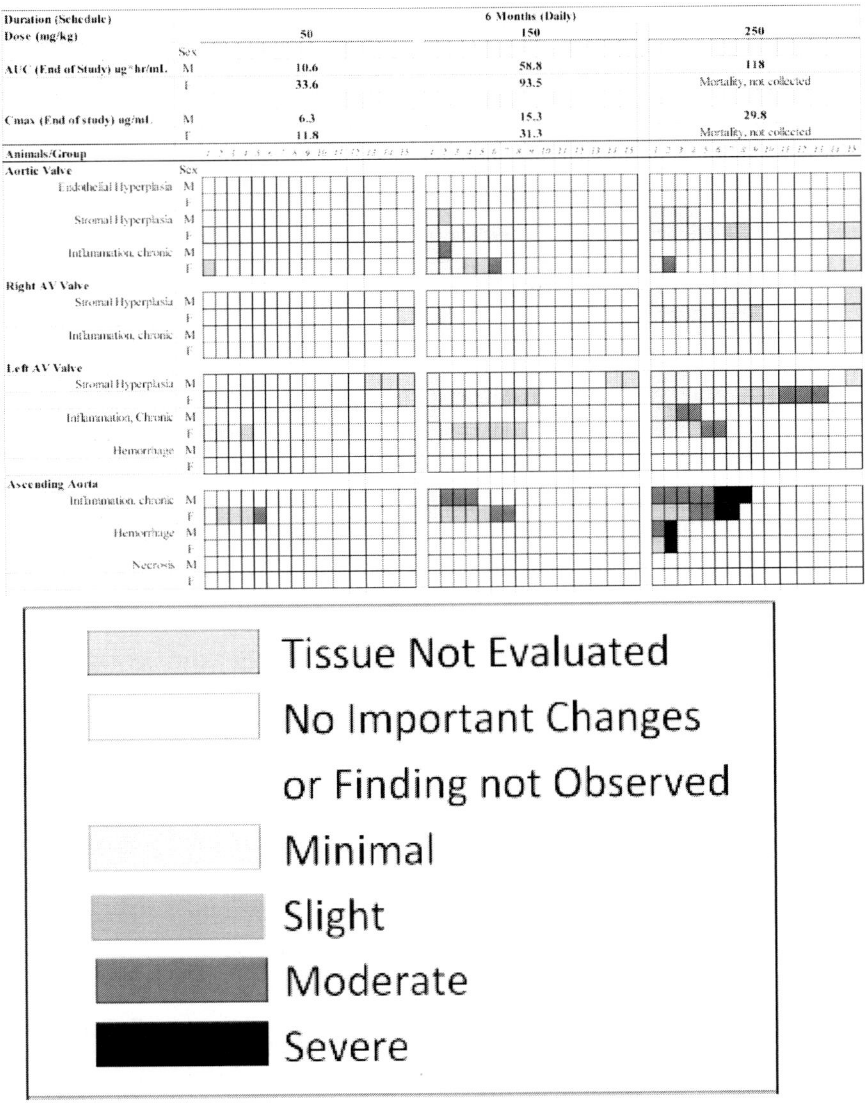

Figure 1: Incidence and severity of cardiac lesions in rats following daily administration of LY2157299 for six months. Exposure was determined at the end of the study in a group of satellite animals receiving compound for six months.

The incidence, severity, and chronicity of valvulopathy were similar in the 3- and 6-month daily dosing studies in rats at all dose levels. In the dog at the end of the 6-month dosing period, most animals in the 20 mg/kg and 60 mg/kg dose groups had at least one valve affected, whereas the 8-mg/kg dose group had fewer affected animals. The valvular lesions were still present in the rat and dog following a treatment-free recovery period. A partial recovery in the rat was observed when rats were dosed with 250 mg/kg LY2157299 daily for 2 months, a dose known to produce virtually 100% incidence of valvular lesions. Rats allowed a 6-week recovery period still had lesions present; however, there was a decrease in severity and incidence, indicating partial reversibility. A chronic end stage lesion characterized by scarring and valvular insufficiency was not recognized microscopically, although valvular function was not assessed in vivo. This could suggest that the onset of the valvular injury occurred at different time points throughout the dosing period.

VASCULAR LESIONS

Vascular lesions occurred when dogs and rats were administered LY2157299 at doses of =8 and =50 mg/kg in the dog and rat, respectively, for prolonged time periods (=3 months). In the rat, the earliest lesions were typically periarteritis and intramural hemorrhage that progressed to transmural inflammation, necrotizing vasculitis, and vascular rupture (Figure 4). Vascular lesions were most prominent in the thoracic cavity and involved the aortic arch, coronary arteries, and other distributing arteries originating from the aorta in the thoracic and cranial abdominal cavities. Lesions were often pronounced at the base of the aorta and where distributing arteries branched from the proximal aorta, a site of high pressure and increased turbulence. In a subset of rats, severe degenerative and inflammatory changes were associated with vascular rupture resulting in fatal hemothorax or hemoabdomen. The intramural hemorrhage was compatible with aortic dissection, although vascular dilation and aneurysms were not apparent. There were no lesions in the standard section of thoracic descending aorta.

Changes in the dog aorta did not occur in the 1-month studies but were noted in the 6-month study. While the valvular lesions in the dog were similar to those described in the rat, the aortic vascular change differed. The changes in the base of the ascending aorta of dogs were characterized by focal to multifocal degeneration and disorganization of the mural elastic lamellae, increased prominence of mucopolysaccharide-rich ground substance, but without accompanying inflammation or intramural hemorrhage (Figure 5). In the most affected animals, irregular separation between the elastic layers occurred. There were no findings in the standard section of thoracic descending aorta. Although the microscopic changes likely compromised aortic structural integrity, there were no changes diagnostic of aneurysm.

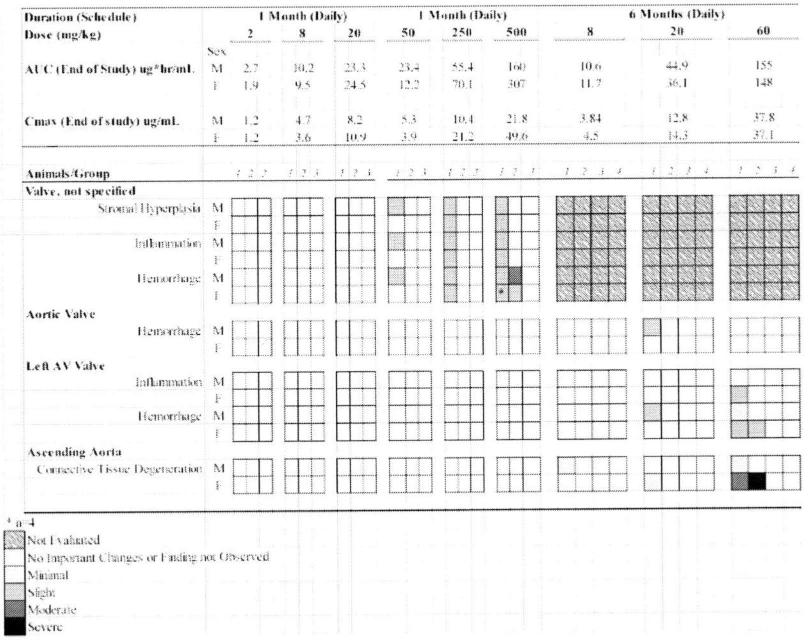

Figure 2: Incidence and severity of cardiac lesions in dogs following daily administration of LY2157299 for one and six months. Exposure was determined at the end of the study.

ADDITIONAL CARDIAC FINDINGS

Myocardial degeneration and necrosis occurred in rats that had the most pronounced valvular and/or aortic changes, and were interpreted as secondary to the valvular and coronary artery changes rather than a primary test article effect.

Multifocal myocardial degeneration and necrosis typically involved the base of the heart, at the base of the valves, and the ventricles, and was generally minimal in severity. Microscopic cardiovascular changes were rarely associated with increases in cardiac troponin I concentrations, most likely due to timing of the serum collection and the relatively mild myocardial degenerative changes.

LY2157299 induces slight-to-moderate decreases in blood pressure and increases in heart rate in dogs. These changes were not considered adverse. No biologically relevant change in QT/QTcf, test-article related arrhythmia, or altered waveforms were identified.

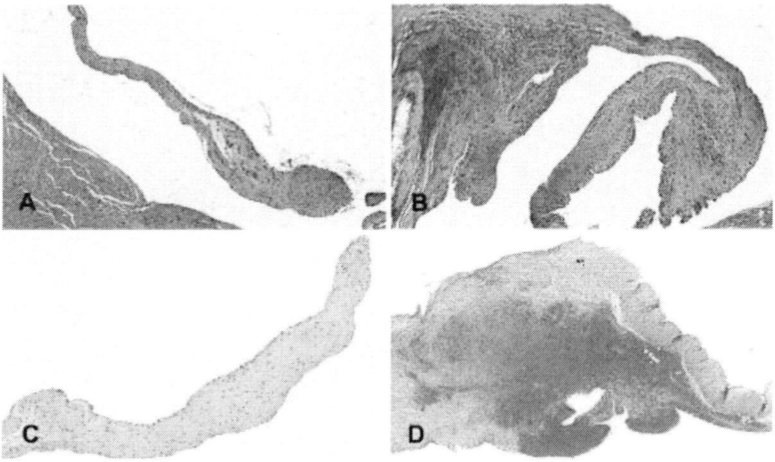

Figure 3: Hematoxylin and eosin stain of atrioventricular heart valves (A) Control rat demonstrating some mild spontaneous myxomatous change at the distal end of the valve. (B) Rat dosed for 3 months with LY2157299. A chronic valvulopathy characterized by expansion of the valve leaflet with increased myxomatous matrix, inflammatory cells, and hemorrhage. (C) Control dog. (D) Valve from a dog dosed for 1 month with LY2157299. An acute valvulopathy characterized by marked expansion due to hemorrhage, edema, and inflammatory cells.

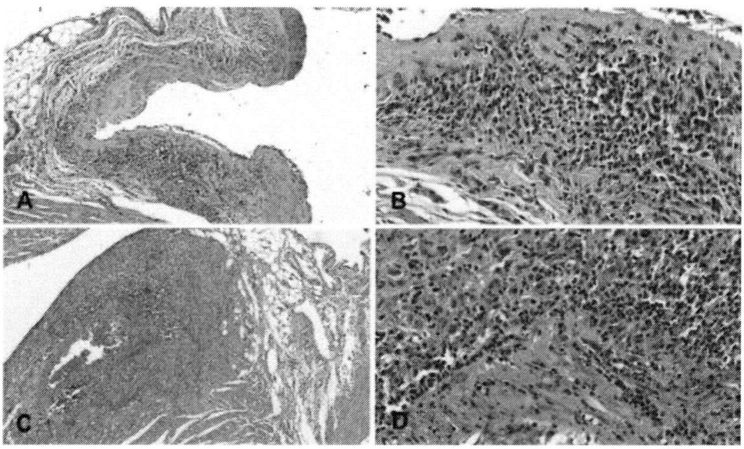

Figure 4: (A and B) Low- and high-magnification images (hematoxylin and eosin stain) of the ascending aorta including the origin of the coronary artery from a rat dosed for 14 weeks with LY2157299. The origin of the coronary artery has disruption of the tunica media with transmural infiltration by mixed inflammatory cells. (C and D) Low- and high-magnification images from a rat dosed for 6 months with LY2157299. The coronary artery has a pronounced perivascular infiltration of mixed inflammatory cells with only limited involvement of the tunica media.

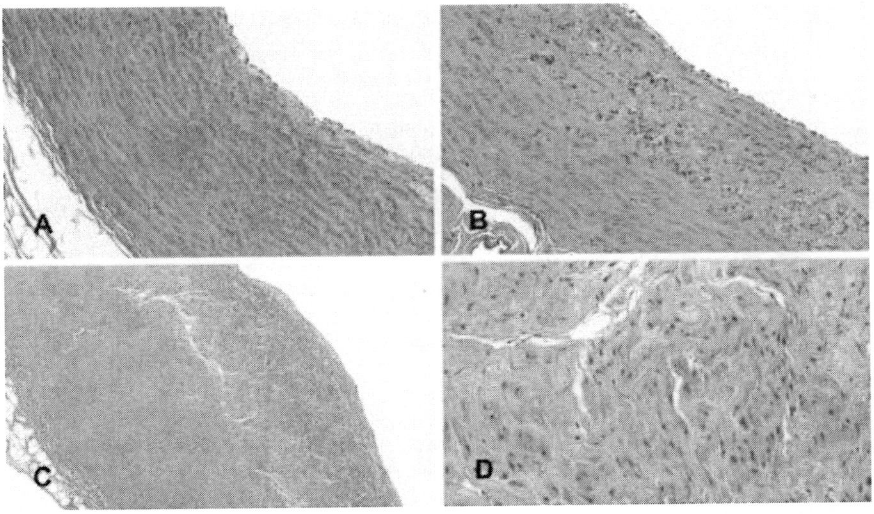

Figure 5: Ascending aorta from dogs dosed for 6 months with LY2157299 (hematoxylin and eosin). (A) Control dog. (B) Focally extensive disruption of the internal elastic laminae within the tunica media. (C and D) Marked disruption of the tunica media resulting in disorganization with loss of the laminar arrangement,increased prominence of mucopolysaccharide-rich ground substance, and cleft formation.

CHARACTERIZING A NON OBSERVED EFFECT LEVEL FOR CARDIAC LESIONS

In order to support the clinical development of LY2157299 with the cardiovascular changes described above, an additional nonclinical toxicology study was conducted in rats to determine if a no observed effect level (NOEL) for cardiovascular changes could be identified using an alternative dosing schedule consisting of 2 weeks on treatment followed by 2 weeks off treatment.

The duration of this study was 3 months and compared daily dosing to the alternative dosing schedule. This duration supports the clinical time interval relevant to patients who are enrolled and then reassessed for disease status 2 months after receiving study drug. Based on the previous 6-month study in rats that identified when early mortality occurred due to cardiac events, a 3-month treatment duration was considered sufficient to induce a high incidence of cardiac lesions with a daily dosing schedule. The doses used for this study were 50 mg/kg, 150 mg/kg, and 250 mg/kg. As expected, mortality due to vascular rupture was observed in rats administered 250 mg/kg daily. In contrast, the rats on the 2-weeks-on/2-

weeks-off dosing schedule at all doses survived until the scheduled termination of the study.

Rats administered 150 mg/kg and 250 mg/kg on a 2-weeks-on/2-weeks-off dosing schedule had cardiovascular lesions similar to those in the daily dosing groups, however, the incidence and often severity were lower in rats administered LY2157299 on the intermittent dosing schedule (Figure 6).

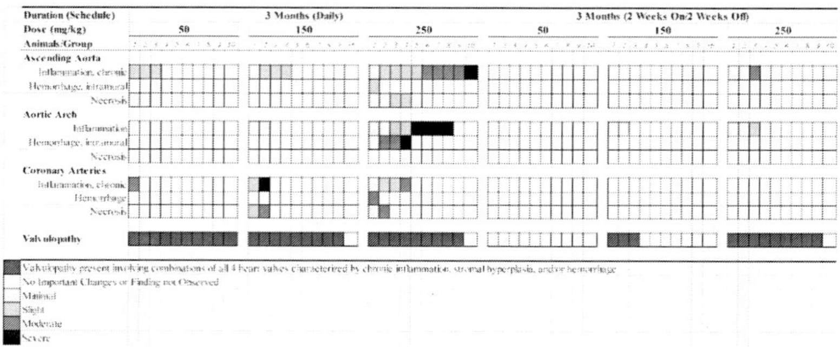

Figure 6: Comparison of incidence and severity of the cardiac lesions in female Fischer 344 rats given LY2157299 in two different dosing schedules for three months.

IMMUNE SYSTEM

Numerous changes occurred in multiple organs of the rat and dog that were consistent with altered immune responses, both proinflammatory and also suggestive of immune dysfunction. These patterns occurred most often in organs such as the gastrointestinal tract and skin that are exposed to environmental antigenic stimulation and host commensal bacteria.

Increased mixed inflammatory cell infiltrates within the lamina propria of the large intestines, and less commonly the small intestines and stomach, occurred in studies of at least 1 month in duration. Additionally, increased inflammatory cell infiltrates occurred in the kidneys, lung, gallbladder, and/or prostate gland depending on the species. Rats in the 6-month toxicity study also had multiple subcutaneous abscesses, and acute inflammation involving the preputial and clitoral glands, suggesting immune dysfunction and/or alterations in the innate immune responses.

GASTROINTESTINAL SYSTEM

Gastrointestinal changes frequently involved the large intestine, but at higher doses also included the small intestines, stomach, and gallbladder. Findings in the large intestine were most apparent in the 6-month studies and included mucosal inflammation in dogs and rats, and mucosal epithelial hyperplasia and neoplasia in rats. In the cecum of the rat, a continuum was identified with numerous rats showing increased mononuclear inflammatory cell infiltrates within the lamina propria of the mucosa, and fewer animals displaying enterocyte hyperplasia, and adenoma formation. In the colon, fewer rats had increased mucosal inflammation, possibly because of higher background infiltrates; however, some rats had epithelial hyperplasia and adenocarcinoma (Figure 7). Some of the adenocarcinomas were quite large and were the cause of adverse clinical signs, early euthanasia and/or death. Similar inflammatory and hyperplastic changes also involved the rectum.

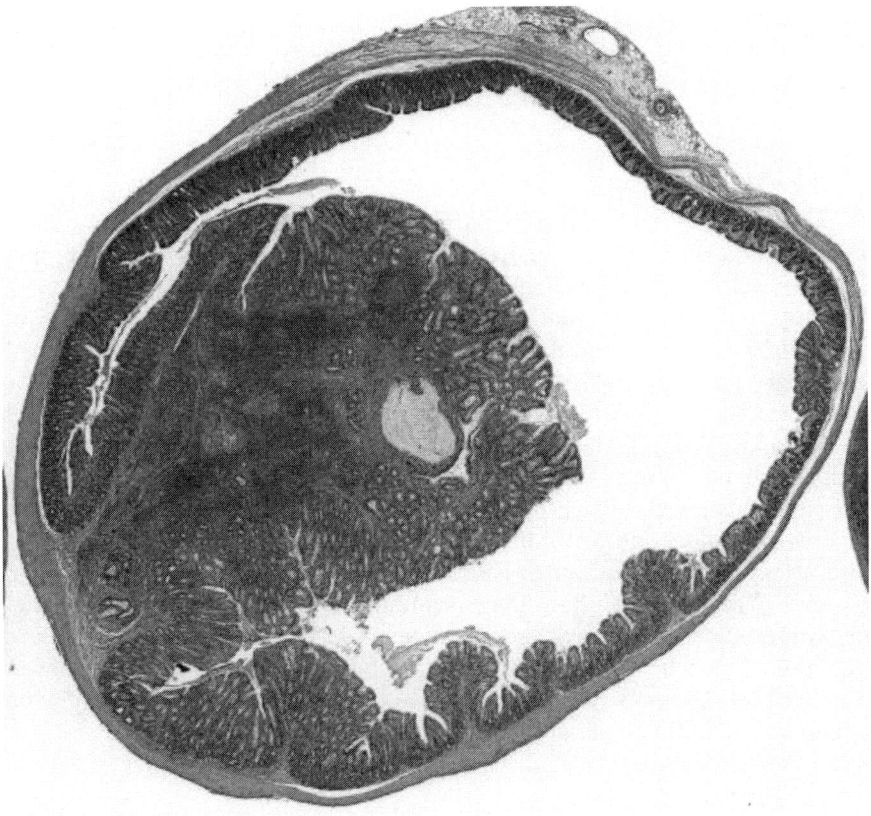

Figure 7: Adenocarcinoma of the colon from a rat that was euthanized after 23 weeks of daily dosing with LY2157299 (hematoxylin and eosin stain).

Dogs in the 1-month toxicity study at high doses (=50 mg/kg) had acute intestinal changes with more pronounced lesions in the small intestine than the large intestine. These microscopic findings included inflammation and hemorrhage, and increased numbers of mucosal cysts. These dogs also had inflammation of the gallbladder that in one dog included necrotizing cholecystitis with rupture that resulted in bile peritonitis.At the highest dose, some dogs had hepatic inflammation and hepatocellular degeneration, changes that may have been secondary to the intestinal changes.Intestinal lesions in dogs in the 6-month toxicity study were uncommon and limited to mild inflammatory cell infiltrates within the large intestine that ranged from acute to chronic, and sometimes included increased numbers of eosinophils and/or plasma cells. These dogs also had chronic inflammation of the gastric mucosa with glandular atrophy and vasculitis that was most prominent in the cardia.Increased amounts of mucus within the gallbladder occurred in some dogs.

SKELETAL SYSTEM

Osseous changes included physeal and subphyseal changes in the femur, tibia, and sternum of rats that occurred after just 14 days of dosing and were similar to those previously described with TGF-β inhibition [12,13]. The zones of maturation and hypertrophy were expanded in the physes of the long bones and endplates of the sternabrae, and the primary and secondary spongiosa were denser with increased connectivity (Figure 8). These changes were orderly and appeared to result in relatively normal endochondral ossification. In animals that had a recovery phase, these osseous changes were characterized by a band of metaphyseal hyperostosis (increased trabecular density) that was separated from the physis by a zone of normal endochondral ossification, presumably the result of an earlier alteration in endochondral ossification that had reversed. Microscopic osseous changes were often more pronounced in rats that had the pectus excavatum defect. Similar physeal changes occurred in the sternebrae of the dog after 6 months of dosing. In addition to the physeal lesions, rats dosed for 6 months had additional changes of multifocal cartilaginous degeneration in the joints of the sternebrae and stifle.

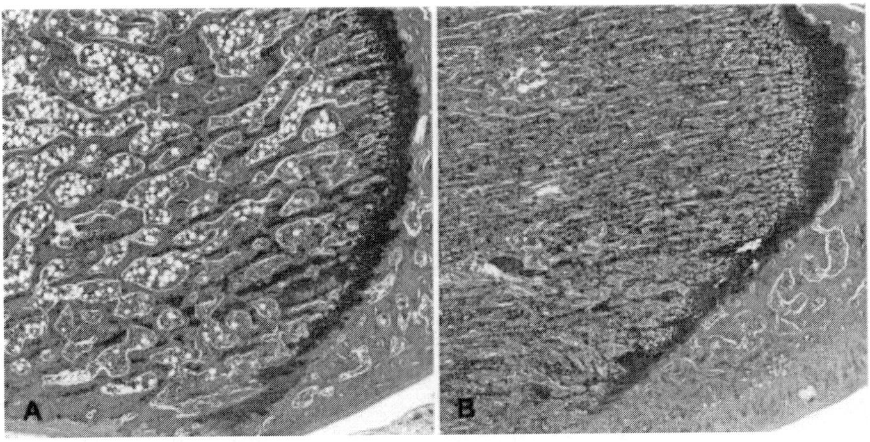

Figure 8: Proximal femur including the physis (hematoxylin and eosin stain). (A) Control rat. (B) Rat dosed for 6 months with LY2157299 had expansion of the physeal zones of maturation and hypertrophy, increased density and interconnectivity of the subphyseal primary and secondary spongiosa, and an orderly increase in the trabecular bone density.

DISCUSSION

TGF-β plays a key role in several disease states, including cancer, cardiovascular disease, and fibrosis, which has made the TGF-β signaling pathway a logical and attractive target of pharmaceutical development. However, TGF-β signaling has posed a challenge with respect to understanding safe therapeutic doses that do not confer toxicities associated with prolonged inhibition of this critical pathway [16]. We characterized the toxicity profile of LY2157299 in nonclinical species for up to 6 months of daily dosing. LY2157299 was well tolerated in the rat and the dog for one month of daily dosing and in the rats for 3 months of intermittent dosing. Long-term daily dosing of =3 months caused alterations in the gastrointestinal, immune, musculoskeletal, renal and cardiovascular systems of animals, which is anticipated based on the integral role for TGF-β signaling for maintaining homeostasis.

Other pharmaceutical companies have published similar valvular toxicities with inhibitors from their development efforts, indicating that the lesions described are likely an on target effect [12,13]. Many explanations have been considered to understand mechanism of this phenomenon; no definitive explanation has been provided at this time [12]. TGF-β is a required mediator for valve structural and functional homeostasis [20]. In response to injury, valvularmyofibroblasts are activated by TGF-β and results in increases in a-SMA expression [21]. This paradoxical inhibition of TGF-β signaling with LY2157299 -treated animals resulting in increased SMA in their heart valves is not understood and awaits further clarification.

We have characterized cardiac effects in nonclinical species that extend beyond the previously reported valvulopathy and, in contrast to the valvulopathy, constituted a pathologic lesion defined as dose limiting within the context of a given study. In both species, the aorta and arteries of the heart are major target organs for toxicity. The vascular lesions in the rat are characterized by inflammation and hemorrhage with associated disruption of the mural organization, sometimes resulting in vessel rupture and hemorrhage into the thoracic cavity. In contrast to the rat, the dog showed changes in the base of the ascending aorta, characterized by degeneration and disorganization of the mural elastic lamellae and often increased prominence of mucopolysaccharide-rich ground substance without accompanying inflammation. The reason for this apparent species difference is unknown.

Our findings support that TGF-β plays a pleiotropic role in cell growth and on the extracellular matrix. It is important in both production of matrix proteins and the degradation of the matrix [22]. Two well-studied genetic diseases Marfan syndrome (MFS) and Loeys-Dietz syndrome (LDS) affect the skeletal, ocular, and cardiovascular systems [23]. Clinically, these syndromes present with overlapping symptoms including aortic aneurysms, dissections and mitral valve prolapse [24]. MFS is caused by mutations in the gene encoding fibrillin-1 that interacts with latent TGF-β binding protein to upregulate TGF-β bioavailability and activity [25]. LDS is an autosomal dominant aortic aneurysm syndrome caused by mutations in the promoter of TGF-β R1 and R2 with increased TGF-β activity [23,26]. In LDS, the microscopic aortic lesion is characterized by fragmentation of elastic fibers, loss of elastin content, and accumulation of amorphous matrix components in the aortic media, changes very similar to those observed in the 6-month dog study with LY2157299 [23]. In cardiac tissue from affected patients and mouse models of MFS, TGF-β signaling is increased, potentially due toaltered receptor processing or alternative pathways responsible for upregulation of TGF-β [27]. Mutations in the TGF-β signaling pathway that should lead to decreased TGF-β signaling resulting in an upregulation of this pathway underscores that in vivo, precise regulation of multiple members is required to maintain homeostasis and that canonical and non-canonical signaling pathways play an important role [28,29].

The findings described in the aorta and cardiac arteries in the toxicology studies in the rat and the dog, based on similarity with genetic syndromes, appear to be more consistent with an upregulation rather than inhibition of the TGF-β pathway. The contributions of the non-canonical TGF-β signaling are not yet understood. Activation of the ERK pathway contributes to progression of aneurysms in *Fbn1*mice. RDEA119, a selective MEK inhibitor, reduced aortic root growth independent of increased *Smad2* activation, suggesting that non-canonical pathway modulation may contribute to progression to aneursyms in *Fbn1*mice [28]. As with MFS and LDS, which are genetic mutations resulting in a clinical phenotype, additional genetic differences may also play a role in susceptibility to cardiac injury observed ·in nonclinical studies with inhibitors of TGF-β signaling. Behmoaraset al. evaluated seven genetically different strains in rats for their differences in composition of the extracellular matrix and noted a differentiation among strains with respect to matrix composition and occurrence of internal elastic lamina rupture [30].

The changes to the heart following long-term treatment of LY2157299 at toxicologically relevant doses are of largest concern with respect to clinical safety, inhibition of TGF-β signaling resulted in changes to several other organ systems in both the rat and dog. LY2157299 resulted in changes to the bone similar to previous reports [12,13]. Reversibility of bone changes was assessed following the 1-month study in rats revealing a band of increased trabecular density that was separated from the normal physis by a zone of normal endochondral ossification. Following 6 months of treatment in the rat, degeneration in the joints of the sternebrum and stifle was reported. Dominant negative mutant TGF-β receptor II mice are reported to have bifurcated xiphoid process and sternum as well as a progressive osteoarthritis-like disease [31]. Dogs used in the 6-month study were between 6-8 months of age and are expected to still have active longitudinal growth in their bones; therefore, alterations in the endochondral cartilage similar to what was described in the rat may be anticipated [32].

TGF-β also plays a critical role in epithelial biology, acting as a tumor suppressor or promoter [33]. Alteration in TGF-β signaling is common in many cancers [34]. Disruption of TGF-β signaling in either epithelial or stromal cells increases inflammatory responses that promote tumor initiation, progression, and metastasis [35]. *TGFβ1-/- Rag2 -/-* mice develop inflammation-associated adenomas and carcinomas through an inability to maintain epithelial tissue organization [36]. Mice genetically deficient in TGF-β R2 signaling have increased susceptibility to cancer [37]. Similarly, inactivation of downstream proteins *Smad3*and *Smad4* results in carcinomas of the intestine [38,10]. The observations of hyperplasia, adenoma, and carcinoma of the intestine in rats treated with LY2157299 for 6 months are consistent with the well-characterized biology.

The use of intermittent dosing is common practice in oncology: clinicians will reduce the dose or permit a drug holiday to allow the patient to recover from toxicities prior to receiving another dose. Studies have shown that anticancer agents were more tolerable and efficacious when administered in intermittent dosing schedules [39,40]. In contrast to a dosing schedule based on clinical findings, clinical dosing with LY2157299 proactively applied an indirect model to support a clinical dosing regimen that would support a desired pharmacologic effect and avoid unwanted toxicity. Buenoet al. have been able to relate plasma concentrations to pSmad inhibition [41]. Tumor growth inhibition was then linked to pSmad inhibition, providing a tool to titrate the dose required for efficacy. Furthermore, the data support the concept of intermittent dosing in maintaining a similar tumor response. The ability to maintain plasma concentrations required for efficacy lower than those observed to cause adverse toxicities in nonclinical species is important, especially when considering drugs targeting signaling pathways such as TGF-β that have been demonstrated to be critical in maintaining normal function. We demonstrate continuous administration of LY2157299 at high doses or durations =3 months in nonclinical species results in adverse toxicities to numerous organ systems including cardiovascular, gastrointestinal, immune, bone/cartilage, reproductive, and renal. The cardiovascular system appears to be most sensitive to abrogation of TGF-β signaling with the small molecule inhibitor LY2157299. Our data have defined a NOEL for cardiovascular lesions of 150 mg/kg and 20 mg/kg, in rat and dog, respectively, after one month of daily dosing and 50 mg/kg for 3 months on a 2-week-on/2-week-off dosing schedule in the rat. An intermittent dosing paradigm demonstrated that the severity and incidence of cardiac lesions could be lessened. The data from the 3-month intermittent dosing study in rats along with the absence of cardiovascular lesions in the 1-month rat and dog study established a nonclinical data package that supports a clinical dosing schedule of 2 weeks on/2 weeks off LY2157299. Successful development of such inhibitors will need to have a thorough characterization of nonclinical toxicity and an understanding of schedule and plasma concentrations needed to demonstrate anti-tumor effect and determine if a sufficient therapeutic window exists to test the clinical efficacy of molecules involved in inhibiting key pathways.

FUNDING

This work was supported by Eli Lilly and Company, Indianapolis, IN, USA.

ACKNOWLEDGEMENTS

We would like to thank Drs. Michael Lahn, Marcus Andrews, Kyla Carroll, and Armando Irizarry for the critical review of this manuscript. We are thankful to Dr. DurisalaDesaiah for writing assistance of this manuscript.

REFERENCES

1. Siegel PM, Massagué J (2003) Cytostatic and apoptotic actions of TGF-beta in homeostasis and cancer. Nat Rev Cancer 3: 807-821.
2. Akhurst RJ, Hata A (2012) Targeting the TGFÎ² signalling pathway in disease. Nat Rev Drug Discov 11: 790-811.
3. Massagué J, Blain SW, Lo RS (2000) TGFbeta signaling in growth control, cancer, and heritable disorders. Cell 103: 295-309.
4. Millan FA, Denhez F, Kondaiah P, Akhurst RJ (1991) Embryonic gene expression patterns of TGF beta 1, beta 2 and beta 3 suggest different developmental functions in vivo. Development 111: 131-143.
5. Feng XH, Derynck R (2005) Specificity and versatility in tgf-beta signaling through Smads. Annu Rev Cell Dev Biol 21: 659-693.
6. Kulkarni AB, Ward JM, Yaswen L, Mackall CL, Bauer SR, et al. (1995) Transforming growth factor-beta 1 null mice. An animal model for inflammatory disorders. Am J Pathol 146: 264-275.
7. Sanford LP, Ormsby I, Gittenberger-de Groot AC, Sariola H, Friedman R, et al. (1997) TGFbeta2 knockout mice have multiple developmental defects that are non-overlapping with other TGFbeta knockout phenotypes. Development 124: 2659-2670.
8. Larsson J, Goumans MJ, Sjöstrand LJ, van Rooijen MA, Ward D, et al. (2001) Abnormal angiogenesis but intact hematopoietic potential in TGF-beta type I receptor-deficient mice. EMBO J 20: 1663-1673.
9. ten Dijke P, Arthur HM (2007) Extracellular control of TGFbetasignalling in vascular development and disease. Nat Rev Mol Cell Biol 8: 857-869.
10. Zhu Y, Richardson JA, Parada LF, Graff JM (1998) Smad3 mutant mice develop metastatic colorectal cancer. Cell 94: 703-714.
11. Li HY, McMillen WT, Heap CR, McCann DJ, Yan L, et al. (2008) Optimization of a dihydropyrrolopyrazole series of transforming growth factor-beta type I receptor kinase domain inhibitors: discovery of an orally bioavailable transforming growth factor-beta receptor type I inhibitor as antitumor agent. J Med Chem 51: 2302-2306.
12. Anderton MJ, Mellor HR, Bell A, Sadler C, Pass M, et al. (2011) Induction of heart valve lesions by small-molecule ALK5 inhibitors. ToxicolPathol 39: 916-924.

13. Frazier K, Thomas R, Scicchitano M, Mirabile R, Boyce R, et al. (2007) Inhibition of ALK5 signaling induces physeal dysplasia in rats. ToxicolPathol 35: 284-295.
14. Stauber AJ, Zimmermann JL, Berridge BR (2006) Pathobiology of a valvulopathy in Fischer 344 rats given a transforming growth factor-ß RI kinase inhibitor. Society of Toxicology, 290.
15. Lonning S, Mannick J, McPherson JM (2011) Antibody targeting of TGF-Î² in cancer patients. Curr Pharm Biotechnol 12: 2176-2189.
16. Garber K (2009) Companies waver in efforts to target transforming growth factor beta in cancer. J Natl Cancer Inst 101: 1664-1667.
17. Gueorguieva I, Cleverly AL, Stauber AJ, Pillay NS, Rodon JA, et al (2013) Defining a therapeutic window for the novel TFG-ß inhibitor LY2157299 monohydrate based on a pharmacokinetic/pharmacodynamic model. Br J ClinPharmacol 77:796-807.
18. Morton D, Sellers RS, Barale-Thomas E, Bolon B, George C, et al. (2010) Recommendations for pathology peer review. ToxicolPathol 38: 1118-1127.
19. Donnelly KB (2008) Cardiac valvular pathology: comparative pathology and animal models of acquired cardiac valvular diseases. ToxicolPathol 36: 204-217.
20. Walker GA, Masters KS, Shah DN, Anseth KS, Leinwand LA (2004) Valvular myofibroblast activation by transforming growth factor-beta: implications for pathological extracellular matrix remodeling in heart valve disease. Circ Res 95: 253-260.
21. Skalli O, Pelte MF, Peclet MC, Gabbiani G, Gugliotta P, et al. (1989) Alpha-smooth muscle actin, a differentiation marker of smooth muscle cells, is present in microfilamentous bundles of pericytes. J HistochemCytochem 37: 315-321.
22. Jones JA, Spinale FG, Ikonomidis JS (2009) Transforming growth factor-beta signaling in thoracic aortic aneurysm development: a paradox in pathogenesis. J Vasc Res 46: 119-137.
23. Loeys BL, Chen J, Neptune ER, Judge DP, Podowski M, et al (2005) A syndrome of altered cardiovascular, craniofacial, neurocognitive and skeletal development caused by mutations in TGF?R1 or TGF?R2. Nat Genet 37: 275-281
24. Ramirez F, Dietz HC (2007) Marfan syndrome: from molecular pathogenesis to clinical treatment. CurrOpin Genet Dev 17: 252-258.
25. Horbelt D, Guo G, Robinson PN, Knaus P (2010) Quantitative analysis of TGFBR2 mutations in Marfan-syndrome-related disorders suggests a correlation between phenotypic severity and Smad signaling activity. J Cell Sci 123: 4340-4350.
26. Loeys BL, Schwarze U, Holm T, Callewaert BL, Thomas GH, et al. (2006) Aneurysm syndromes caused by mutations in the TGF-beta receptor. N Engl J Med 355: 788-798.
27. Carta L, Smaldone S, Zilberberg L, Loch D, Dietz HC, et al. (2009) p38 MAPK is an early determinant of promiscuous Smad2/3 signaling in the aortas of fibrillin-1 (Fbn1)-null mice. J BiolChem 284: 5630-5636.
28. Holm TM, Habashi JP, Doyle JJ, Bedja D, Chen Y, et al. (2011) Noncanonical TGFÎ² signaling contributes to aortic aneurysm progression in Marfan syndrome mice. Science 332: 358-361.
29. Ng CM, Cheng A, Myers LA, Martinez-Murillo F, Jie C, et al. (2004) TGF-beta-dependent pathogenesis of mitral valve prolapse in a mouse model of Marfan syndrome. J Clin Invest 114: 1586-1592.
30. Behmoaras J, Osborne-Pellegrin M, Gauguier D, Jacob MP (2005) Characteristics of the aortic elastic network and related phenotypes in seven inbred rat strains. Am J Physiol Heart CircPhysiol 288: H769-777.
31. Serra R, Johnson M, Filvaroff EH, LaBorde J, Sheehan DM, et al. (1997) Expression of a truncated, kinase-defective TGF-beta type II receptor in mouse skeletal tissue promotes terminal chondrocyte differentiation and osteoarthritis. J Cell Biol 139: 541-552.
32. Yonamine H, Ogi N, Ishikawa T, Ichiki H (1980) Radiographic studies on skeletal growth of the pectoral limb of the beagle. Nihon JuigakuZasshi 42: 417-425.

33. Dumont N, Arteaga CL (2003) Targeting the TGF beta signaling network in human neoplasia. Cancer Cell 3: 531-536.
34. Derynck R, Akhurst RJ, Balmain A (2001) TGF-beta signaling in tumor suppression and cancer progression. Nat Genet 29: 117-129.
35. Achyut BR, Yang L (2011) Transforming growth factor-Î² in the gastrointestinal and hepatic tumor microenvironment. Gastroenterology 141: 1167-1178.
36. Engle SJ, Hoying JB, Boivin GP, Ormsby I, Gartside PS, et al. (1999) Transforming growth factor beta1 suppresses nonmetastatic colon cancer at an early stage of tumorigenesis. Cancer Res 59: 3379-3386.
37. Lu SL, Herrington H, Reh D, Weber S, Bornstein S, et al. (2006) Loss of transforming growth factor-beta type II receptor promotes metastatic head-and-neck squamous cell carcinoma. Genes Dev 20: 1331-1342.
38. Xu X, Brodie SG, Yang X, Im YH, Parks WT, et al. (2000) Haploid loss of the tumor suppressor Smad4/Dpc4 initiates gastric polyposis and cancer in mice. Oncogene 19: 1868-1874.
39. Boss DS, Schwartz GK, Middleton MR, Amakye DD, Swaisland H, et al. (2010) Safety, tolerability, pharmacokinetics and pharmacodynamics of the oral cyclin-dependent kinase inhibitor AZD5438 when administered at intermittent and continuous dosing schedules in patients with advanced solid tumours. Ann Oncol 21: 884-894.
40. Wang X, Zhang L, Goldberg SN, Bhasin M, Brown V, et al. (2011) High dose intermittent sorafenib shows improved efficacy over conventional continuous dose in renal cell carcinoma. J Transl Med 9: 220.
41. Bueno L, de Alwis DP, Pitou C, Yingling J, Lahn M, et al. (2008) Semi-mechanistic modelling of the tumour growth inhibitory effects of LY2157299, a new type I receptor TGF-beta kinase antagonist, in mice. Eur J Cancer 44: 142-150.

CHAPTER 19

Debromoaplysiatoxin as the Causative Agent of Dermatitis in a Dog after Exposure to Freshwater in California

Birgit Puschner[1]*, Adrienne C. Bautista[2] and Chris Wong[3]

[1]*Department of Molecular Biosciences, School of Veterinary Medicine, University of California, Davis, CA, USA*

[2]*California Animal Health and Food Safety Laboratory System, School of Veterinary Medicine, University of California, Davis, CA, USA*

[3]*VCA Sacramento Veterinary Referral Center, Sacramento, CA, USA*

ABSTRACT

Contamination of recreational waters with cyanobacterial toxins continues to increase and presents a risk to animals and humans. Although cases of acute hepato- and neurotoxicoses in dogs following cyanotoxin exposure exist, no reports of skin-related reactions in dogs exist. A 5-year-old female spayed 34 kg Bracco Italiano was initially presented for rapid onset of severe pruritus and urticaria. Marked excoriation and erythema were noted over the chest and neck, while urticaria was noted in the inguinal regions and ventral abdomen. Initial basic dermatology work-up excluded parasitic, fungal, and bacterial organisms. Due to the severity and progression of urticaria, the dog received IV dexamethasone and IM diphenhydramine. Improvement of the urticaria and the dog's clinical status was noted over the next 45 min. Assessment of the dog's environment revealed access to a lake on the property with visible algal bloom. Water from the lake was submitted for toxicology testing and revealed the presence of debromoaplysiatoxin. Access to the lake was discontinued and follow-up evaluation over the next few weeks revealed a complete resolution of the skin irritation. To the authors' knowledge, this is the first case report of debromoaplysiatoxin exposure in a dog after swimming in cyanobacteria-contaminated water. Veterinarians should recognize the potential harm that contaminated waters may cause in terms of dermal, hepatic, and neurological conditions. In addition, more prudent oversight of contaminated recreational waters is recommended for animals and humans to prevent adverse events and intoxications.

Keywords: dog, cyanobacteria, cyanotoxins, debromoaplysiatoxin, *Lyngbya* sp., skin irritation, toxicosis

INTRODUCTION

Cyanobacteria poisoning in dogs is not a new phenomenon. Most reports document exposure to hepatotoxic microcystins (1–4) and neurotoxic anatoxin-a (5, 6). In contrast, documentation of the irritant and allergenic effects of cyanotoxins is limited with most reports referring to studies in laboratory animals (7, 8) or epidemiological studies in humans (9–12). Although skin irritation after exposure to recreational waters is routinely mentioned as a health risk, a majority of effort has been focused on the association between total coliform, fecal coliform, and *E. coli* and skin-related symptoms (13) and not cyanobacteria. Dermatotoxins produced by cyanobacteria have been linked to outbreaks of skin irritation and include aplysiatoxins, debromoaplysiatoxins, and lyngbyatoxins. However, cyanobacterial dermatotoxins were not widely recognized until the late 1950s when an epidemic of acute contact dermatitis was reported in bathers along the beaches of Oahu, Hawaii (14). Exposure to *Lyngbya majuscula* was suspected and studies using purified extracts from *L. majuscula* positively identified exposure to this cyanobacterium as the inciting cause (15). Nevertheless, it would take more than a decade before the exact toxin, debromoaplysiatoxin, was isolated and identified (16). Unfortunately our understanding of cyanobacteria-associated dermatologic illnesses from exposure to freshwater is lacking with most acute dermatitis studies focusing on marine cyanobacteria such as the Hawaii epidemic.

A 2006 review of anecdotal and case reports and epidemiological studies of recreational and occupational exposure to freshwater cyanobacteria found that the true incidence of acute cyanobacteria-associated illness is unknown, likely due to under-recognition and under-diagnosis by healthcare providers (10). As our understanding of the toxins associated with cyanobacteria increase and our ability to detect the various toxins in biologic samples expand, linking exposure to illness will become easier. In addition, given the association between increasing temperatures and increased frequency of cyanobacteria blooms around the world, the incidence of cyanobacteria-associated illness will likely increase (17). Changing climates and anthropogenic activities are also likely to have an impact; therefore, healthcare providers and veterinarians will have to be vigilant when it comes to evaluating all aspects of adverse outcomes to cyanotoxins from recreational waters.

CASE PRESENTATION

Within 24 h of swimming in a lake in Northern California, a 5-year-old, 34-kg, spayed female Bracco Italiano developed severe pruritus, urticaria, and malaise and presented to the VCA Sacramento Veterinary Referral Center (SVRC) for evaluation. On evaluation, a large area of excoriation with erythema was noted over the cranial chest and at the base of the neck. In addition, the pinnae were erythematous and there was frequent head shaking by the dog. On otoscopic examination, the tympanic membranes were intact with only a small amount of cerumin noted; no erythema was seen within the canals. The dog's vulva was also thickened and erythematous with urticaria noted on the skin adjacent. Urticaria was evident in the inguinal regions and ventral abdomen and appeared to progress during the physical evaluation. A skin scraping of the cranial thorax revealed no evidence of parasites. An impression smear, also of the cranial thorax, revealed degenerate neutrophils, a few eosinophils, and no bacteria or fungal organisms. No fungal organisms or bacteria were found on cytology of the ear.

Due to the progression of urticaria over the inguinal region, ventral abdomen, and shoulders during the initial evaluation, a hypersensitivity reaction to an unknown stimulus was suspected. The dog was given dexamethasone SP (0.02 mg/kg IV) and diphenhydramine (2 mg/kg IM) and kept for observation for the next 45 min. During that time, the urticaria appeared to be resolving, although there was still some residual erythema along the cranial thorax and the head shaking continued. Concern by the owner for the possibility of blue-green algae intoxication prompted the collection of blood for selected chemistries, CBC, and venous blood gas analysis. Abnormalities were only identified on the CBC and consisted of a mild neutrophilia (12,940 cells/μL; ref range: 3,000–10,500 cells/μL). The dog was discharged on cephalexin (1000 mg PO q 12 h × 14 days), prednisone (30 mg PO q 12 h for 7 days, followed by tapering to q 48 h for 7 days), and diphenhydramine (75 mg PO q 8–12 h PRN × 5 days). Referral to a dermatology clinic was also recommended if the erythema and head shaking continued. Overnight, the dog vomited several times and developed diarrhea. She was presented to VCA SVRC the next morning and continued to vomit bile and was inappetent. On physical examination, the dog was alert, eupneic, dehydrated, and the abdomen palpated as non-painful. Due to the recent history of hypersensitivity reaction, GI shock was considered the most likely cause for vomiting and diarrhea; however, other underlying causes were also considered especially because of the dog's previous history of eosinophilic and lymphocytic-plasmacytic enteritis. The owner elected inpatient care consisting of the administration of LRS (on a 5% rehydration rate over 6 h, IV), famotidine (1 mg/kg, IV, q 24 h), metronidazole (10 mg/kg, IV, q 12 h), dolasetron (0.6 mg/kg, IV, q 24 h), and diphenhydramine (2 mg/kg, IM, TID). She was also placed on NPO before being referred to the VCA critical care unit for further monitoring, where clinical improvement was noted over the next 48 h. The dog was sent home after 2 days of hospitalization. Follow-up evaluation after 1 week revealed a complete resolution of clinical signs.

Careful evaluation of the dog's environment revealed access to several lakes with visible algal bloom (Figures 1 and 2). Water from the lake the dog had access to prior to developing illness was submitted for phycological and toxicological evaluation. Microscopic observation of the water sample (Green Water Laboratories/Cyanolab, Palatka, FL, USA) showed a dominant presence of a tiny vascular plant *Wolffia columbiana* (common name: watermeal), a high density of motile bacterial rods and bacterial filaments, and an abundant filamentous cyanophyte called *Microchaete cf. uberrima*. The water sample also contained green algae (*Chlorophyta*) and diatoms (*Bacillariophyta*). Specific to cyanobacteria, the water sample contained *Microchaete cf. uberrima, Lyngbya* sp. (Figure 3), *Pseudanabaena* sp., *Anabaena/Trichormus* sp. (Figure 4), *Microchaete* sp., *Scytonema* sp., *Calothrix* sp., and *Aphanothece elabens*. Amongst the cyanobacteria identified, only *Anabaena /Trichormus* sp. and *Lyngbya* sp. were of toxicological significance. Unfortunately, positive species identification for *Anabaena/Trichormus* sp. and *Lyngbya* sp. could not be achieved from collected specimens.

The water samples were also analyzed for cyanotoxins by LC-MS/MS (Green Water Laboratories/Cyanolab, Palatka, FL, USA and the California Animal Health and Food Safety Laboratory, University of California, Davis, CA, USA). The water contained 3.8 μg/g debromoaplysiatoxin [limit of detection (LOD) 0.3 μg/g]. The water also contained a low concentration of anatoxin-a that was estimated to be 1 μg/L. The water contained no aplysiatoxin (LOD 0.5 μg/g), lyngbyatoxin-a (LOD 0.2 μg/g), or cylindrospermopsin (LOD 1 μg/L).

Figure 1. Mat-forming algal bloom along the shore of the lake the dog accessed.

FIGURE 2. LAKE SURFACE COVERED WITH A DENSE GROWTH OF *WOLFFIA COLUMBIANA*. *W. columbiana* is a small, free-floating aquatic plant commonly known as watermeal and native to California. Watermeal is non-toxic and thrives in nutrient-rich water, conditions suitable for cyanbacteria growth and toxin production.

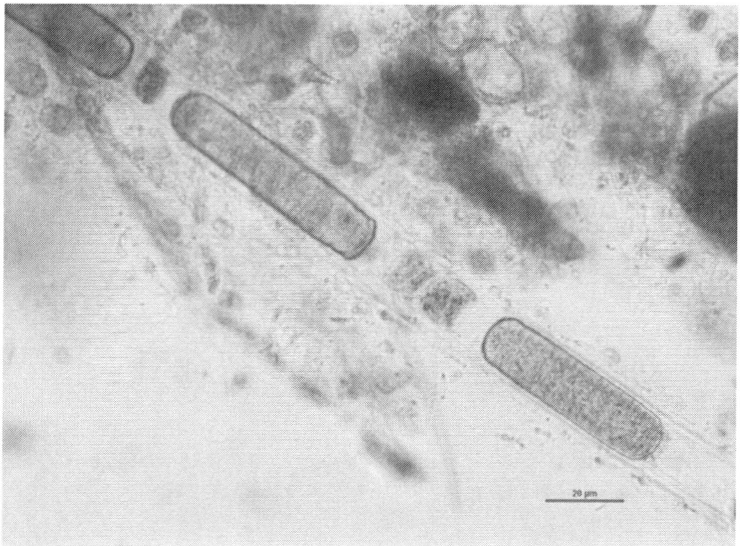

FIGURE 3. *LYNGBYA* SP. IDENTIFIED IN THE LAKE WATER. Filamentous cyanobacteria with a thin, colorless sheath. Wet mount of lake water, ×400 (scale bar = 20 μm).

FIGURE 4. *ANABAENA/TRICHORMUS* SP. IDENTIFIED IN THE LAKE WATER. Only one colony of filaments was observed in the sample. Akinetes, important characteristics morphological features were not present which prevented further identification to the lowest possible taxonomic level. Wet mount of lake water, ×400 (scale bar = 20 μm).

DISCUSSION

The current case study confirms a case of debromoaplysiatoxin poisoning in a dog following access to contaminated lake water and highlights the potential for serious skin irritation after cyanotoxin exposure. While most of the toxic cyanobacteria are freshwater species, the first documentation of algal toxin-induced dermatitis dates back to 1958 when swimmers along the beaches of Oahu, Hawaii, developed acute, vesicular contact dermatitis (14). A similar outbreak in 1976 again on the shores of Oahu, Hawaii was referred to as “swimmers' itch” (18). In the past decade, numerous outbreaks in swimmers have occurred in Australia and Hawaii (19, 20). Clinical findings in people include gradual itching and burning progressing to reddening and swelling of the skin with blister formation especially in areas covered by a bathing suit, where the seaweed/toxins are trapped. Desquamation of blisters can lead to skin erosions. The dermatitis typically resolves completely within 1 week. In more severe cases, symptomatic treatment with analgesics, antihistamine, and steroids is provided. *Lyngbya* dermatitis can develop after only minutes of exposure to contaminated waters; airborne contact dermatitis on faces has also been noted during weather patterns with strong winds. The dog in this case developed severe pruritus and urticaria within 24 h of swimming in contaminated lake water. Prompt treatment with dexamethasone and diphenhydramine prevented progression to a more severe dermatitis. Within 1 week, the dog's skin condition had resolved completely.

Isolation of the actual dermatotoxins from *L. majuscula* and from the digestive gland of the sea hare *Stylocheilus longicauda* occurred in the early 1970s (21, 22). Subsequent toxicity studies with debromoaplysiatoxin in laboratory animals and human volunteers confirmed this compound to be a potent primary skin irritant (23). Mice and rabbits developed erythema, edema, and sloughing of skin within 6 h of topical application of a 0.05% debromoaplysiatoxin solution. Human volunteers described tingling and burning sensation after application and developed an irritant pustular folliculitis between 6 and 12 h of topical administration that required up to 10 days to resolve. Debromoaplysiatoxin was subsequently isolated from other tropical marine cyanobacteria including *Lyngbya gracilis, Oscillatoria nigroviridis*, and *Schizothrix calcicola* (16). In the freshwater environment, the filamentous, benthic cyanobacterium *Lyngbya wollei* is now the cause for common proliferations in lakes and rivers from the St. Lawrence/Great Lakes basin to Florida (24). In the lake sample collected for this case work-up, the presence of *Lyngbya* sp. was confirmed by phycological evaluation. Unfortunately, only one colony of filaments was observed in the presence, which did not allow for species identification. The finding of *Lyngbya* sp. in the lake water is not surprising, considering *L. wollei* has been recorded in three sites of California streams (25). *L. wollei* has the ability to produce a wide range of toxins including paralytic shellfish toxins, and hepatotoxic cylindrospermopsin and its derivatives (26). Currently, it is unclear whether *L. wollei* is capable of producing dermatotoxins. Interestingly, most recent review articles provide in-depth information on the neuro- and hepatotoxicity of cyanotoxins but highlight the need for further examination of dermatotoxins because they bear major potential public health consequences for recreational users, while there are very limited toxicity and geographic distribution data available (27–29). In addition to debromoaplysiatoxin, aplysiatoxin and lyngbyatoxins have been associated with skin irritation (30). A debromoaplysiatoxin-producing *Lyngbya* bloom that occurred in the Homosassa River in 2008 was hypothesized as the cause for ulcerative dermatitis in manatees (31). Thus, while debromoaplysiatoxin is predominantly detected in marine water, one report documents its occurrence in freshwater ecosystems in Florida. Our report adds additional occurrence data of debromoaplysiatoxin in freshwater and is the first documentation of debromoaplysiatoxin in freshwater from a Northern California lake resulting in skin irritation in a dog. Debromoaplysiatoxin produced dermatitis on the murine ear at 0.005 nmol (equivalent to 2.7 ng) per ear (32). Considering an estimated biomass of 30 mg/L dry weight in the lake water, 1 L of water would have contained 114 ng of debromoaplysiatoxin. Based on the limited toxicity data available for topical exposure, the concentrations found are considered to be capable of resulting in dermal irritation. The lake water contained no detectable amounts of aplysiatoxin or lyngbyatoxin-a. Debromoaplysiatoxin, aplysiatoxin, and lyngbyatoxins exert their toxic effects by binding and activating protein kinase C isozymes (33). PKC belongs to a family of serine/threonine kinases and plays important roles in cellular signaling transduction for proliferation, differentiation, and apoptosis (34); activation of PKC can induce cutaneous inflammation (35). Debromoaplysiatoxin and its brominated analog aplysiatoxin can also cause diarrhea in animals and humans (36, 37). The dog developed diarrhea and vomiting approximately 48 h after swimming in the lake. These gastrointestinal signs could have been a direct toxic insult of debromoaplysiatoxin on the intestine similar to what has been confirmed after oral administration of aplysiatoxin to mice (38). However, because of the dog's previous history of eosinophilic and lymphocytic-plasmacytic enteritis, other underlying causes were also considered.

It is important to consider the role of anatoxin-a in the clinical presentation of this dog since anatoxin-a was detected in the lake water sample. Anatoxin-a is a potent cholinergic agonist at nicotinic acetylcholine receptors in neurons and at the neuromuscular junctions and can lead to muscle tremors and, in severe cases, seizures and respiratory paralysis (6, 39). The toxin is produced by many different cyanobacteria including as *Anabaena, Plantkothrix, Oscillatoria, Microcystis, Aphanizomenon, Cylindrospermum*, and *Phormidium* sp. Anatoxin-a has been found in rivers and lakes in CA (39). *Anabaena/Trichormus* sp. was identified in the water sample; due to lack of akinetes in the sample, further identification to the species level was not possible. However, the fact that anatoxin-a was identified in the water sample is a clear indication of the presence of a toxin producer. Fortunately, the dog did not develop neurologic signs, which is likely attributable to exposure to a dose below the toxic threshold. In other reported anatoxin-a poisoning cases, values of anatoxin-a ranging from 0.01 mg/L to 1 mg/L were detected (6, 39). In contrast, an estimated concentration of 1 μg/L was found in this case. This finding highlights the importance of toxicological evaluation of suspect water sample. While phycological examination of a water sample aids in the diagnostic approach of a suspect cyanotoxins exposure, it is important to note that the production of toxins by cyanobacteria is strain specific, and morphological observations alone cannot predict the hazard level. Thus, detection of toxins is needed to confirm a suspect intoxication.

CONCLUDING REMARKS

This study adds an additional toxin, debromoaplysiatoxin, to the list of freshwater toxins resulting in acute illness in animals. The presence of debromoaplysiatoxin in recreational waters has health implications for pets and humans using environments containing high concentrations of this cyanotoxin. Our investigation offers insight into the dermatotoxic risk from cyanotoxin exposure in companion animals, as well as the risk for co-exposure to multiple toxins and a complex clinical presentation. To the authors' knowledge, this is the first documentation of debromoaplysiatoxin-induced skin irritation in a dog. Owners and veterinarians should recognize the potential harm in allowing dogs to access cyanobacteria-contaminated waters. If adverse effects are noted, access to water must be denied, and supportive care should be initiated. In addition, more prudent oversight of recreational waters is recommended to prevent adverse events/intoxications not only for companion animals but also for humans.

AUTHOR CONTRIBUTIONS

AB was responsible for the toxicology case work-up and interpretation of test results. CW performed the clinical work-up and care of the dog. BP drafted the manuscript, mentored AB with the toxicological work-up, and acquired photographic materials. All authors read and approved the final manuscript.

CONFLICT OF INTEREST STATEMENT

The authors declare that they have no competing interests. All data generated or analyzed during this study are included in this published article.

ACKNOWLEDGMENTS

The authors would like to thank Mrs. Sue Watkins for providing photographs and video material, Dr. Steven Gallego for assistance during the case investigation, and Ms. Elizabeth Tor for anatoxin-a analysis.

ABBREVIATIONS

IM, intramuscular; IV, intravenous; NPO, nil *per os*, withhold oral intake of food and fluids; PO, orally; PRN, pre re nata, when necessary; Q, every.

REFERENCES

1. DeVries SE, Galey FD, Namikoshi M, Woo JC. Clinical and pathologic findings of blue-green algae (*Microcystis aeruginosa*) intoxication in a dog. *J Vet Diagn Invest* (1993) 5(3):403–8.
2. Sebbag L, Smee N, van der Merwe D, Schmid D. Liver failure in a dog following suspected ingestion of blue-green algae (*Microcystis* spp.): a case report and review of the toxin. *J Am Anim Hosp Assoc* (2013) 49(5):342–6.
3. Bautista AC, Moore CE, Lin Y, Cline MG, Benitah N, Puschner B. Hepatopathy following consumption of a commercially available blue-green algae dietary supplement in a dog. *BMC Vet Res* (2015) 11:136.
4. Moore CE, Juan J, Lin Y, Gaskill CL, Puschner B. Comparison of protein phosphatase inhibition assay with LC-MS/MS for diagnosis of microcystin toxicosis in veterinary cases. *Mar Drugs* (2016) 14(3):E54.
5. Gugger M, Lenoir S, Berger C, Ledreux A, Druart JC, Humbert JF, et al. First report in a river in France of the benthic cyanobacterium *Phormidium favosum* producing anatoxin-a associated with dog neurotoxicosis. *Toxicon* (2005) 45(7):919–28.
6. Puschner B, Pratt C, Tor ER. Treatment and diagnosis of a dog with fulminant neurological deterioration due to anatoxin – a intoxication. *J Vet Emerg Crit Care (San Antonio)* (2010) 20(5):518–22.

7. Torokne A, Palovics A, Bankine M. Allergenic (sensitization, skin and eye irritation) effects of freshwater cyanobacteria—experimental evidence. *Environ Toxicol* (2001) 16(6):512–6.
8. Stewart I. *Recreational Exposure to Freshwater Cyanobacteria: Epidemiology, Dermal Toxicity and Biological Activity of Cyanobacterial Lipopolysaccharides.* St. Lucia, QLD: The University of Queensland, School of Population Health (2005). 417 p.
9. Williamson M, Corbett S. Investigating health risks from riverine blooms of blue green algae. *NSW Public Health Bull* (1993) 4(3):27–9.
10. Stewart I, Webb PM, Schluter PJ, Fleming LE, Burns JW, Gantar M, et al. Epidemiology of recreational exposure to freshwater cyanobacteria–an international prospective cohort study. *BMC Public Health* (2006) 6(1):1.
11. Lévesque B, Gervais M-C, Chevalier P, Gauvin D, Anassour-Laouan-Sidi E, Gingras S, et al. Prospective study of acute health effects in relation to exposure to cyanobacteria. *Sci Total Environ* (2014) 466:397–403.
12. Lin CJ, Wade TJ, Sams EA, Dufour AP, Chapman AD, Hilborn ED. A prospective study of marine phytoplankton and reported illness among recreational beachgoers in Puerto Rico, 2009. *Environ Health Perspect* (2016) 124(4):477–83.
13. Yau V, Wade TJ, de Wilde CK, Colford JM. Skin-related symptoms following exposure to recreational water: a systematic review and meta-analysis. *Water Qual Expo Health* (2009) 1(2):79–103.
14. Grauer F. Dermatitis escharotica caused by a marine alga. *Hawaii Med J* (1959) 19(1):32–4.
15. Grauer FH, Arnold HL Jr. Seaweed dermatitis. First report of a dermatitis-producing marine alga. *Arch Dermatol* (1961) 84:720–32.
16. Mynderse JS, Moore RE, Kashiwagi M, Norton TR. Antileukemia activity in the Oscillatoriaceae: isolation of debromoaplysiatoxin from *Lyngbya*. *Science* (1977) 196(4289):538–40.
17. Paerl HW, Gardner WS, Havens KE, Joyner AR, McCarthy MJ, Newell SE, et al. Mitigating cyanobacterial harmful algal blooms in aquatic ecosystems impacted by climate change and anthropogenic nutrients. *Harmful Algae* (2016) 54:213–22.
18. Serdula M, Bartolini G, Moore RE, Gooch J, Wiebenga N. Seaweed itch on windward Oahu. *Hawaii Med J* (1982) 41(7):200–1
19. Osborne NJ, Shaw GR. Dermatitis associated with exposure to a marine cyanobacterium during recreational water exposure. *BMC Dermatol* (2008) 8:5.
20. Werner KA, Marquart L, Norton SA. Lyngbya dermatitis (toxic seaweed dermatitis). *Int J Dermatol* (2012) 51(1):59–62.
21. Moikeha SN, Chu GW, Berger LR. Dermatitis-producing alga *Lygnbya majuscula* Gomont in Hawaii. I. Isolation and chemical characterization of the toxic factor 1,2. *J Phycol* (1971) 7(1):4–8.
22. Kato Y, Scheuer PJ. Aplysiatoxin and debromoaplysiatoxin, constituents of the marine mollusk *Stylocheilus longicauda*. *J Am Chem Soc* (1974) 96(7):2245–6.
23. Solomon AE, Stoughton RB. Dermatitis from purified sea algae toxin (debromoaplysiatoxin). *Arch Dermatol* (1978) 114(9):1333–5.
24. Hudon C, De Sève M, Cattaneo A. Increasing occurrence of the benthic filamentous cyanobacterium *Lyngbya wollei*: a symptom of freshwater ecosystem degradation. *Freshwater Science* (2014) 33(2):606–18.
25. Fetscher AE, Howard MDA, Stancheva R, Kudela RM, Stein ED, Sutula MA, et al. Wadeable streams as widespread sources of benthic cyanotoxins in California, USA. *Harmful Algae* (2015) 49:105–16.
26. Foss AJ, Phlips EJ, Aubel MT, Szabo NJ. Investigation of extraction and analysis techniques for *Lyngbya wollei* derived Paralytic shellfish toxins. *Toxicon* (2012) 60(6):1148–58.
27. van Apeldoorn ME, van Egmond HP, Speijers GJ, Bakker GJ. Toxins of cyanobacteria. *Mol Nutr Food Res* (2007) 51(1):7–60.
28. Merel S, Walker D, Chicana R, Snyder S, Baurès E, Thomas O. State of knowledge and concerns on cyanobacterial blooms and cyanotoxins. *Environ Int* (2013) 59:303–27.

29. Corbel S, Mougin C, Bouaïcha N. Cyanobacterial toxins: modes of actions, fate in aquatic and soil ecosystems, phytotoxicity and bioaccumulation in agricultural crops. *Chemosphere* (2014) 96:1–15.
30. Rzymski P, Poniedziałek B. Dermatotoxins synthesized by blue-green algae (Cyanobacteria). *Post Dermatol Alergol* (2012) 29:47–50.
31. Harr KE, Szabo NJ, Cichra M, Phlips EJ. Debromoaplysiatoxin in *Lyngbya*-dominated mats on manatees (*Trichechus manatus* latirostris) in the Florida King's Bay ecosystem. *Toxicon* (2008) 52(2):385–8.
32. Kaneshima H, Hiai H, Fujiki H, Oguro YB, Iijima S, Sugimura T, et al. Tumor promoter-dependent mouse leukemia cell line. *Cancer Res* (1983) 43(10):4676.
33. Shimomura K, Mullinix MG, Kakunaga T, Fujiki H, Sugimura T. Bromine residue at hydrophilic region influences biological activity of aplysiatoxin, a tumor promoter. *Science* (1983) 222(4629):1242–4.
34. Mackay HJ, Twelves CJ. Targeting the protein kinase C family: are we there yet? *Nat Rev Cancer* (2007) 7(7):554–62.
35. Cataisson C, Joseloff E, Murillas R, Wang A, Atwell C, Torgerson S, et al. Activation of cutaneous protein kinase Cα induces keratinocyte apoptosis and intraepidermal inflammation by independent signaling pathways. *J Immunol* (2003) 171(5):2703–13.
36. Centers for Disease Control and Prevention (CDC). Outbreak of gastrointestinal illness associated with consumption of seaweed – Hawaii, 1994. *MMWR Morb Mortal Wkly Rep* (1995) 44(39):724.
37. Nagai H, Yasumoto T, Hokama Y. Aplysiatoxin and debromoaplysiatoxin as the causative agents of a red alga *Gracilaria coronopifolia* poisoning in Hawaii. *Toxicon* (1996) 34(7):753–61.
38. Ito E, Nagai H. Morphological observations of diarrhea in mice caused by aplysiatoxin, the causative agent of the red alga *Gracilaria coronopifolia* poisoning in Hawaii. *Toxicon* (1998) 36(12):1913–20.
39. Puschner B, Hoff B, Tor ER. Diagnosis of anatoxin-a poisoning in dogs from North America. *J Vet Diagn Invest* (2008) 20(1):89–92.

CHAPTER 20

Bioequivalence Study of Two Long-Acting Formulations of Oxytetracycline Following Intramuscular Administration in Bovines

Nora Mestorino[1*], María Laura Marchetti[1], Mariana Florencia Lucas[1], Pilar Modamio[2],Pedro Zeinsteger[1], Cecilia Fernández Lastra[2], Ignacio Segarra[2] and Eduardo Luis Mariño[2]

[1]*Laboratory of Pharmacological and Toxicological Studies (LEFyT), Faculty of Veterinary Science, Universidad Nacional de La Plata, La Plata, Argentina*

[2]*Clinical Pharmacy and Pharmacotherapy Unit, Department of Pharmacy and Pharmaceutical Technology, Faculty of Pharmacy, University of Barcelona, Barcelona, Spain*

ABSTRACT

The aim of this study was to evaluate the bioequivalence of two commercial long-acting formulations based on oxytetracycline (OTC) hydrochloride between the reference formulation (Terramycin LA, Pfizer) and a test formulation (Cyamicin LA, Fort Dodge Saude Animal). Both formulations were administered in a single intramuscular route at a dose of 20 mg OTC/kg of body weight in clinically healthy bovines. The study was carried out according to a one-period parallel design. Plasma samples were analyzed by high-pressure liquid chromatography. The limit of quantitation was 0.050 μg/mL with an accuracy of 101.67% with a coefficient of variation of 13.15%. Analysis of variance and 90% confidence interval tests were used to compare the bioavailability parameters (maximum plasma concentration, C_{max}, and the area under the concentration-versus-time curve extrapolated to infinity, $AUC_{0-\infty}$) of both products. In the case of the time to maximum concentration (T_{max}), non-parametric tests based on Wilcoxon's signed rank test were preferred. The comparison of the mean $AUC_{0-\infty}$ values did not reveal any significant differences (311.40 ± 93.05 µg h/mL and 287.71 ± 45.31 µg h/mL, respectively). The results were similar for the T_{max} (3.58 ± 0.90 h versus 3.42 ± 0.51 h). However, when comparing the mean C_{max} some significant differences were found (8.73 ± 3.66 µg/mL and 10.43 ± 3.84 µg/mL, respectively). The 90% confidence intervals for the ratio of $AUC_{0-\infty}$ and T_{max} values for the reference and test product are within the interval 80–125%, but the 90% confidence intervals for the ratio of C_{max} falls outside the proposed interval. It was concluded that C_{max} of test product are not within the 20% of those of the reference, thus suggesting that test OTC is not bioequivalent to the reference formulation.

Keywords: oxytetracycline, pharmacokinetics, AUC, C_{max}, T_{max}, bioequivalence, bovines

INTRODUCTION

The determination of the bioequivalence of veterinary drug formulations has become an increasingly important issue for both the European Union and the United States of America. Bioequivalence determination guidelines have been established by authorities of both regulatory bodies. For the American guideline, "two products are considered to be bioequivalent when they are equally bioavailable; that is, equal in the rate and extent to which the active ingredients(s) or therapeutic ingredient(s) is (are) absorbed and become(s) available at the site(s) of drug action" (1, 2). European guideline affirms that "bioequivalence exists between veterinary medicinal products or between routes of administration if, under identical and appropriate experimental conditions, the bioavailability of the active ingredient differs within acceptable limits" (3). In other words, both definitions state that equivalent rate and extent of absorption will lead to same plasma drug concentration–time profiles and, therefore, essentially the same magnitude of therapeutic or toxic effects.

The goal of the bioequivalence trial is to demonstrate that, with controlled risk to the patient, two formulations are bioequivalent and, therefore, the practitioner may use them interchangeably.

The determination of drug product bioequivalence is based on a statistical comparison of selected pharmacokinetic parameters of the two formulations of the same drug where the differences in rate and extent of drug absorption are shown to be no more than ±20% with a 90% degree of confidence.

"*In vivo*" experiments are the best way to compare the bioavailability of formulations with theoretically similar effects. In general terms, two formulations are compared, the new formulation (test formulation) versus a reference formulation efficacy of which has been clinically determined. According to the guidelines of the EMEA (3), the pharmacokinetic parameters compared between reference and test formulation in bioequivalence studies after a single dose are the area under the plasma concentration time (Cpt) curve to last concentration ($AUC_{0\text{-}t}$) or the area under the Cpt curve extrapolate to infinity ($AUC_{0-\infty}$), the peak maximum plasma concentration (C_{max}), and the time to maximum concentration (T_{max}). Non-compartmental analysis methods should be used for determination of the pharmacokinetic parameters in bioequivalence studies. If these three parameters deviate within the acceptance range, then the test formulation is considered to be bioequivalent to the reference product.

Briefly, the width of the confidence interval is determined by the within subject variance (between subject variance for parallel group studies) and the number of subjects in the study. In general, the confidence interval for untransformed data should be 80–120% (the confidence interval should lie within ±20% of the mean of the reference product). For logarithmically transformed data, the confidence interval is generally 80–125%.

In veterinary medicine, oxytetracycline (OTC) as well as the other major tetracyclines, are widely used mainly to treat respiratory, gastrointestinal, skin, locomotive organs, and genito-urinary bacterial infections, as well as systemic infections and sepsis (4).

Oxytetracycline is one of the cheapest classes of tetracycline antibiotics available, and this is due to modern manufacturing processes. Such conditions make it particularly attractive for its use in developing countries. The molecule has been available for veterinary medicine for the last half-century.

Due to the difficulties that daily OTC injections represent, long-acting (LA) alternatives were developed to achieve fast and high blood levels and to provide greater effective plasma concentrations during several days (5–7). An ideal OTC formulation for cattle should have some characteristics, such as the ability to maintain antimicrobial concentrations in tissues above the bacterial minimum inhibitory concentration for a long period of time, be easy to administrate, and with minimal tissue irritation or damage.

Consequently, parenteral LA injections are used as an alternative for the therapy of diseases that normally require several daily parenteral treatments to provide sustained concentrations of the antibiotic at the site of infection. The formulation provides prolonged circulating antibacterial concentration of the active agent, without the profile produced by repeated injections that may lead to the concentration of the agent in the blood and tissue falling below effective values. Such preparations have been particularly popular in cattle and swine because of the convenience of a single injection (4, 5, 8). Licensed 20 and 30% formulations of OTC have persistent actions because of the high strength and high dosage used (20 or 30 mg/kg), leading to sustained absorption from the reservoir site at the intramuscular injection.

Many OTC pharmacokinetic studies have been conducted in cattle. For generic medicinal products, the purpose of establishing bioequivalence is to demonstrate equivalence in pharmacokinetic parameters between the generic medicinal product (test product) and a reference one (3). Consequently, bioequivalence studies are important for the development of new pharmaceutical formulations (9).

The purpose of this study was to demonstrate the bioequivalence of two LA commercial OTCs, a test formulation (Cyamicin 20% LA) and the reference formulation (Terramycin 20% LA) after intramuscular administration to bovine.

MATERIALS AND METHODS

Chemicals and Reagents

Oxytetracycline hydrochloride of high purity (>95%) was purchased from Sigma Aldrich Chemicals and high-performance liquid chromatography (HPLC) grade solvents were obtained from J.T. Baker®.

Drugs

Two commercial products of OTC, containing 200 mg of OTC hydrochloride were compared. A reference formulation (Terramycin 20% LA, Pfizer) and a test formulation (Cyamicin 20% LA, Fort Dodge Saude Animal) were used. In order to confirm OTC concentrations, the two formulations were previously analyzed.

Animals

Twenty four young castrated male Aberdeen Angus bovines weighing 150–280 kg and aged between 6 and 12 months were used in this study. Animals were clinically evaluated to assure satisfactory health status prior to the beginning of the study and were identified with plastic numbered ear-tags.

Experimental Design

Animals were divided into two groups (group A and group B) of 12 bovines each, according VICH guidelines (VICH GL 52 – Bioequivalence, 2015) (10). The animals were ranked by weight and allocated by Greek guard to one of the two groups, after random allocation of the first animal. The scale was a digital Ezy-weight, made in New Zealand.

Dosage way and route of administration were done following the label instructions of each antibiotic. Only one treatment was carried out per bovine; thus, the study was designed as a one-period parallel study. Parallel designs do not need a washout period between treatments and they are often used for bioequivalence studies conducted in patients or for drugs with a long half-life where crossover studies are difficult or impossible to perform. If the bioequivalence study is problematic, single-dose parallel designs can be an alternative choice, because it is easy to organize, easy to analyze, and easy to interpret. The intramuscular (i.m.) injection was applied on the gluteal area.

- Animals of group A were given a single i.m. dose of 20 mg/kg of the reference (R) product (Terramycin LA, Pfizer).
- Animals of group B received the test (T) formulation (Cyamicin LA, Fort Dodge) at the same dose and by the same route.

The 24 animals were housed in a grass paddock. Feeding was on natural pasture and water was available *ad libitum*. The protocol followed the "Guide for the Care and Use of Agricultural Animals in Agricultural Research and Teaching" (Federation of Animal Science Societies, FASS) and was approved by the Experimental Ethics Committee of the Faculty of Veterinary Science, UNLP, Argentina.

Animals were sampled during the following 5 days after the OTC application. Blood samples were obtained from each animal from the jugular vein at the following post-treatment hours: 1, 2, 3, 4, 6, 8, 10, 12, 15, 18, 24, 36, 48, 60, 72, 96, and 120 to measure the level of active ingredient. Bleeding was done with 10 mL plastic disposal heparinized syringes and 21Gx11/2″ gage. Plasma was obtained by centrifugation and each tube was properly identified and was stored at −20°C until assayed.

Oxytetracycline was assayed by HPLC. Pharmacokinetics and comparative bioavailability were determined.

Oxytetracycline Assay

Plasma samples were analyzed for OTC determination following an analytical method validated at LEFyT (SOP: OTC-PL-VMA-12/05).

Oxytetracycline was assayed by HPLC with UV detection. For this, 0.1 mL plasma samples were deproteinized by adding 0.2 mL of acetonitrile and vigorously mixing for 5 min. After centrifugation of the samples for 10 min, the supernatant was evaporated to dryness at 45°C under nitrogen stream. The residue was reconstituted with 100% of mobile phase and then a 50 μL aliquot was injected directly into the HPLC system.

The chromatographic system consisted of an isocratic pump (Gilson Inc. 307), an automatic injector (Gilson Inc. 234) and a UV-VIS detector (Gilson Inc. 155) set at a wavelength of 254 nm and an octadecylsilane column (Luna C18, 4.6 mm × 150 mm, 5 μm; Phenomenex, Torrance, CA, USA). The sample was eluted with a mixture of 0.05 M phosphate buffer (pH 4.5) and acetonitrile (85:15, v/v) at a flow rate of 1.5 mL/min and at room temperature. The calibration curve from the standard samples was linear over the concentration range of 0.05–10 μg/mL.

Precision of the System

One standard solution containing 1 μg/mL of OTC was prepared and the precision of the system was evaluated after the placement of 20 injections in the chromatographic system. Thereby, the efficiencies of the column and of the system were evaluated. After 20 injections a coefficient of variation (CV) of 8.59% was determined.

Calibration and Validation

Assay was validated by measuring concentrations of known amounts of OTC in plasma of cattle. Linearity, precision, accuracy, recovery, and specificity were determined ($n = 6$).

Linearity of Standard Curves

The ratio between different concentrations was determined. Calibration curve was obtained for OTC (concentrations ranging between 0.05 and 10 μg/mL).

Precision and Accuracy

The inter-day precision was determined to estimate the run-to-run extraction and chromatographic variation in the method. Inter-day variation was measured during three (3) consecutive working days for plasma samples. Precision was expressed as %CV and accuracy as %ER (11, 12).

Accuracy is defined as the extent to which the experimental values agree with the true values. Accuracy of the method was determined by the differences between desired and calculated concentration results divided by desired concentration, and expressed as the relative error (% RE) (11, 12).

Lower Limit of Quantitation

The lower limit of quantitation (LLOQ) was calculated ($n = 12$) as the lowest drug concentration on the standard curve that could be quantitated with CV and relative error not exceeding 20%, and recovery between 20 and 80% of nominal value (11, 12).

Pharmacokinetic and Statistical Analysis

Pharmacokinetic analyses were performed using WinNonlin Professional version 6.4 (Pharsight Corp.) software. The linear trapezoidal rule was used to calculate the $AUC_{0\text{-last}}$. In addition, the elimination rate constant (k_{el}) determined from the terminal slope by log-linear regression, $T_{1/2}$, and $AUC_{0-\infty}$ calculated as the addition of $AUC_{0\text{-last}}$ and C_{last}/k_{el}, were determined for the analysis.

The one-way analysis of variance (ANOVA) for parallel design was used to assess the effect of formulations on the raw (untransformed) and logarithmically transformed data of $AUC_{0-\infty}$, T_{max} and C_{max} according VICH GL52 (Bioequivalence, August 2015) (10). In the case of T_{max}, non-parametric tests based on Wilcoxon's signed rank test were preferred (13). Parametric 90% confidence intervals based on the ANOVA of the mean Test/Reference ratios of $AUC_{0-\infty}$, and C_{max} were computed under the assumption of multiplicative model using log-transformed data. Confidence intervals were determined by the method of Westlake (14).

RESULTS

Assay Linearity

This assay exhibited a linear dynamic range between 0.05 and 10 µg/mL with r^2 value >0.999. A linear relationship was obtained across one dynamic range, as determined for the plasma spiked curves shown in Figure 1.

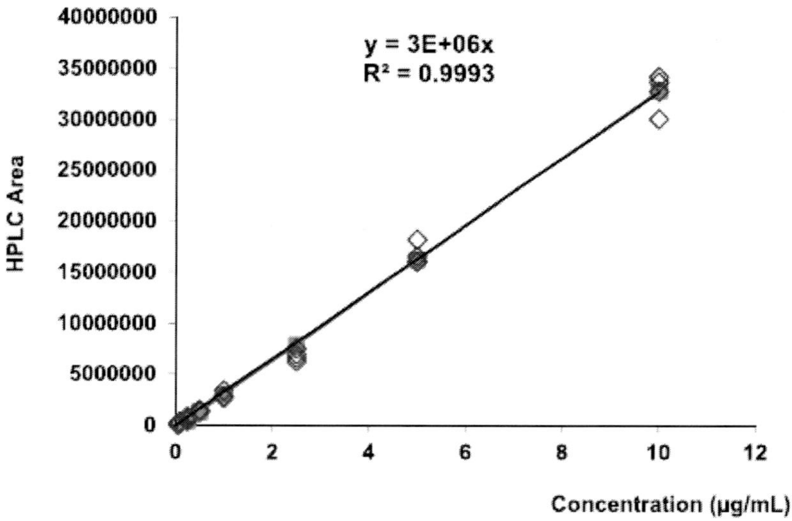

Figure 1. Calibration curve corresponding to otc standard.

Limit of Detection

The limit of detection (LOD) is the smallest detectable but not quantifiable quantity of analyte. It was estimated by means of the analysis of 20 aliquots of control plasma (free of antibiotic). The noise of the base-line was measured; the average and the SD were calculated. The LOD corresponds to three of those SDs, which in this case allowed to detect levels of 0.013 µg/mL with a recovery of >85%.

Intra-Day and Inter-Day Accuracy and Precision

To assess the inter-day (over 3 days) assay accuracy and precision, three sets of plasma samples were spiked with OTC at 0.05, 0.25, 5, and 10 µg/mL concentrations.

To determine the intra-day accuracy and precision, six replicates at each four concentrations were analyzed along with duplicate standard calibration curves prepared from two separate stock solutions.

The intra-day and inter-day variation recovery, precision, and accuracy of the method are illustrated in Tables 1 and 2, respectively. The mean recovery was within the range of 85–115%, and precision and accuracy for the 0.25, 5.00, and 10.00 µg/mL were, respectively, ≤15 and ±15%.

Table 1. Intra-day recovery, precision, and accuracy.

Desired (µg/mL)	Calculated (µg/mL)	Recovery (%)	Mean	SD	Precision (% CV)	Accuracy (% RE)
0.05	0.05	100.00				
0.05	0.04	80.00				
0.05	0.06	120.00				
0.05	0.05	100.00				
0.05	0.05	100.00				
0.05	0.05	100.00	100.00	12.65	12.65	0.00
0.25	0.2	80.00				
0.25	0.26	104.00				
0.25	0.2	80.00				
0.25	0.22	88.00				
0.25	0.2	80.00				
0.25	0.26	104.00	89.33	11.78	13.18	−12.00
5	4.14	82.80				
5	4.12	82.40				
5	5.13	102.60				
5	4.22	84.40				
5	5.08	101.60				
5	3.82	76.40	88.37	10.98	12.43	−11.60
10	8.26	82.60				
10	8.54	85.40				
10	7.41	74.10				
10	8.47	84.70				
10	10.93	109.30				
10	8.91	89.10	87.53	11.78	13.46	−12.5

% CV = (SD/calculated concentration) × 100; % RE = (calculated concentration − desired concentration)/(desired concentration × 100).

Table 2. Inter-day recovery, precision, and accuracy.

Desired (µg/mL)	Day 1 (calculate)	% Recovery	Day 2 (calculate)	% Recovery	Day 3 (calculate)	% Recovery	Mean (% Recovery)	SD (% Recovery)	%CV (% Recovery)	Accuracy (%RE)
0.05	0.05	100.00	0.06	120.00	0.04	84.00	101.33	18.04	17.80	
0.05	0.04	80.00	0.05	100.00	0.06	116.00	98.67	18.04	18.28	
0.05	0.06	120.00	0.05	100.00	0.06	120.00	113.33	11.55	10.19	
0.05	0.05	100.00	0.05	100.00	0.06	120.00	106.67	11.55	10.83	
0.05	0.05	100.00	0.06	120.00	0.06	120.00	113.33	11.55	10.19	
0.05	0.05	100.00	0.04	80.00	0.04	80.00	86.67	11.55	13.32	
Mean	0.05	100.00	0.05	103.33	0.05	106.67	103.33			
SD	0.01	12.65	0.01	15.06	0.01	19.21	10.15			
CV%	12.65	12.65	14.57	14.57	18.01	18.01	9.82			0.00
0.25	0.20	80.00	0.23	92.00	0.28	112.00	94.67	16.17	17.08	
0.25	0.26	104.00	0.23	92.00	0.24	96.00	97.33	6.11	6.28	
0.25	0.20	80.00	0.22	88.00	0.24	96.00	88.00	8.00	9.09	
0.25	0.22	88.00	0.21	84.00	0.25	100.00	90.67	8.33	9.18	
0.25	0.20	80.00	0.23	92.00	0.23	92.00	88.00	6.93	7.87	
0.25	0.26	104.00	0.24	96.00	0.24	96.00	98.67	4.62	4.68	
Mean	0.22	89.33	0.23	90.67	0.25	98.67	92.89			
SD	0.03	11.78	0.01	4.13	0.02	7.00	4.67			
CV%	13.18	13.18	4.56	4.56	7.10	7.10	5.03			−8.00
5	4.14	82.80	4.72	94.40	4.18	83.6	86.93	6.48	7.45	
5	4.12	82.40	4.21	84.20	3.98	79.6	82.07	2.32	2.82	
5	5.13	102.60	5.08	101.60	4.01	80.20	94.80	12.65	13.35	
5	4.22	84.40	4.87	97.40	4.87	97.40	93.07	7.51	8.06	
5	5.08	101.60	3.95	79.00	3.98	79.60	86.73	12.88	14.85	
5	3.82	76.40	4.14	82.80	4.21	84.20	81.13	4.16	5.13	
Mean	4.42	88.37	4.50	89.90	4.21	84.10	87.46			
SD	0.55	10.98	0.46	9.11	0.34	6.82	5.57			
CV%	12.43	12.43	10.14	10.14	8.11	8.11	6.37			−12.40
10	8.26	82.60	8.82	88.20	8.09	80.90	83.90	3.82	4.55	
10	8.54	85.40	8.78	87.80	8.99	89.90	87.70	2.25	2.57	
10	7.41	74.10	10.93	109.30	8.77	87.70	90.37	17.75	19.64	
10	8.47	84.70	10.37	103.70	8.93	89.30	92.57	9.91	10.71	
10	10.93	109.30	9.23	92.30	9.21	92.10	97.90	9.87	10.08	
10	8.91	89.10	8.97	89.70	8.94	89.40	89.40	0.30	0.34	
Mean	8.75	87.53	9.52	95.17	8.82	88.22	90.31			
SD	1.18	11.78	0.91	9.09	0.39	3.85	4.72			
CV%	13.46	13.46	9.56	9.56	4.37	4.37	5.23			−9.70

% CV = (SD/calculated concentration) × 100; % RE = (calculated concentration − desired concentration)/(desired concentration × 100).

Specificity

Six different samples from control plasma (free of antibiotics) and six plasma samples loaded with OTC were analyzed by HPLC and the corresponding chromatograms were compared. With this technique, no interferences in the retention time (3.85 ± 0.14 min) for OTC were found (Figure 2).

Lower Limit of Quantitation

The lower limit of quantitation was 0.050 µg/mL with a recovery of 103.33%, and a precision and accuracy, respectively, ≤20 and ±20%.

Pharmacokinetic Analysis

The mean ± SD OTC plasma concentration–time profiles after the intramuscular administration of each formulation (R and T) to bovines are presented in Figure 3, whereas the intramuscular pharmacokinetic data for Terramycin LA and Cyamicin LA, are presented in Table 3.

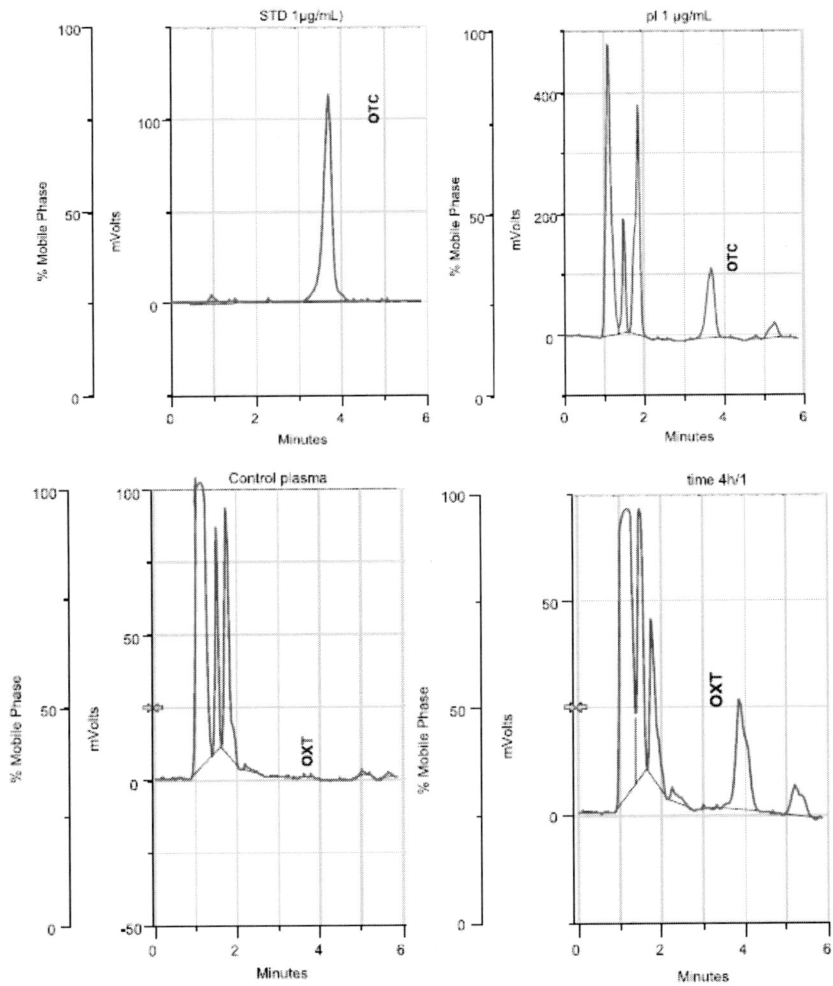

FIGURE 2. HPLC CHROMATOGRAMS (UV DETECTOR SET AT 254 NM). **(A)** Chromatogram of standard OTC solution (1 µg/mL); **(B)** chromatogram of control plasma sample (free of OTC) after extraction procedure; **(C)** chromatogram after extraction of plasma spiked with OTC (1 µg/mL); and **(D)** chromatogram after extraction of problem sample (animal 1 after 4 h OTC formulation administration).

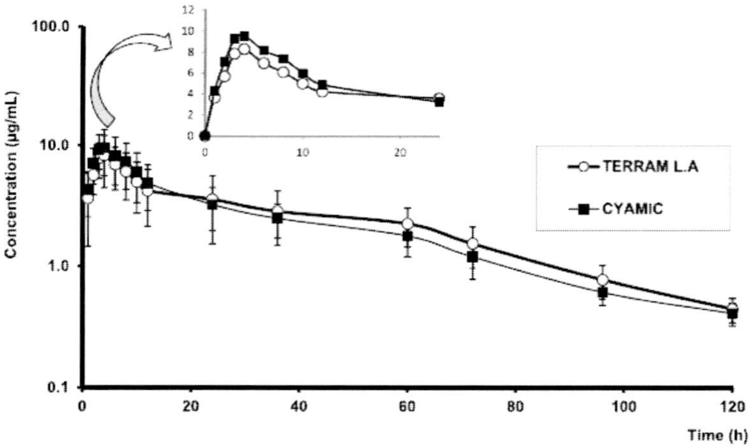

FIGURE 3. SEMILOGARITHMIC COMPARATIVE CURVES (MEAN ± SD) OF OXYTETRACYCLINE PLASMA PROFILES OBTAINED FOLLOWING INTRAMUSCULAR ADMINISTRATION OF BOTH FORMULATIONS TO BOVINES. Small inserted figure: arithmetic comparative curves (0–24 h after each administration).

Table 3. Comparison of the pharmacokinetic parameters (mean ± sd) of the reference (terramycin la) and test (cyamicin la) formulations at a single dose (20 mg/kg) administration to bovines.

		Terramycin		Cyamicin	
Parameter	**Unit**	**Mean**	**SD**	**Mean**	**SD**
λ	h^{-1}	0.020	0.005	0.022	0.004
$T½\lambda$	h	36.88	11.50	32.61	5.14
C_{max}	µg/mL	8.73	3.66	10.43	3.84
T_{max}	h	3.58	0.90	3.42	0.51
$AUC_{0-\infty}$	µg h/mL	311.40	93.05	287.71	45.31
AUC_{0-24}	µg h/mL	118.29	45.46	131.15	44.56

λ, rate constant associated with the terminal elimination phase; $T½\lambda$, elimination half-life; C_{max}, maximum concentration; T_{max}, time to reach maximum concentration; $AUC_{0-\infty}$, area under the concentration–time curve from 0 to infinity; AUC_{0-24}, area under the concentration time curve from 0 to 24 h.

Oxytetracycline plasma concentrations evolved following known features for 20% formulations. Maximum plasma concentrations (C_{max}) of 8.73 ± 3.66 and 10.43 ± 3.84 μg/mL were obtained for R and T formulations, respectively. Both formulations reached the C_{max} between 3 to 4 h. The terminal elimination half-lives (T½λ) of OTC were 36.88 ± 11.50 h and 32.61 ± 5.14 h after administration of Terramycin LA and Cyamicin LA, respectively (Table 3).

Bioequivalence Analysis

The bioavailability of the Test product (Cyamicin L.A) relative to the Reference product (Terramycin L.A) was compared on the basis of the Log-transformed and untransformed study data. Differences between the R and T products were statistically evaluated by means of confidence intervals. C_{max}, T_{max}, and $AUC_{0-\infty}$ were analyzed by ANOVA. The criterion for product bioequivalence requires for C_{max}, T_{max}, and $AUC_{0-\infty}$that the 90% confidence intervals about the difference in product means (test minus reference values) be within ±20% of the reference mean.

Tables 4 and 5 displays the statistic ANOVA comparison for log-transformation and untransformed data and the Westlake's 90% CI interval, respectively. Using this criterion, the test product did not demonstrate bioequivalence with the reference product. The parameters T_{max} and $AUC_{0-\infty}$ of the test formulation were not significantly different from the reference formulation. However, in the case of C_{max}, T product presented a significantly higher value compared to R, which exceeded the bioequivalence criterion.

Table 4. Statistical comparison between the pharmacokinetic parameters obtained for terramycin l.a (r) and cyamicin (t) by anova test with log transformed and untransformed data.

Parameter	Reference	Test	ANOVA	ANOVA (log)
AUC	311.40 ± 93.05	287.71 ± 45.31	>0.30	>0.30
C_{max}	8.73 ± 3.66	10.43 ± 3.84	0.22	0.21
T_{max}*	3.58 ± 0.90	3.42 ± 0.51	>0.30	>0.30

T_{max}: bioequivalence evaluation by Steinijans and Diletti's non-parametric test.*
Arithmetic mean of individual differences = −1.66667E−01.
90% confidence interval (Test vs. Ref.) = 86.0–114.0% vs. mean Ref.

Table 5. Westlake's 90% interval calculated iteratively for log untransformed and transformed data.

Parameter	Reference	Test	Westlake, interval for untransformed 90% (USA)	Westlake, interval after ln-transformation 90% (log) (CEE)
AUC	311.40 ± 93.05	287.71 ± 45.31	80.58–119.42%	85.89–114.11%
C_{max}	8.73 ± 3.66	10.43 ± 3.84	60.13–139.87%	65.04–134.97%
T_{max}	3.58 ± 0.90	3.42 ± 0.51	83.29–116.71%	86.28–113.72%

DISCUSSION

The analytical method showed good specificity, sensitivity, linearity, precision, and accuracy for the quantitation of OTC in plasma samples, thus allowing its use in bioequivalence assays.

The average plasma profile obtained for the reference (Terramycin LA, Pfizer) and test (Cyamicin LA, Fort Dodge) products were similar, as well as the pharmacokinetic parameters.

The area under the concentration-versus-time curve (AUC) is the parameter that indicates the exposure to the drug in base to the fraction of the dose reaching the systemic circulation and systemic clearance of the drug. In our study, the area under the OTC curve extrapolate to infinity ($AUC_{0-\infty}$) was 311.40 ± 93.05 and 287.71 ± 45.31 μg h/mL for R and T formulation, respectively. These values were high; this makes sense due to the prolonged elimination slope. While AUC_{0-24} values were smaller (118.29 ± 45.46 and 131.15 ± 44.56 μg h/mL respectively) compared to those reported by Craigmill et al. (15) in calves receiving the same dose of OTC LA (168 ± 14.6 μg h/mL), and with the area under the curve calculated by Achenbach (16) which was 256.73 μg h/mL, both authors performed these assays on sera samples, while our study was conducted using plasma.

To determine if two formulations would yield the same efficacy behavior, their areas under the respective curves, their maximum concentrations, and the times at which the C_{max} is reached, should be determined and compared statistically. If the differences among such parameters are not significant, the drugs are bioequivalent, but if the differences are significant, the formulations are bioinequivalent. In the present work, these parameters were compared through ANOVA, since the populations were normal, using untransformed and log-transformed values. No significant differences were encountered for areas under the curves, and times to the maximum concentrations for the values found were inside the confidence intervals set by the European Union to determine bioequivalence. However, the maximum concentration (C_{max}) of the test product (Cyamicin LA) was significantly higher than that of the reference formulation and exceeded the bioequivalence interval. These significant differences were found in the untransformed and the log-transformed methodologies and these values were outside the confidence intervals.

The results obtained with i.m. administration of OTC LA in this study for C_{max} were similar to those of Ozdemir and Yildirim (17) who reported that C_{max} was 8.02 μg/mL at 2.75 h. Although this was a study of the bioequivalence of two LA OTC formulations in sheep, the authors obtained similar results. They mentioned that the multivariate analysis, accomplished through ANOVA for pharmacokinetic parameters (after log-transformation of the data), showed no statistically significant differences between the two products for the parameters C_{max} and AUC except between two periods and two groups for AUC. By contrast, much lower C_{max} values of 4, 3.89 ± 1.48, and 5.7 ± 0.32 were reported in cattle by Toutain and Raynaud (6), Mestorino et al. (5), and Kumar and Malik (18).

As a final conclusion, the formulations behaved similarly but were not bioequivalent, although T_{max} and AUC fell within the bioequivalence criterion. In the case of C_{max}, a parameter classically considered of importance regarding antimicrobial efficacy, the value obtained after administration of Cyamicin was statistically significantly higher. The 90% confidence interval of T/R was 65.04–134.97% for log-transformation C_{max}. This interval is not included in the equivalence interval (80–125%), therefore, cannot be considered bioequivalent. The test product showed an extent of absorption appreciably larger than the reference product following administration of the same dose; therefore, suprabioavailability was found. Probably equivalent AUC and C_{max} are achieved following administration of a lower dose of the test product as compared to the reference formulation. So could be expected that the two products will have similar systemic efficacy and safety although administered at different doses, but a new bioequivalence assay must be performed.

AUTHOR CONTRIBUTIONS

Substantial contributions to the design of the work: NM. Conduction of the field work: animal management and blood sampling: MM and ML (Arg. team). HPLC assay: NM, MM, and ML (Arg. team). Acquisition and analysis of data for the work: NM, MM, and ML (Arg. team). Interpretation of data:

NM, PM, IS, CL, and EM. Wrote or contributed to the writing of the manuscript: NM, MM, ML, PM, PZ, CL, IS, and EM. Revised the manuscript: NM, PM, PZ, CL, IS, and EM. The manuscript was critically reviewed and the final version approved by all authors.

CONFLICT OF INTEREST STATEMENT

None of the authors of this paper have a financial or personal relationship with other people or organizations that could inappropriately influence or bias the content of the paper.

FUNDING

This research was partially supported by the University of La Plata (Project n° V11/202) and by the Agencia Nacional de Promoción Científica y Tecnológica (PICT N° 535).

REFERENCES

1. Food and Drug Administration, U.S. Department of Health and Human Services, Center for Veterinary Medicine (CVM). *Guidance for Industry # 35: Bioequivalence Guidance*. Silver Spring, MD: FDA (2006).
2. Food and Drug Administration. *Bioavailability and Bioequivalence Requirements. Title 21*. (Vol. 5). Silver Spring, MD: FDA (2015).
3. European Medicines Agency. *Guideline on the Conduct of Bioequivalence Studies for Veterinary Medicinal Products. EMA/CVMP/016/00-Rev.2 Committee for Medicinal Products for Veterinary Use (CVMP)*. London: EMA (2011).
4. Giguère S, Prescott JF, Dowling PM. *Antimicrobial Therapy in Veterinary Medicine*. 5th ed. New York City, NY: John Wiley & Sons, Inc (2013).
5. Mestorino N, Mariño Hernández E, Marchetti ML, Errecalde JO. Pharmacokinetics and tissue residues of an oxytetracycline/diclofenac combination in cattle. *Rev Sci Tech* (2007) 26(3):679–90.
6. Toutain PL, Raynaud JP. Pharmacokinetics of oxytetracycline in young cattle: comparison of conventional vs long-acting formulations. *Am J Vet Res* (1983) 44:1203–9.
7. Michalova E, Novotna P, Schelegelova J. Tetracyclines in veterinary and bacterial resistance to them. *Vet Med Czech* (2004) 49(3):79–100.
8. Escudero E, Cárceles CM, Serrano JM. Pharmacokinetics of oxytetracycline in goats: modifications induced by a long-acting formulation. *Vet Rec* (1994) 135:548–52.
9. Vetchý D, Frýbortová K, Rabisková M, Danecková H. Bioequivalence studies of pharmaceutical preparations. *Cas Lek Cesk* (2007) 146(5):431–3.
10. VICH GL 52. *Bioequivalence: Blood Level Bioequivalence Study*. Brussels: International Cooperation on Harmonisation of Technical Requirements for Registration of Veterinary Medicinal Products (2015).

11. Snyder L, Kirkland J, Glajch J. *Practical HPLC Method Development*. 2nd ed. New York, NY: Wiley Interscience (1997).
12. Chandran S, Singh RSP. Comparison of various international guidelines for analytical method validation. *Pharmazie* (2007) 62:4–14.
13. Steinijans VW, Diletti E. Statistical analysis of bioequivalence studies: parametric and non-parametric confidence intervals. *Eur J Clin Pharmacol* (1983) 24:127.
14. Westlake WJ. Symmetrical confidence intervals for bioequivalence trials. *Biometrics* (1976) 32:741–4.
15. Craigmill AL, Holland RE, Robinson D, Wetzlich S, Arndt T. Serum pharmacokinetics of oxytetracycline in sheep and calves and tissue residues in sheep following a single intramuscular injection of a long acting preparation. *J Vet Pharm Therap* (2000) 23:345–52.
16. Achenbach TE. *Physiological and Classical Pharmacokinetic Models of Oxytetracycline in Cattle*. Master's thesis. Simon Fraser University, Canada (2000).
17. Ozdemir N, Yildirim M. Bioequivalence study of two long-acting oxytetracycline formulations in sheep. *Vet Res Commun* (2006) 30:929–34.
18. Kumar N, Malik JK. Some pharmacokinetic parameters and dosage regimens for a long-acting formulation of oxytetracycline in 6- to 8-month-old male calves. *Vet Res Commun* (1998) 22(8):533.

Index